DUALISABILITY

Unary Algebras and Beyond

Advances in Mathematics

VOLUME 9

DUALISABILITY

Unary Algebras and Beyond

By

JANE PITKETHLY
La Trobe University, Victoria, Australia

BRIAN DAVEY
La Trobe University, Victoria, Australia

Springer

Library of Congress Cataloging-in-Publication Data

Pitkethly, Jane.
Dualisability : unary algebras and beyond / by Jane Pitkethly, Brian Davey.
p. cm. — (Advances in mathematics ; v. 9)
Includes bibliographical references and index.

1. Unary algebras. 2. Quasivarieties (Universal algebra) 3. Duality theory (Mathematics)
I. Davey, B.A. II. Title. III. Advances in mathematics (Springer Science+Business Media) ; v. 9.

QA251.P58 2005
511.3′3—dc22

2005050606

ISBN 978-1-4419-3901-2 e-ISBN 978-0-387-27570-3

AMS Subject Classifications: 08–02, 08C15, 08A05, 08A60, 18A40

Printed in the United States of America.

9 8 7 6 5 4 3 2 1

springeronline.com

Contents

Preface

In mathematics, there are many methods for translating a difficult problem into an easier problem in a completely different setting. For example:

- a logarithm function translates the difficult problem of multiplying two positive numbers into the easy problem of adding two different numbers;
- a Galois connection translates the problem of solving an equation over an infinite field into the problem of finding all the subgroups of a finite group.

A natural duality provides another method for translating difficult problems into easier ones. A natural duality can take a difficult, hard-to-visualise problem in a class of algebras and translate it into an easier, pictorial problem in a different class of mathematical structures. For example, Priestley's duality [58] can be used to translate problems about distributive lattices into problems about ordered topological spaces. Over the past two decades, the theory of natural dualities has developed into a practical tool for studying algebras.

There are 'classical' examples of dualities dating from the 1930s and beyond: Pontryagin's duality for abelian groups [55], Stone's duality for Boolean algebras [63] and Priestley's duality for distributive lattices [56, 57]. The number of known dualities began to escalate in the early 1980s, when Davey and Werner [29] set out the general theory of natural dualities. This general theory encompassed nearly all previously known dualities and introduced methods for finding new dualities.

While natural dualities have found varied applications in both algebra [8] and logic [1], the theory of natural dualities is a beautiful area of mathematics in its own right. Accordingly, there has been ongoing interest in understanding natural dualities at the theoretical level.

Primary theme: dualisability

A finite algebra is said to be *dualisable* if it can be used to create a natural duality. The most fundamental problem in the theory of natural dualities is the *Dualisability Problem*, which asks: 'Which finite algebras are dualisable?' At present, the Dualisability Problem seems to be unsolvable. Indeed, it may be formally undecidable. The holy grail for natural-duality theoreticians is the *Decidability Problem for Dualisability*, which asks: 'Is there an algorithm for deciding whether or not any given finite algebra is dualisable?'

There are algorithms for deciding dualisability within certain special classes of algebras. For example, the dualisable two-element algebras have been characterised by the second author [16, 8]. The dualisable commutative rings with identity have been completely described by Clark, Idziak, Sabourin, Szabó and Willard [14]. Davey, Idziak, Lampe and McNulty [23] have characterised dualisability within the class of graph algebras. The considerable effort and the different methods required to obtain these partial characterisations lead to the expectation that there is no general algorithm for deciding dualisability.

There is a large class of algebras for which there is a simple description of the dualisable algebras, but for which there is no known algorithm for identifying them. Amongst the finite algebras that generate congruence-distributive varieties, the dualisable algebras are precisely those with a near-unanimity term. (One half of this result was proved by Davey and Werner [29], and the other by Davey, Heindorf and McKenzie [22].) At present, it is not known whether there is an algorithm that can identify algebras with a near-unanimity term.

In this text, we illustrate further the complexity of dualisability. We do this by studying dualisability amongst the simplest algebras of all: unary algebras. A *unary algebra* is an algebra all of whose operations are unary. The simplicity of unary algebras, and their pictorial nature, makes them easy to work with. Nevertheless, from a duality-theory viewpoint, unary algebras are not particularly well behaved. Perhaps surprisingly, we shall find that the class of unary algebras appears to reflect much of the complexity of dualisability.

Secondary theme: strong dualisability

A finite algebra is said to be *strongly dualisable* if it can be used to create a natural duality that is a special sort of dual category equivalence. The *Strong Dualisability Problem*— 'Which finite algebras are strongly dualisable?'—and the corresponding *Decidability Problem for Strong Dualisability* are unsolved, even within the class of dualisable algebras. However, much is known about strong dualisability, and a vast array of algebras have been shown to be strongly dualisable.

The richness of dualisability and strong dualisability for unary algebras is strikingly illustrated by a discovery of Hyndman and Willard [41]. They answered an important question in duality theory by exhibiting an algebra that is dualisable but not strongly dualisable. Their algebra has just three elements and two unary operations! We shall see that the class of unary algebras is, in fact, an ideal place to search for pathological examples in duality theory.

Relevant background

This research-level text is a sequel to the foundational duality-theory text by Clark and Davey [8]. Nevertheless, this text can be used as a stand-alone introduction to the theory of natural dualities, with an emphasis on developments that have occurred since their text was published. A specially focused overview of the basic universal algebra, topology and category theory needed here can be found in the first chapter and appendices of the Clark–Davey text [8]. Comprehensive treatments of these background topics, well beyond what we require, may be found in the texts on universal algebra by G. Grätzer [34], Burris and Sankappanavar [5] and McKenzie, McNulty and Taylor [48], on topology by J. Dugundji [31] and J. L. Kelley [44], and on category theory by S. Mac Lane [46]. The notation we use largely follows that of Clark and Davey [8]. All specialised notation is listed in the notation index.

Chapter by chapter

In Chapter 1, we briefly introduce finitely generated quasi-varieties and the theory of natural dualities. We also provide motivation for our study of the dualisability of unary algebras, by showing that the quasi-variety generated by a small unary algebra can be complicated.

In this text, we will find that unary algebras are a rich source of examples and counterexamples for the study of some longstanding questions in duality theory. We shall develop techniques for creating dualisable and non-dualisable unary algebras, and then use these techniques to build examples to answer such questions as: 'Can a product of two dualisable algebras be non-dualisable?'

Most of the tools that we will use to create examples are developed in Chapters 2 and 3. In Chapter 2, we investigate ways in which certain binary homomorphisms of a finite algebra can guarantee its dualisability. In particular, we study binary homomorphisms that are lattice, flat-semilattice or group operations. We develop some general tools that we use to prove the dualisability of a large number of unary algebras. For example, we show that the endomorphisms of a finite cyclic group are the operations of a dualisable unary algebra.

In Chapter 3, we completely solve the Dualisability Problem within the class of three-element unary algebras. The dualisable and non-dualisable three-element unary algebras are tightly entangled. Indeed, on the three-element set, there is a chain of six unary clones such that the corresponding unary algebras are alternately dualisable and non-dualisable. The intricacy of the characterisation of dualisability for three-element unary algebras suggests that the Dualisability Problem for unary algebras is difficult. The chapter also includes a proof that every finite *unar* (unary algebra with a single operation) is dualisable.

Chapter 4 builds on the results in Chapter 3 by characterising strong dualisability within the class of three-element unary algebras. Amongst the dualisable three-element unary algebras, the strongly dualisable algebras can be characterised in two alternative ways: as those satisfying a weak injectivity condition, or as those avoiding three particular obstacles. This chapter also makes a contribution to the *Full versus Strong Problem.* A finite algebra is said to be *fully dualisable* if it can be used to create a natural duality that is a dual category equivalence. While full dualisability is formally weaker than strong dualisability, we presently have no example that can differentiate them. The Full versus Strong Problem concerns the existence of such examples. We shall prove that full dualisability and strong dualisability are equivalent for three-element unary algebras. We will also extend the example of Hyndman and Willard [41], by showing that there are many dualisable three-element unary algebras that are not strongly dualisable. Chapter 4 is our most technical chapter. We return to the consideration of strong dualisability for unary algebras in Chapter 7.

In the remainder of the text, we use the experience gained in the earlier chapters to solve some general problems in duality theory. In Chapter 5, we show that there are many natural algebraic constructions under which dualisability is not always preserved. In particular, we find two dualisable unary algebras whose product is not dualisable. We also show that dualisability is not always preserved by taking homomorphic images or coproducts. In addition, this chapter includes a characterisation of dualisability for p-semilattices. This allows us to give examples of dualisable algebras that are retracts of non-dualisable algebras.

In Chapter 6, we solve several clone-theoretic problems in duality theory. We build on our example, from Chapter 3, of a chain of six unary clones that determine alternately dualisable and non-dualisable algebras. We show that, for any natural number n, there is a chain of n unary clones for which the corresponding algebras are alternately dualisable and non-dualisable. We also give an example of a non-dualisable algebra that can be obtained by adding a nullary operation to a dualisable algebra, and we find a non-dualisable entropic algebra.

Chapter 7 contains another alternating-chains result. Beginning from any finite unary algebra with at least two fundamental operations, there is an infinite ascending chain of finite algebras (under the subalgebra order) that are alternately dualisable and non-dualisable. We obtain this result while characterising the finite algebras (of arbitrary type) that can be embedded into a non-dualisable algebra. The chapter concludes with a proof that every finite linear unary algebra is strongly dualisable. Linear unary algebras form an important class that includes all unars. Consequently, this result generalises J. Hyndman's result that all finite unars are strongly dualisable [38].

In the appendix, we prove several important general theorems related to strong dualisability, including two results used extensively in Chapters 4 and 7. The appendix also gives a new approach to R. Willard's technical concept of rank [65], which is a finitary sufficient condition for strong dualisability.

Acknowledgements

This text is a modified version of the first author's PhD thesis [52], which was written under the supervision of the second author. The text provides a coherent treatment of a line of research within duality theory, much of which has appeared in a series of papers.

Chapters 2 and 3 are based on papers by the authors and D. M. Clark [12, 13]. Much of the rest of the text is based on papers by the authors [26, 50, 51, 53, 54]. The presentation of Chapter 4 has been influenced by a paper written by the first author and J. Hyndman [40]. Section 5.1 is largely new. The results presented in Sections 6.1 and 6.2, on chains of alternately dualisable and non-dualisable clones, have not appeared elsewhere in print, nor has the proof in Section 7.3 that linear unary algebras are strongly dualisable. Most of the appendix is also new. In particular, the concept of height is introduced here for the first time.

The authors' joint work with David Clark has shaped much of the research reported here. We would like to thank David for his friendship and for the many fruitful collaborations. Thanks also to Jennifer Hyndman, who visited La Trobe University and showed us how she viewed 'enough algebraic operations' and 'rank' for unary algebras. We must thank Ralph McKenzie, George McNulty and Ross Willard, whose positive reports on the first author's PhD thesis encouraged the authors to write this text. Finally, we wish to acknowledge the ongoing support of the Mathematics Department at La Trobe University.

Jane Pitkethly and Brian Davey

14 May, 2005

1

Unary algebras and dualisability

Dualisability for unary algebras is surprisingly complicated. Even though individual unary algebras are relatively simple and easy to work with, we shall see that as a class they have a rich and complex entanglement with dualisability. This combination of local simplicity and global complexity ensures that, for the study of natural duality theory, unary algebras are an excellent source of examples and counterexamples.

The dualisability of an algebra depends on the structure of the quasi-variety it generates. We begin this preliminary chapter by briefly introducing the concept of a quasi-variety. Then, to illustrate the complexity that can be hidden inside a small unary algebra, we present two three-element unary algebras and explore the structure of the very different quasi-varieties they generate. It will be helpful to look back on these two concrete examples throughout the rest of this text, when we are working more generally.

In the last three sections of this chapter, we introduce the notions of dualisability, full dualisability and strong dualisability. These sections provide a brief overview of the areas of duality theory that we use in this text.

1.1 Quasi-varieties generated by finite algebras

Many familiar classes of algebras can be formed as the quasi-variety generated by a finite algebra: for instance, Boolean algebras, bounded distributive lattices, semilattices and abelian groups of a given finite exponent. The theory of natural dualities provides a method for studying such classes.

In this section, we give a quick introduction to quasi-varieties generated by finite algebras. (See the text by Clark and Davey [8] for a more detailed introduction, and the text by V. A. Gorbunov [33] for an in-depth study of quasi-varieties in general.) We shall start from the more familiar notion of a variety. The variety $\mathrm{Var}(\underline{\mathbf{M}})$ generated by a finite algebra $\underline{\mathbf{M}}$ can be described in various ways. The variety $\mathrm{Var}(\underline{\mathbf{M}})$ is simultaneously:

- the class of all algebras (of the same type as $\underline{\mathbf{M}}$) that satisfy all the equations satisfied by $\underline{\mathbf{M}}$;
- the smallest class of algebras that contains $\underline{\mathbf{M}}$ and is closed under forming homomorphic images, subalgebras and products;
- the class $\mathbb{HSP}(\underline{\mathbf{M}})$ consisting of all homomorphic images of subalgebras of powers of $\underline{\mathbf{M}}$.

The quasi-variety generated by $\underline{\mathbf{M}}$ has analogous descriptions.

To give a syntactic description of the quasi-variety generated by $\underline{\mathbf{M}}$, we replace the notion of an equation with the notion of a quasi-equation. The Cancellative Law for semigroups,

$$xz \approx yz \implies x \approx y,$$

is an example of a quasi-equation. In general, a **quasi-equation** is an implication of the form

$$(\sigma_1 \approx \tau_1)\ \&\ \cdots\ \&\ (\sigma_n \approx \tau_n) \implies \sigma \approx \tau,$$

for some $n \in \omega$ and terms $\sigma_1, \dots, \sigma_n, \tau_1, \dots, \tau_n, \sigma, \tau$ of a given type. Since we can take $n = 0$ in this definition, an equation is a special sort of quasi-equation. An algebra **satisfies a quasi-equation** (of the same type as the algebra) if the implication holds in the algebra for all possible assignments of the variables in the terms.

We can now define the **quasi-variety generated by $\underline{\mathbf{M}}$** to be the class of all algebras (of the same type as $\underline{\mathbf{M}}$) that satisfy all the quasi-equations satisfied by $\underline{\mathbf{M}}$. To give semantic descriptions of the quasi-variety generated by $\underline{\mathbf{M}}$, we require more definitions.

The standard operators $\mathbb{I}$, $\mathbb{S}$ and $\mathbb{P}$ are defined as follows. For each class $\mathcal{K}$ of algebras of the same type:

- $\mathbb{I}(\mathcal{K})$ is the class of all isomorphic copies of algebras in $\mathcal{K}$;
- $\mathbb{S}(\mathcal{K})$ is the class of all subalgebras of algebras in $\mathcal{K}$;
- $\mathbb{P}(\mathcal{K})$ is the class of all products of algebras in $\mathcal{K}$.

For each class $\mathcal{K}$ of algebras, the class $\mathbb{P}(\mathcal{K})$ contains the empty-indexed product, which is a one-element algebra of the same type as $\mathcal{K}$.

Consider a set X of maps from a set A to a set B, and let C be a subset of A. We say that X **separates (the elements of)** C if, for all $c, d \in C$ with $c \neq d$, there is a map $x : A \to B$ in X such that $x(c) \neq x(d)$. In particular, an algebra $\mathbf{A}$ is **separated by homomorphisms into $\underline{\mathbf{M}}$** if the set of all homomorphisms from $\mathbf{A}$ to $\underline{\mathbf{M}}$ separates the elements of A. Note that a one-element algebra of the same type as $\underline{\mathbf{M}}$ is vacuously separated by homomorphisms into $\underline{\mathbf{M}}$.

We are now able to give several descriptions of the quasi-variety generated by a finite algebra.

1.1.1 $\mathbb{ISP}$ Theorem *Let $\underline{\mathbf{M}}$ be a finite algebra. For every algebra $\mathbf{A}$ of the same type as $\underline{\mathbf{M}}$, the following are equivalent:*

(i) $\mathbf{A}$ *satisfies all the quasi-equations satisfied by* $\underline{\mathbf{M}}$;

(ii) $\mathbf{A}$ *can be obtained from* $\underline{\mathbf{M}}$ *by repeated applications of* $\mathbb{I}$, $\mathbb{S}$ *and* $\mathbb{P}$;

(iii) $\mathbf{A}$ *belongs to* $\mathbb{ISP}(\underline{\mathbf{M}})$;

(iv) $\mathbf{A}$ *is separated by homomorphisms into* $\underline{\mathbf{M}}$.

In particular, the quasi-variety generated by $\underline{\mathbf{M}}$ *is the class* $\mathbb{ISP}(\underline{\mathbf{M}})$.

Proof The equivalence of (i), (ii) and (iii) is a special case of a theorem due to A. I. Mal'cev [47]. The equivalence of (iii) and (iv) is an easy exercise. We have restricted ourselves to the case in which $\underline{\mathbf{M}}$ is finite, as that is what is needed here. For an infinite algebra $\underline{\mathbf{M}}$, we need to add the ultraproduct operator $\mathbb{P}_u$ to the description. A direct proof of the theorem as stated here may be found in the Clark–Davey text [8, 1.3.1 and 1.3.4]. ∎

Each of the quasi-varieties mentioned at the start of this section has many possible generators. The smallest generators are as follows.

- Define $\underline{\mathbf{B}} = \langle\{0,1\}; \vee, \wedge, ', 0, 1\rangle$ to be the two-element Boolean algebra. Then the variety of Boolean algebras is equal to $\mathbb{ISP}(\underline{\mathbf{B}})$.
- Let $\underline{\mathbf{D}} = \langle\{0,1\}; \vee, \wedge, 0, 1\rangle$ denote the two-element bounded lattice. Then the variety of bounded distributive lattices is equal to $\mathbb{ISP}(\underline{\mathbf{D}})$.
- Define $\underline{\mathbf{S}} = \langle\{0,1\}; \wedge, 1\rangle$ to be the two-element meet semilattice with 1. The variety of meet semilattices with 1 is equal to $\mathbb{ISP}(\underline{\mathbf{S}})$.
- Let $m > 0$ and define $\underline{\mathbf{Z}}_m = \langle\mathbb{Z}_m; +, ^-, 0\rangle$ to be the cyclic group of order m. Then the variety of abelian groups of exponent m is equal to $\mathbb{ISP}(\underline{\mathbf{Z}}_m)$.

Next, we begin our study of quasi-varieties of unary algebras.

1.2 Two examples

In this section, we consider the two three-element unary algebras $\underline{\mathbf{R}}$ and $\underline{\mathbf{Q}}$, shown in Figure 1.1. The algebras $\underline{\mathbf{R}}$ and $\underline{\mathbf{Q}}$ may seem equally benign, but

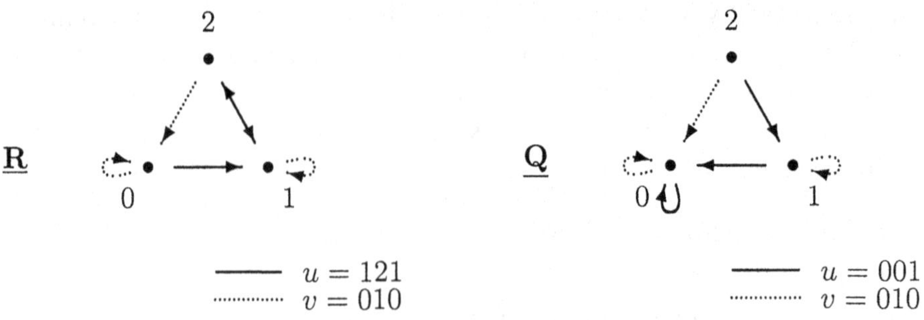

Figure 1.1 Two unary algebras

they are quite different. The algebra $\underline{\mathbf{R}}$ generates a very simple quasi-variety. The non-trivial members of $\mathbb{ISP}(\underline{\mathbf{R}})$ can all be built from two basic types of components. In contrast, the quasi-variety generated by $\underline{\mathbf{Q}}$ is complicated. We shall show that $\mathbb{ISP}(\underline{\mathbf{Q}})$ is equivalent to a category of directed graphs.

Throughout this text, we write unary operations as strings. For each $n \in \omega$, we denote a unary operation

$$u : \{0, \dots, n\} \to \{0, \dots, n\}$$

by the string $u(0) \cdots u(n)$.

1.2.1 Definition Define the unary algebra

$$\underline{\mathbf{R}} := \langle \{0, 1, 2\}; 121, 010 \rangle$$

as in Figure 1.1. Take the type of $\underline{\mathbf{R}}$ to be $\{u, v\}$, where $u^{\underline{\mathbf{R}}} := 121$ and $v^{\underline{\mathbf{R}}} := 010$. To help with our description of the quasi-variety $\mathbb{ISP}(\underline{\mathbf{R}})$, we will introduce two classes of unary algebras. For all (possibly empty) sets I and J, define the 'triangular' unary algebra $\mathbf{Tr}_I$ and the 'square' unary algebra $\mathbf{Sq}_{IJ}$, both of type $\{u, v\}$, as in Figure 1.2.

The overall structure of a unary algebra can be captured by a directed graph. Consider any unary algebra $\mathbf{A} = \langle A; F \rangle$. The algebra $\mathbf{A}$ determines a directed graph

$$G(\mathbf{A}) = \langle A; E_{\mathbf{A}} \rangle, \text{ where } E_{\mathbf{A}} := \{ (a, u(a)) \mid a \in A \text{ and } u \in F \}.$$

The algebra $\mathbf{A}$ is said to be **connected** if the graph $G(\mathbf{A})$ is connected. A **connected component of A** is a maximal connected subalgebra of $\mathbf{A}$.

We can completely describe the quasi-variety $\mathbb{ISP}(\underline{\mathbf{R}})$ by first exhibiting all the algebras that can occur as connected components of algebras in $\mathbb{ISP}(\underline{\mathbf{R}})$,

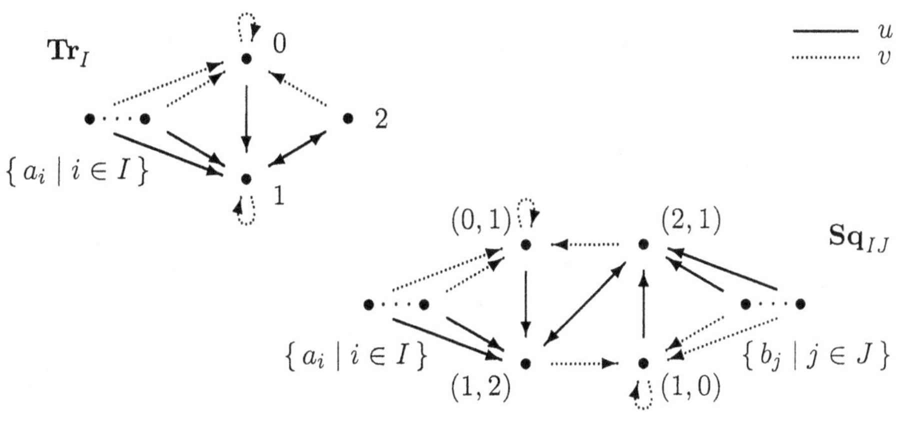

Figure 1.2 The 'triangular' and 'square' algebras

and then saying exactly which combinations of these connected components form an algebra in $\mathbb{ISP}(\underline{\mathbf{R}})$.

1.2.2 Lemma *Invoke the definitions in 1.2.1 and let* $\mathbf{A} = \langle A; u, v \rangle$ *be a non-trivial unary algebra. Then* $\mathbf{A}$ *belongs to* $\mathbb{ISP}(\underline{\mathbf{R}})$ *if and only if*

(i) *each connected component of* $\mathbf{A}$ *is isomorphic to* $\mathbf{Tr}_I$*, for some set* I*, or to* $\mathbf{Sq}_{IJ}$*, for some sets* I *and* J*, and*

(ii) *there is at most one connected component of* $\mathbf{A}$ *that, for some set* I*, is isomorphic to* $\mathbf{Tr}_I$*.*

Proof First assume that $\mathbf{A} \in \mathbb{ISP}(\underline{\mathbf{R}})$. As $\mathbf{A}$ is non-trivial, there is a non-empty set S such that $\mathbf{A}$ is isomorphic to a subalgebra of $\underline{\mathbf{R}}^S$. We will show that conditions (i) and (ii) hold with $\mathbf{A}$ replaced by $\underline{\mathbf{R}}^S$. It will then follow that (i) and (ii) hold for $\mathbf{A}$.

For all $r \in \{0,1,2\}$, let $\widehat{r}$ denote the constant function in R^S with value r. Define the subset C_Δ of R^S by

$$C_\Delta := \{0,2\}^S \cup \{\widehat{1}\}.$$

Using Figure 1.3, it is easy to check that C_Δ forms a connected subalgebra $\mathbf{C}_\Delta$ of $\underline{\mathbf{R}}^S$. In $\underline{\mathbf{R}}^S$, we have

$$121^{-1}(C_\Delta) = 121^{-1}(\widehat{1}) \cup 121^{-1}(\widehat{2}) = \{0,2\}^S \cup \{\widehat{1}\} = C_\Delta$$

and

$$010^{-1}(C_\Delta) = 010^{-1}(\widehat{0}) \cup 010^{-1}(\widehat{1}) = \{0,2\}^S \cup \{\widehat{1}\} = C_\Delta.$$

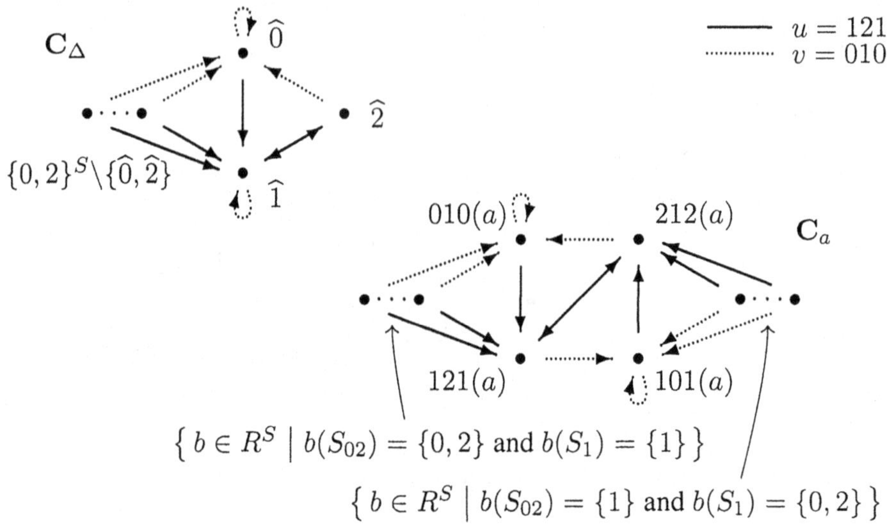

Figure 1.3 Connected components of $\underline{\mathbf{R}}^S$

So $\mathbf{C}_\Delta$ is a connected component of $\underline{\mathbf{R}}^S$. The component $\mathbf{C}_\Delta$ is isomorphic to $\mathbf{Tr}_I$, where $I := \{0,2\}^S\backslash\{\widehat{0},\widehat{2}\}$.

Now choose some $a \in R^S\backslash C_\Delta$. We want to show that there are sets I and J such that the connected component of $\underline{\mathbf{R}}^S$ containing a is isomorphic to $\mathbf{Sq}_{IJ}$. Since $\mathbf{C}_\Delta$ is a connected component of $\underline{\mathbf{R}}^S$, we must have $010(a) \notin C_\Delta$. So $010(a) \in \{0,1\}^S\backslash\{\widehat{0},\widehat{1}\}$, which implies that the subsets

$$S_{02} := a^{-1}(\{0,2\}) \quad \text{and} \quad S_1 := a^{-1}(1)$$

of S are non-empty. Therefore $\{S_{02}, S_1\}$ is a partition of S. Define the subset C_a of R^S by

$$C_a := \{\, b \in R^S \mid b(S_{02}) \subseteq \{0,2\} \text{ and } b(S_1) = \{1\}\,\}$$
$$\cup \{\, b \in R^S \mid b(S_{02}) = \{1\} \text{ and } b(S_1) \subseteq \{0,2\}\,\}.$$

Then, by Figure 1.3, the set C_a forms a connected subalgebra of $\underline{\mathbf{R}}^S$. Since

$$121^{-1}(C_a) = C_a \quad \text{and} \quad 010^{-1}(C_a) = C_a,$$

the algebra $\mathbf{C}_a$ is a connected component of $\underline{\mathbf{R}}^S$. The component $\mathbf{C}_a$ is isomorphic to $\mathbf{Sq}_{IJ}$, where I is the set of non-empty proper subsets of S_{02} and J is the set of non-empty proper subsets of S_1.

We have established that conditions (i) and (ii) hold with $\underline{\mathbf{R}}^S$ in place of $\mathbf{A}$. For each set I, every subalgebra of $\mathbf{Tr}_I$ is of the form $\mathbf{Tr}_K$, for some $K \subseteq I$. Similarly, for all sets I and J, every subalgebra of $\mathbf{Sq}_{IJ}$ is of the form $\mathbf{Sq}_{KL}$, for some $K \subseteq I$ and some $L \subseteq J$. Since $\mathbf{A}$ is isomorphic to a subalgebra of $\underline{\mathbf{R}}^S$, it follows that (i) and (ii) hold.

Now assume that (i) and (ii) are satisfied. By the $\mathbb{ISP}$ Theorem, 1.1.1, we can prove that $\mathbf{A}$ belongs to $\mathbb{ISP}(\underline{\mathbf{R}})$ by showing that $\mathbf{A}$ is separated by homomorphisms into $\underline{\mathbf{R}}$. We begin by proving that, for all sets I and J,

(a) both $\mathbf{Tr}_I$ and $\mathbf{Sq}_{IJ}$ are separated by homomorphisms into $\underline{\mathbf{R}}$,

(b) for each $a \in Sq_{IJ}$, there exist homomorphisms $x, y : \mathbf{Sq}_{IJ} \to \underline{\mathbf{R}}$ with $x(a) \neq y(a)$.

To do this, let I and J be sets. For each subset K of I, we can define the homomorphism $x_K : \mathbf{Tr}_I \to \underline{\mathbf{R}}$ by

$$x_K(r) = r, \quad \text{for all } r \in \{0, 1, 2\},$$

and

$$x_K(a_i) = \begin{cases} 2 & \text{if } i \in K, \\ 0 & \text{otherwise,} \end{cases} \qquad \text{for all } i \in I.$$

The homomorphisms in $\{\, x_K \mid K \subseteq I \,\}$ separate the elements of $\mathbf{Tr}_I$. Now, for each subset K of I, define the homomorphism $y_K : \mathbf{Sq}_{IJ} \to \underline{\mathbf{R}}$ by

$$y_K\big((r,s)\big) = r, \quad \text{for all } (r,s) \in \big\{(0,1), (1,2), (2,1), (1,0)\big\},$$

$$y_K(b_j) = 1, \quad \text{for all } j \in J,$$

and

$$y_K(a_i) = \begin{cases} 2 & \text{if } i \in K, \\ 0 & \text{otherwise,} \end{cases} \qquad \text{for all } i \in I.$$

By symmetry, for each subset L of J, we can define $z_L : \mathbf{Sq}_{IJ} \to \underline{\mathbf{R}}$ by

$$z_L\big((r,s)\big) = s, \quad \text{for all } (r,s) \in \big\{(0,1), (1,2), (2,1), (1,0)\big\},$$

$$z_L(a_i) = 1, \quad \text{for all } i \in I,$$

and

$$z_L(b_j) = \begin{cases} 2 & \text{if } j \in L, \\ 0 & \text{otherwise,} \end{cases} \qquad \text{for all } j \in J.$$

It is easy to check that the elements of $\mathbf{Sq}_{IJ}$ are separated by the maps in

$$\{\, y_K \mid K \subseteq I \,\} \cup \{\, z_L \mid L \subseteq J \,\}.$$

Furthermore, for every $a \in Sq_{IJ}$, we have $y_{\varnothing}(a) \neq z_{\varnothing}(a)$. Hence (a) and (b) hold for all sets I and J.

To check that $\mathbf{A} \in \mathbb{ISP}(\underline{\mathbf{R}})$, let $a, b \in A$ with $a \neq b$. We can define homomorphisms from $\mathbf{A}$ to $\underline{\mathbf{R}}$ independently on each of the connected components of $\mathbf{A}$. So, since (i) holds for $\mathbf{A}$, we can use (a) to find at least one homomorphism $x : \mathbf{A} \to \underline{\mathbf{R}}$. First assume that a and b belong to the same connected component $\mathbf{C}$ of $\mathbf{A}$. By (a), there is a homomorphism $y : \mathbf{C} \to \underline{\mathbf{R}}$ such that $y(a) \neq y(b)$. Thus

$$x{\restriction}_{A \backslash C} \cup y : \mathbf{A} \to \underline{\mathbf{R}}$$

is a homomorphism separating a and b. Now assume that a and b belong to different connected components of $\mathbf{A}$. Since (ii) holds, we can assume that the connected component $\mathbf{D}$ of $\mathbf{A}$ that contains a is isomorphic to $\mathbf{Sq}_{IJ}$, for some sets I and J. By (b), there is a homomorphism $z : \mathbf{D} \to \underline{\mathbf{R}}$ such that $z(a) \neq x(b)$. So the homomorphism

$$x{\restriction}_{A \backslash D} \cup z : \mathbf{A} \to \underline{\mathbf{R}}$$

separates a and b. By the $\mathbb{ISP}$ Theorem, 1.1.1, it follows that $\mathbf{A} \in \mathbb{ISP}(\underline{\mathbf{R}})$. ∎

Later, we will show that the algebra $\underline{\mathbf{R}}$ is dualisable. The way that we prove this exploits the 'finiteness' and simplicity of the quasi-variety $\mathbb{ISP}(\underline{\mathbf{R}})$. In fact, the dualisability of $\underline{\mathbf{R}}$ will follow from two separate general results: Theorem 3.2.10 and Theorem 7.2.12. Each of these general results applies to finite unary algebras that generate a quasi-variety that is in some way similar in its simplicity to $\mathbb{ISP}(\underline{\mathbf{R}})$.

Now we will show that the quasi-variety generated by a small unary algebra can be complicated.

1.2.3 Definition Define the unary algebra

$$\underline{\mathbf{Q}} := \langle \{0, 1, 2\}; 001, 010 \rangle$$

as in Figure 1.1. Take the type of $\underline{\mathbf{Q}}$ to be $\{u, v\}$, where $u^{\underline{\mathbf{Q}}} := 001$ and $v^{\underline{\mathbf{Q}}} := 010$. We will represent algebras in $\mathbb{ISP}(\underline{\mathbf{Q}})$ as graphs. Define a $\underline{\mathbf{Q}}$-graph to be a directed graph $\mathbf{G} = \langle V; E \rangle$ with a vertex $0 \in V$ such that $(0, \nu) \in E$, for all $\nu \in V$, and $(0, 0)$ is the only loop in E.

The $\mathbb{ISP}$ Theorem, 1.1.1, tells us that the class $\mathbb{ISP}(\underline{\mathbf{Q}})$ is determined by some set of quasi-equations. The following lemma gives a finite list of quasi-equations that suffice.

1.2.4 Lemma *Define the unary algebra* $\underset{\sim}{\mathbf{Q}}$ *as in 1.2.3, and define the unary term* $\underline{0}(x) := u(u(x))$. *Then the quasi-variety* $\mathbb{ISP}(\underset{\sim}{\mathbf{Q}})$ *is determined by the quasi-equations*

(i) $\underline{0}(x) \approx \underline{0}(y)$,

(ii) $u(v(x)) \approx \underline{0}(x)$,

(iii) $v(u(x)) \approx u(x)$,

(iv) $v(v(x)) \approx v(x)$,

(v) $u(x) \approx v(x) \implies u(x) \approx \underline{0}(x)$,

(vi) $u(x) \approx u(y) \;\&\; v(x) \approx v(y) \implies x \approx y$.

Proof It is easy to check that $\underset{\sim}{\mathbf{Q}}$ satisfies quasi-equations (i) to (vi). So assume that the unary algebra $\mathbf{A} = \langle A; u, v\rangle$ satisfies (i) to (vi). We want to show that $\mathbf{A} \in \mathbb{ISP}(\underset{\sim}{\mathbf{Q}})$. First we shall show that the elements of $v(A)$ are separated by the homomorphisms from $\mathbf{A}$ into $\underset{\sim}{\mathbf{Q}}$.

Let $c, d \in v(A)$ with $c \neq d$. By equation (i), the term function $\underline{0}^{\mathbf{A}}$ of $\mathbf{A}$ is constant. Let $0^{\mathbf{A}}$ denote the value of $\underline{0}^{\mathbf{A}}$ in $\mathbf{A}$. We can assume that $c \neq 0^{\mathbf{A}}$. For all $a \in A$ such that $u(a) = v(a)$, we have $u(a) = 0^{\mathbf{A}} \neq c$, by (v). So we can define the map $x : A \to \{0, 1, 2\}$ by

$$x(a) = \begin{cases} 2 & \text{if } u(a) = c, \\ 1 & \text{if } v(a) = c, \\ 0 & \text{otherwise.} \end{cases}$$

To see that $x : \mathbf{A} \to \underset{\sim}{\mathbf{Q}}$ is a homomorphism, let $a \in A$. As $u \circ u(a) = 0^{\mathbf{A}} \neq c$, we must have $x(u(a)) \neq 2$. So

$$u(x(a)) \in \{0, 1\} \quad \text{and} \quad x(u(a)) \in \{0, 1\}.$$

By (iii), we have

$$\begin{aligned} u(x(a)) = 1 \iff x(a) = 2 \iff u(a) = c \\ \iff v(u(a)) = c \iff x(u(a)) = 1. \end{aligned}$$

So it follows that $u(x(a)) = x(u(a))$. Using (ii), we must have

$$v(x(a)) \in \{0, 1\} \quad \text{and} \quad x(v(a)) \in \{0, 1\},$$

and, using (iv), we must have

$$\begin{aligned} v(x(a)) = 1 \iff x(a) = 1 \iff v(a) = c \\ \iff v(v(a)) = c \iff x(v(a)) = 1. \end{aligned}$$

Therefore $v(x(a)) = x(v(a))$. Thus $x : \mathbf{A} \to \underset{\sim}{\mathbf{Q}}$ is a homomorphism. Since $c, d \in v(A)$, we have $v(c) = c$ and $v(d) = d$, by (iv). So $x(c) = 1 \neq x(d)$, and $v(A)$ is separated by the homomorphisms from $\mathbf{A}$ into $\underset{\sim}{\mathbf{Q}}$.

By the $\mathbb{ISP}$ Theorem, 1.1.1, to show that $\mathbf{A} \in \mathbb{ISP}(\underset{\sim}{\mathbf{Q}})$, it is enough to prove that $\mathbf{A}$ is separated by homomorphisms into $\underset{\sim}{\mathbf{Q}}$. Let $a, b \in A$ such that $a \neq b$. Then $u(a) \neq u(b)$ or $v(a) \neq v(b)$, by (vi). We have $u(a), u(b) \in v(A)$, by (iii). The elements of $v(A)$ are separated by the homomorphisms from $\mathbf{A}$ to $\underset{\sim}{\mathbf{Q}}$. So there is a homomorphism $y : \mathbf{A} \to \underset{\sim}{\mathbf{Q}}$ such that

$$y(u(a)) \neq y(u(b)) \quad \text{or} \quad y(v(a)) \neq y(v(b)).$$

In either case, we have $y(a) \neq y(b)$. ∎

1.2.5 Lemma *Invoking the definitions in 1.2.3, there is a category equivalence between the quasi-variety $\mathbb{ISP}(\underset{\sim}{\mathbf{Q}})$ and the category of all $\underset{\sim}{\mathbf{Q}}$-graphs.*

Proof Let $\mathcal{G}$ denote the category of all $\underset{\sim}{\mathbf{Q}}$-graphs. Each directed graph $\mathbf{G}$ in $\mathcal{G}$ has a unique looped vertex $0^{\mathbf{G}}$, and there is an edge from $0^{\mathbf{G}}$ to every other vertex of $\mathbf{G}$. Now define the quasi-variety $\mathcal{A} := \mathbb{ISP}(\underset{\sim}{\mathbf{Q}})$. We will be using the quasi-equational basis for $\mathcal{A}$ established in the previous lemma. For each $\mathbf{A} \in \mathcal{A}$, there is a distinguished element $0^{\mathbf{A}}$ of $\mathbf{A}$ corresponding to the constant term function $\underline{0}^{\mathbf{A}}$.

We shall set up a pair of functors

$$\Gamma : \mathcal{A} \to \mathcal{G} \quad \text{and} \quad \Delta : \mathcal{G} \to \mathcal{A}.$$

First let $\mathbf{A} = \langle A; u, v \rangle$ be an algebra in $\mathcal{A}$. Since $u(A) \subseteq v(A)$, by quasi-equation (iii), we can define the directed graph

$$\Gamma(\mathbf{A}) = \langle v(A); \Xi_{\mathbf{A}} \rangle, \quad \text{where } \Xi_{\mathbf{A}} := \{\, (u(a), v(a)) \mid a \in A \,\}.$$

(See Figure 1.4 for an example with $\mathbf{A} \leqslant \underset{\sim}{\mathbf{Q}}^4$.) The only looped vertex of $\Gamma(\mathbf{A})$ is $0^{\mathbf{A}}$, by (v). For all $a \in A$, we have

$$\big(0^{\mathbf{A}}, v(a)\big) = \big(u \circ v(a), v \circ v(a)\big) \in \Xi_{\mathbf{A}},$$

by (ii) and (iv). So there is an edge from $0^{\mathbf{A}}$ to every other vertex of $\Gamma(\mathbf{A})$. Thus the directed graph $\Gamma(\mathbf{A})$ belongs to $\mathcal{G}$.

Let $\varphi : \mathbf{A} \to \mathbf{B}$ be a homomorphism in $\mathcal{A}$. We want to define the graph homomorphism

$$\Gamma(\varphi) : \Gamma(\mathbf{A}) \to \Gamma(\mathbf{B}) \quad \text{by} \quad \Gamma(\varphi) := \varphi{\restriction}_{v(A)}.$$

The map $\Gamma(\varphi)$ is well defined, as

$$\varphi(v(a)) = v(\varphi(a)) \in v(B),$$

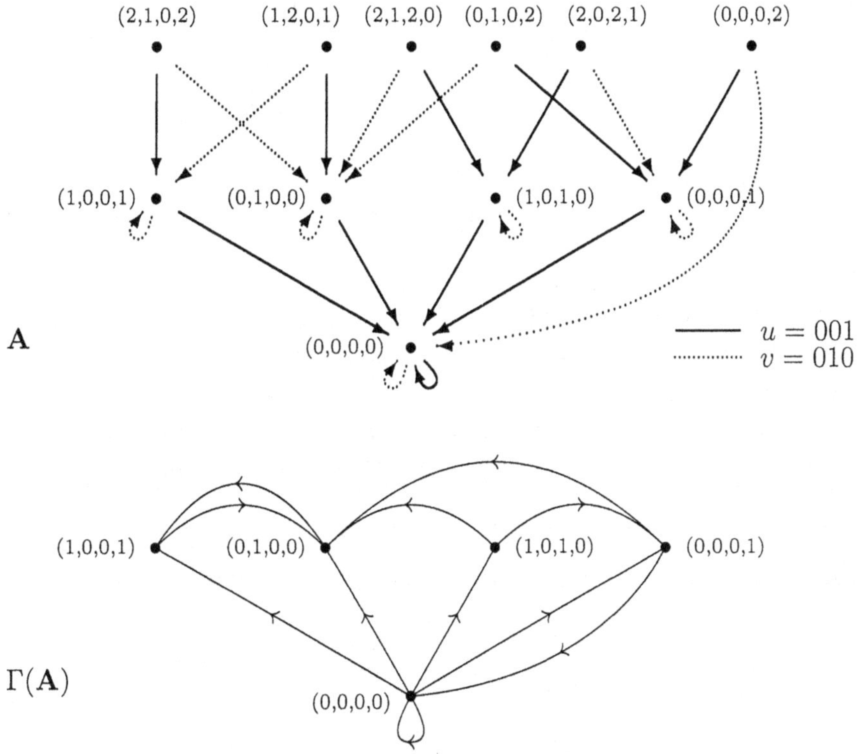

Figure 1.4 Representing an algebra in $\mathbb{ISP}(\underline{\mathbf{Q}})$ as a graph

for all $a \in A$. The map $\Gamma(\varphi)$ is a graph homomorphism, since, for all $a \in A$, we have

$$\begin{aligned}\big(\Gamma(\varphi)(u(a)), \Gamma(\varphi)(v(a))\big) &= \big(\varphi(u(a)), \varphi(v(a))\big) \\ &= \big(u(\varphi(a)), v(\varphi(a))\big) \in \Xi_{\mathbf{B}}.\end{aligned}$$

It is easy to see that $\Gamma : \mathcal{A} \to \mathcal{G}$ is a functor.

Now let $\mathbf{G} = \langle V; E\rangle$ be a directed graph in $\mathcal{G}$. There is an edge from $0^{\mathbf{G}}$ to every vertex of $\mathbf{G}$. So we can define the unary algebra

$$\Delta(\mathbf{G}) = \langle E; u, v\rangle, \text{ where } u\big((\mu,\nu)\big) = (0^{\mathbf{G}}, \mu) \text{ and } v\big((\mu,\nu)\big) = (0^{\mathbf{G}}, \nu),$$

for all $(\mu,\nu) \in E$. Using Lemma 1.2.4, it is easy to verify that the algebra $\Delta(\mathbf{G})$ belongs to $\mathcal{A}$. Consider a graph homomorphism $\psi : \mathbf{G} \to \mathbf{H}$ in $\mathcal{G}$. We wish to define

$$\Delta(\psi) : \Delta(\mathbf{G}) \to \Delta(\mathbf{H}) \text{ by } \Delta(\psi)\big((\mu,\nu)\big) = (\psi(\mu), \psi(\nu)),$$

for each edge $(\mu, \nu) \in E$. The map $\Delta(\psi)$ is well defined, since ψ is a graph homomorphism. For each edge $(\mu, \nu) \in E$, we have

$$\begin{aligned} u \circ \Delta(\psi)\big((\mu,\nu)\big) &= u\big((\psi(\mu), \psi(\nu))\big) = (0^{\mathbf{H}}, \psi(\mu)) \\ &= \big(\psi(0^{\mathbf{G}}), \psi(\mu)\big) = \Delta(\psi) \circ u\big((\mu,\nu)\big) \end{aligned}$$

and

$$\begin{aligned} v \circ \Delta(\psi)\big((\mu,\nu)\big) &= v\big((\psi(\mu), \psi(\nu))\big) = (0^{\mathbf{H}}, \psi(\nu)) \\ &= \big(\psi(0^{\mathbf{G}}), \psi(\nu)\big) = \Delta(\psi) \circ v\big((\mu,\nu)\big). \end{aligned}$$

So $\Delta(\psi)$ is a homomorphism. Thus $\Delta : \mathcal{G} \to \mathcal{A}$ is a functor.

We now want to show that Γ and Δ give us a category equivalence between $\mathcal{A}$ and $\mathcal{G}$. We do this by setting up a pair of natural isomorphisms $\eta : \mathrm{id}_{\mathcal{A}} \to \Delta\Gamma$ and $\zeta : \mathrm{id}_{\mathcal{G}} \to \Gamma\Delta$.

First let $\mathbf{A} \in \mathcal{A}$. The universe of the algebra $\Delta\Gamma(\mathbf{A})$ is the edge set of the graph $\Gamma(\mathbf{A})$, which is $\Xi_{\mathbf{A}} := \{\, (u(a), v(a)) \mid a \in A \,\}$. Define

$$\eta_{\mathbf{A}} : \mathbf{A} \to \Delta\Gamma(\mathbf{A}) \quad \text{by} \quad \eta_{\mathbf{A}}(a) = (u(a), v(a)).$$

Then $\eta_{\mathbf{A}}$ is a homomorphism since, for all $a \in A$, we have

$$u(\eta_{\mathbf{A}}(a)) = u\big((u(a), v(a))\big) = (0^{\Gamma(\mathbf{A})}, u(a)) = (0^{\mathbf{A}}, u(a)) = \eta_{\mathbf{A}}(u(a)),$$

by (iii), and

$$v(\eta_{\mathbf{A}}(a)) = v\big((u(a), v(a))\big) = (0^{\Gamma(\mathbf{A})}, v(a)) = (0^{\mathbf{A}}, v(a)) = \eta_{\mathbf{A}}(v(a)),$$

by (ii) and (iv). The homomorphism $\eta_{\mathbf{A}}$ is clearly surjective, and $\eta_{\mathbf{A}}$ is injective by (vi). So $\eta_{\mathbf{A}} : \mathbf{A} \to \Delta\Gamma(\mathbf{A})$ is an isomorphism.

Now let $\varphi : \mathbf{A} \to \mathbf{B}$ be a homomorphism in $\mathcal{A}$. To see that η is a natural transformation, we need to show that the square below commutes.

$$\begin{array}{ccc} \mathbf{A} & \xrightarrow{\ \eta_{\mathbf{A}}\ } & \Delta\Gamma(\mathbf{A}) \\ {\scriptstyle\varphi}\big\downarrow & & \big\downarrow{\scriptstyle\Delta\Gamma(\varphi)} \\ \mathbf{B} & \xrightarrow[\ \eta_{\mathbf{B}}\ ]{} & \Delta\Gamma(\mathbf{B}) \end{array}$$

For each $a \in A$, we have

$$\begin{aligned} \Delta\Gamma(\varphi) \circ \eta_{\mathbf{A}}(a) &= \Delta\Gamma(\varphi)\big((u(a), v(a))\big) = \Delta(\varphi{\restriction_{v(A)}})\big((u(a), v(a))\big) \\ &= \big(\varphi(u(a)), \varphi(v(a))\big) = \big(u(\varphi(a)), v(\varphi(a))\big) \\ &= \eta_{\mathbf{B}} \circ \varphi(a). \end{aligned}$$

Thus $\eta : \mathrm{id}_{\mathcal{A}} \to \Delta\Gamma$ is a natural isomorphism.

Let $\mathbf{G} = \langle V; E\rangle$ be a directed graph in $\mathcal{G}$. We want to define the isomorphism

$$\zeta_{\mathbf{G}} : \mathbf{G} \to \Gamma\Delta(\mathbf{G}) \text{ by } \zeta_{\mathbf{G}}(\nu) = (0^{\mathbf{G}}, \nu),$$

for each $\nu \in V$. The vertex set of the graph $\Gamma\Delta(\mathbf{G})$ is $v(E)$. For each edge $(\mu, \nu) \in E$, we have $v\big((\mu,\nu)\big) = (0^{\mathbf{G}}, \nu)$ in the algebra $\Delta(\mathbf{G})$. For each vertex $\nu \in V$, we know that $(0^{\mathbf{G}}, \nu) \in E$, and therefore

$$(0^{\mathbf{G}}, \nu) = v\big((0^{\mathbf{G}}, \nu)\big) \in v(E).$$

So the vertex set of $\Gamma\Delta(\mathbf{G})$ is

$$v(E) = \big\{\, (0^{\mathbf{G}}, \nu) \;\big|\; \nu \in V \,\big\},$$

and it follows that $\zeta_{\mathbf{G}}$ is a well-defined bijection. The edge set of the graph $\Gamma\Delta(\mathbf{G})$ is

$$\begin{aligned}\Xi_{\Delta(\mathbf{G})} &:= \big\{\, \big(u((\mu,\nu)), v((\mu,\nu))\big) \;\big|\; (\mu,\nu) \in E \,\big\}\\ &= \big\{\, \big((0^{\mathbf{G}}, \mu), (0^{\mathbf{G}}, \nu)\big) \;\big|\; (\mu,\nu) \in E \,\big\}.\end{aligned}$$

The map $\zeta_{\mathbf{G}}$ is a graph homomorphism since, for each $(\mu, \nu) \in E$, the pair

$$\big(\zeta_{\mathbf{G}}(\mu), \zeta_{\mathbf{G}}(\nu)\big) = \big((0^{\mathbf{G}}, \mu), (0^{\mathbf{G}}, \nu)\big)$$

is an edge of $\Gamma\Delta(\mathbf{G})$. To finish proving that $\zeta_{\mathbf{G}}$ is an isomorphism, assume that $(\zeta_{\mathbf{G}}(\mu), \zeta_{\mathbf{G}}(\nu))$ is an edge of $\Gamma\Delta(\mathbf{G})$, for some $\mu, \nu \in V$. Then there exists $(\mu', \nu') \in E$ such that

$$\big((0^{\mathbf{G}}, \mu), (0^{\mathbf{G}}, \nu)\big) = \big((0^{\mathbf{G}}, \mu'), (0^{\mathbf{G}}, \nu')\big).$$

Thus $(\mu, \nu) = (\mu', \nu')$, and (μ, ν) is an edge of $\mathbf{G}$. We have shown that the map $\zeta_{\mathbf{G}}$ is an isomorphism.

It remains to show that $\zeta : \mathrm{id}_{\mathcal{G}} \to \Gamma\Delta$ is a natural transformation. Let $\psi : \mathbf{G} \to \mathbf{H}$ be a graph homomorphism in $\mathcal{G}$, where $\mathbf{G} = \langle V; E\rangle$. For each $\nu \in V$, we get

$$\begin{aligned}\Gamma\Delta(\psi) \circ \zeta_{\mathbf{G}}(\nu) &= \Gamma\Delta(\psi)\big((0^{\mathbf{G}}, \nu)\big) = \Delta(\psi){\restriction_{v(E)}}\big((0^{\mathbf{G}}, \nu)\big)\\ &= \big(\psi(0^{\mathbf{G}}), \psi(\nu)\big) = \big(0^{\mathbf{H}}, \psi(\nu)\big)\\ &= \zeta_{\mathbf{H}} \circ \psi(\nu).\end{aligned}$$

So ζ is a natural isomorphism. Hence $\mathcal{A}$ and $\mathcal{G}$ are equivalent as categories. ∎

As we shall later see, it seems as though a finite unary algebra that generates a very simple quasi-variety is likely to be dualisable. On the other hand, a finite unary algebra that generates a complicated quasi-variety is not necessarily non-dualisable. It turns out that the algebra $\underline{\mathbf{Q}} = \langle\{0,1,2\}; 001, 010\rangle$ is non-dualisable (Theorem 3.0.1). But the quasi-variety generated by a dualisable unary algebra can be just as complicated as $\mathbb{ISP}(\underline{\mathbf{Q}})$. For instance, the unary algebra $\langle\{0,1,2,3\}; 0011, 0101\rangle$ is dualisable, even though it has $\underline{\mathbf{Q}}$ as a subalgebra (Example 5.3.2). The algebra $\langle\{0,1,2\}; 001, 010, 002\rangle$ is also dualisable, and yet it has $\underline{\mathbf{Q}}$ as a reduct (Theorem 3.0.1).

1.3 An introduction to dualisability

The remainder of this chapter is meant to serve as a brief introduction to the theory of natural dualities for the uninitiated, and as a quick refresher for the enlightened. We shall concentrate on the aspects of duality theory that will be required in this text. The text by Clark and Davey [8] gives a more thorough treatment of the theory of natural dualities, and fills in all details missing here.

Roughly speaking, a finite algebra is said to be dualisable if it is possible to set up a representation for the quasi-variety it generates in a special, natural way. To make this more precise, let $\underline{\mathbf{M}}$ be a finite algebra and define $\mathcal{A}$ to be the quasi-variety $\mathbb{ISP}(\underline{\mathbf{M}})$ generated by $\underline{\mathbf{M}}$. We want to represent each algebra in $\mathcal{A}$ as an algebra of continuous structure-preserving maps. Our construction is split into four main steps.

Step 1: An alter ego for the generator First we will give names to the different types of structure on the set M that are compatible with the algebra $\underline{\mathbf{M}}$:

- an **algebraic operation on $\underline{\mathbf{M}}$** is a homomorphism $g : \underline{\mathbf{M}}^n \to \underline{\mathbf{M}}$, for some $n \in \omega$;
- an **algebraic partial operation on $\underline{\mathbf{M}}$** is a homomorphism $h : \mathbf{D} \to \underline{\mathbf{M}}$, where $\mathbf{D}$ is a subalgebra of $\underline{\mathbf{M}}^n$, for some $n \in \omega$;
- an **algebraic relation on $\underline{\mathbf{M}}$** is a subset r of M^n, for some $n \in \omega\backslash\{0\}$, such that r forms a subalgebra of $\underline{\mathbf{M}}^n$.

We shall say that an operation, partial operation or relation on M is **algebraic over $\underline{\mathbf{M}}$** if it is an algebraic operation, partial operation or relation on $\underline{\mathbf{M}}$, respectively.

We now choose a topological structure $\underset{\sim}{\mathbf{M}}$, with the same underlying set as $\underline{\mathbf{M}}$, of the form

$$\underset{\sim}{\mathbf{M}} = \langle M; G, H, R, \mathcal{T}\rangle,$$

where

- G is a set of algebraic operations on $\underline{\mathbf{M}}$,
- H is a set of algebraic partial operations on $\underline{\mathbf{M}}$,

- R is a set of algebraic relations on $\underline{\mathbf{M}}$, and
- $\mathcal{T}$ is the discrete topology on M.

We call $\underset{\sim}{\mathbf{M}}$ an **alter ego of $\underline{\mathbf{M}}$**. The algebra $\underline{\mathbf{M}}$ and the alter ego $\underset{\sim}{\mathbf{M}}$ will form the basis of our construction.

Note that we distinguish *total* operations from partial operations in the type of $\underset{\sim}{\mathbf{M}}$, even though a total operation is just special sort of partial operation. We do this because structures without partial operations are more familiar and easier to handle. Moreover, some of the fundamental theorems of the theory depend upon there being no proper partial operations in the type of $\underset{\sim}{\mathbf{M}}$.

Step 2: The dual category We shall use $\underset{\sim}{\mathbf{M}}$ to define a class $\mathfrak{X}$ of structures that we hope will mirror the class $\mathcal{A}$ of algebras. Define $\mathfrak{X}$ to be the topological quasi-variety $\mathbb{IS}_c\mathbb{P}^+(\underset{\sim}{\mathbf{M}})$ consisting of all isomorphic copies of (topologically) closed substructures of non-zero powers of $\underset{\sim}{\mathbf{M}}$. Note that, if the type of $\underset{\sim}{\mathbf{M}}$ includes no nullary operations, then the empty structure is a closed substructure of $\underset{\sim}{\mathbf{M}}$, and so belongs to $\mathfrak{X}$.

We shall show explicitly how non-zero powers of $\underset{\sim}{\mathbf{M}}$ are constructed. Let S be a non-empty set. Then the operations, partial operations and relations in the type of $\underset{\sim}{\mathbf{M}}$ are lifted pointwise to $\underset{\sim}{\mathbf{M}}^S$, as follows.

- Let $g : M^n \to M$ be an operation in G, for some $n \in \omega$. Then we define the operation $g^{\underset{\sim}{\mathbf{M}}^S} : (M^S)^n \to M^S$ of $\underset{\sim}{\mathbf{M}}^S$ by

$$g^{\underset{\sim}{\mathbf{M}}^S}(a_0, \dots, a_{n-1})(s) := g(a_0(s), \dots, a_{n-1}(s)),$$

  for all $a_0, \dots, a_{n-1} \in M^S$ and $s \in S$.

- Let $h : \mathrm{dom}(h) \to M$ be a partial operation in H, with its domain, $\mathrm{dom}(h)$, a subset of M^n, for some $n \in \omega$. Then the domain $\mathrm{dom}(h^{\underset{\sim}{\mathbf{M}}^S})$ of the partial operation $h^{\underset{\sim}{\mathbf{M}}^S}$ is defined to be the following subset of $(M^S)^n$:

$$\{\, (a_0, \dots, a_{n-1}) \in (M^S)^n \mid (\forall s \in S)\, (a_0(s), \dots, a_{n-1}(s)) \in \mathrm{dom}(h) \,\}.$$

  The partial operation $h^{\underset{\sim}{\mathbf{M}}^S} : \mathrm{dom}(h^{\underset{\sim}{\mathbf{M}}^S}) \to M^S$ is given by

$$h^{\underset{\sim}{\mathbf{M}}^S}(a_0, \dots, a_{n-1})(s) := h(a_0(s), \dots, a_{n-1}(s)),$$

  for all $(a_0, \dots, a_{n-1}) \in \mathrm{dom}(h^{\underset{\sim}{\mathbf{M}}^S})$ and $s \in S$.

- Now let r be an n-ary relation in R, for some $n \in \omega\backslash\{0\}$. The relation $r^{\underset{\sim}{\mathbf{M}}^S}$ is defined to be the subset

$$\{\, (a_0, \dots, a_{n-1}) \in (M^S)^n \mid (\forall s \in S)\, (a_0(s), \dots, a_{n-1}(s)) \in r \,\}$$

  of $(M^S)^n$.

The usual definition of the product topology is used to lift the topology on $\underset{\sim}{\mathbf{M}}$ up to $\underset{\sim}{\mathbf{M}}^S$, as follows.

- The topology on $\underset{\sim}{\mathbf{M}}^S$ is determined by the clopen subbasis consisting of all sets of the form
$$U_{s,m} := \{\, a \in M^S \mid a(s) = m \,\},$$
for some $s \in S$ and $m \in M$.

If S is a finite set, then the topology on $\underset{\sim}{\mathbf{M}}^S$ will be discrete. However, if S is infinite, then the topology on $\underset{\sim}{\mathbf{M}}^S$ will be compact and Hausdorff. Indeed, all the structures in $\mathcal{X} := \mathbb{IS}_c\mathbb{P}^+(\underset{\sim}{\mathbf{M}})$ have an underlying *boolean* topology—they are compact spaces whose elements are separated by clopen subsets.

A substructure of $\underset{\sim}{\mathbf{M}}^S$ must be closed under each operation in G and also closed under each partial operation in H, where it is defined.

The morphisms between the structures in $\mathcal{X}$ are the continuous maps that preserve the operations, partial operations and relations in the natural sense. In particular, for a map to preserve a partial operation, it must preserve the domain of the partial operation, and preserve the partial operation where it is defined.

Step 3: The contravariant functors Because the operations, partial operations and relations in the type of $\underset{\sim}{\mathbf{M}}$ are algebraic over the algebra $\underline{\mathbf{M}}$, we are able to define a natural pair of contravariant hom-functors

$$\mathrm{D} : \mathcal{A} \to \mathcal{X} \quad \text{and} \quad \mathrm{E} : \mathcal{X} \to \mathcal{A}$$

between the category of algebras $\mathcal{A} := \mathbb{ISP}(\underline{\mathbf{M}})$ and the category of topological structures $\mathcal{X} := \mathbb{IS}_c\mathbb{P}^+(\underset{\sim}{\mathbf{M}})$.

- For each algebra $\mathbf{A} \in \mathcal{A}$, the **dual of A** is the closed substructure $\mathrm{D}(\mathbf{A})$ of $\underset{\sim}{\mathbf{M}}^A$ formed by the set $\mathcal{A}(\mathbf{A}, \underline{\mathbf{M}})$ of all homomorphisms from $\mathbf{A}$ to $\underline{\mathbf{M}}$.
- For each structure $\mathbf{X} \in \mathcal{X}$, the **dual of X** is the subalgebra $\mathrm{E}(\mathbf{X})$ of $\underline{\mathbf{M}}^X$ formed by the set $\mathcal{X}(\mathbf{X}, \underset{\sim}{\mathbf{M}})$ of all morphisms from $\mathbf{X}$ to $\underset{\sim}{\mathbf{M}}$.
- For each homomorphism $\varphi : \mathbf{A} \to \mathbf{B}$, where $\mathbf{A}, \mathbf{B} \in \mathcal{A}$, the morphism $\mathrm{D}(\varphi) : \mathrm{D}(\mathbf{B}) \to \mathrm{D}(\mathbf{A})$ is given by $\mathrm{D}(\varphi)(x) := x \circ \varphi$, for $x \in \mathcal{A}(\mathbf{B}, \underline{\mathbf{M}})$.
- For each morphism $\psi : \mathbf{X} \to \mathbf{Y}$, where $\mathbf{X}, \mathbf{Y} \in \mathcal{X}$, the homomorphism $\mathrm{E}(\psi) : \mathrm{E}(\mathbf{Y}) \to \mathrm{E}(\mathbf{X})$ is given by $\mathrm{E}(\psi)(\alpha) := \alpha \circ \psi$, for $\alpha \in \mathcal{X}(\mathbf{Y}, \underset{\sim}{\mathbf{M}})$.

Step 4: The natural evaluations Consider an algebra $\mathbf{A} \in \mathcal{A}$. Then there is a natural evaluation homomorphism

$$e_{\mathbf{A}} : \mathbf{A} \to \mathrm{ED}(\mathbf{A}),$$

given by $e_{\mathbf{A}}(a)(x) := x(a)$, for all $a \in A$ and all $x \in \mathcal{A}(\mathbf{A}, \underline{\mathbf{M}})$. The map $e_{\mathbf{A}}$ is automatically an embedding, since $\mathbf{A}$ must be separated by homomorphisms

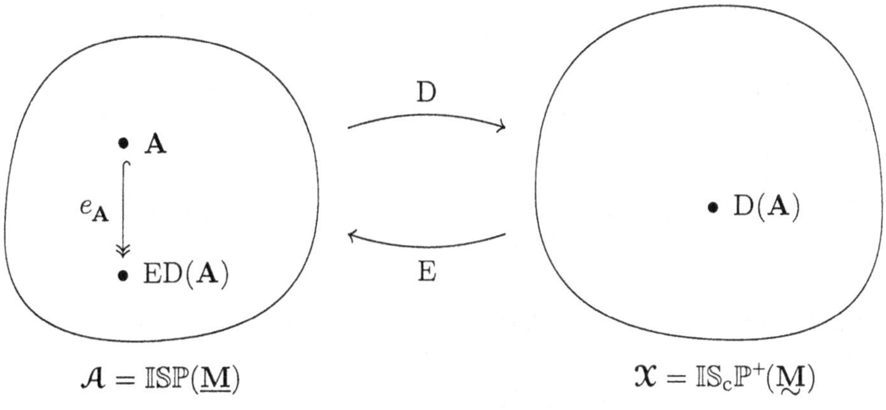

Figure 1.5 A natural duality

into $\underline{\mathbf{M}}$; see the $\mathbb{ISP}$ Theorem, 1.1.1. If the map $e_{\mathbf{A}}$ is an isomorphism, then we say that $\underset{\sim}{\mathbf{M}}$ **yields a duality on A** or, alternatively, that $G \cup H \cup R$ **yields a duality on A**.

The previous step finishes our construction. We now say that the alter ego $\underset{\sim}{\mathbf{M}}$ **yields a duality on $\mathcal{A}$ (based on $\underline{\mathbf{M}}$)** if $\underset{\sim}{\mathbf{M}}$ yields a duality on each algebra in $\mathcal{A}$. In this case, we have a representation for $\mathcal{A}$: each algebra $\mathbf{A}$ in $\mathcal{A}$ is isomorphic to the algebra $\mathrm{ED}(\mathbf{A})$ of all continuous structure-preserving maps from $\mathrm{D}(\mathbf{A})$ to $\underset{\sim}{\mathbf{M}}$. (See Figure 1.5.)

If the structure $\underset{\sim}{\mathbf{M}}$ yields a duality on $\mathcal{A}$ and we want to emphasise the role of the generating algebra $\underline{\mathbf{M}}$, then we can say that $\underset{\sim}{\mathbf{M}}$ is a **dualising structure for $\underline{\mathbf{M}}$** or, more briefly, that **$\underset{\sim}{\mathbf{M}}$ dualises $\underline{\mathbf{M}}$**. The algebra $\underline{\mathbf{M}}$ is called **dualisable** if it is dualised by some alter ego $\underset{\sim}{\mathbf{M}}$.

A duality that arises from the construction described above is referred to as a **natural duality**. Particularly from the standpoint of an algebraist, natural dualities are indeed natural:

- the structure on the alter ego $\underset{\sim}{\mathbf{M}}$—the operations, partial operations and relations—is compatible with the original algebra $\underline{\mathbf{M}}$;
- the duality between the quasi-variety $\mathcal{A}$ and the topological quasi-variety $\mathcal{X}$ is given by the naturally defined hom-functors D and E;
- products in the dual category $\mathcal{X}$ are just the natural cartesian products with the structure extended pointwise;
- for each non-empty set S, the dual of the S-generated free algebra in $\mathcal{A}$ is the power $\underset{\sim}{\mathbf{M}}^S$ of the alter ego $\underset{\sim}{\mathbf{M}}$.

1.3.1 Examples There are many well-known natural dualities. Below we list four of the most familiar, and refer to Clark and Davey's text [8] for a host of other examples.

- M. H. Stone's [63] duality for Boolean algebras can be built from the two-element Boolean algebra $\underline{\mathbf{B}} = \langle \{0,1\}; \vee, \wedge, ', 0, 1\rangle$ and the discretely topologised set $\underset{\sim}{\mathbf{B}} = \langle \{0,1\}; \mathcal{T}\rangle$.
- H. A. Priestley's [56, 57] duality for bounded distributive lattices is determined by the two-element lattice $\underline{\mathbf{D}} = \langle \{0,1\}; \vee, \wedge, 0, 1\rangle$ and the ordered discrete space $\underset{\sim}{\mathbf{D}} = \langle \{0,1\}; \leqslant, \mathcal{T}\rangle$, where $0 \leqslant 1$.
- The Hofmann–Mislove–Stralka [37] duality for semilattices is determined by the two-element semilattice $\underline{\mathbf{S}} = \langle \{0,1\}; \wedge, 1\rangle$ and the discrete topological semilattice $\underset{\sim}{\mathbf{S}} = \langle \{0,1\}; \wedge, 1, \mathcal{T}\rangle$.
- Davey and Werner [29] showed that L. S. Pontryagin's [55] duality restricted to abelian groups of exponent m, for any finite $m > 0$, can be obtained from the cyclic group $\underline{\mathbf{Z}}_m = \langle \mathbb{Z}_m; +, ^-, 0\rangle$ and the discrete topological cyclic group $\underset{\sim}{\mathbf{Z}}_m = \langle \mathbb{Z}_m; +, ^-, 0, \mathcal{T}\rangle$.

Throughout this text, we shall meet many examples of dualisable unary algebras. For instance, every finite unary algebra with just one fundamental operation is dualisable, 3.5.1.

1.3.2 Examples There are many seemingly simple algebras that are not dualisable. The **two-element implication algebra** $\underline{\mathbf{I}} = \langle \{0,1\}; \rightarrow\rangle$ has one binary operation, given by the table below.

$\rightarrow$	0	1
0	1	1
1	0	1

The algebra $\underline{\mathbf{I}}$ was the first known example of a non-dualisable algebra (Davey and Werner [29]). We shall see that $\underline{\mathbf{I}}$ satisfies a condition even stronger than non-dualisability, in Lemma 7.1.4.

The first family of non-dualisable unary algebras was discovered by L. Heindorf [35]: a finite unary algebra $\langle M; F\rangle$, with $\{0,1,2\} \subseteq M$, must be non-dualisable if F contains every operation in $\{0,1\}^M$. We will generalise this result later, in Theorem 3.4.5.

1.3.3 Remark Natural duality theory can be extended to encompass quasi-varieties generated by infinite algebras that have a compatible compact topology. The best known example is certainly the circle group $\underline{\mathbf{T}} = \langle T; \cdot, ^{-1}, 1\rangle$, where $T := \{\, z \in \mathbb{C} \mid |z| = 1 \,\}$. The class $\mathbb{ISP}(\underline{\mathbf{T}})$ is the variety of abelian groups.

Now choose $\underset{\sim}{\mathbf{T}} = \langle T; \cdot, ^{-1}, 1, \mathcal{T}\rangle$, where $\mathcal{T}$ is the relative topology from $\mathbb{C}$. As shown by Davey and Werner [29], an extended version of natural duality theory can be used to establish Pontryagin's duality between the variety of abelian groups and the class $\mathbb{IS}_c\mathbb{P}^+(\underset{\sim}{\mathbf{T}})$ of compact topological abelian groups [55]. The paucity of other infinitely based examples, and the richness and scope of the finitely based theory, has led researchers to concentrate on the latter. (The general Pontryagin duality for *locally* compact groups seems to be a beautiful but isolated example, and is outside the theory of natural dualities. Indeed, the locally compact case is definitely a theorem of *topological* algebra; whereas the finitely based dualities, even though they involve topology, can be obtained via methods that are purely algebraic.)

Duality theory has many applications within algebra. For examples, see the text by Clark and Davey [8]. Throughout this text, our main aim is not to use duality theory to study algebras, but to use algebras to study duality theory.

1.4 A dualisability toolkit

Having introduced the idea behind dualisability, we now present a tightly focused set of results that will be our basic tools of trade. The rest of the text will be built using just the few tools described in this and the next section.

While dualisability is defined as a property of an algebra, it is really more a property of a quasi-variety. The following theorem, proved independently by M. J. Saramago [61, 2.5] and Davey and Willard [30], shows that different finite algebras that generate the same quasi-variety must share dualisability or non-dualisability.

1.4.1 Independence Theorem [61, 30] *Let* $\underline{\mathbf{M}}$ *and* $\underline{\mathbf{N}}$ *be finite algebras such that* $\mathbb{ISP}(\underline{\mathbf{M}}) = \mathbb{ISP}(\underline{\mathbf{N}})$. *If* $\underline{\mathbf{M}}$ *is dualisable, then* $\underline{\mathbf{N}}$ *is dualisable.*

There are several fundamental theorems that provide conditions under which a given alter ego dualises a finite algebra. For example, both the Second Duality Theorem [8, 2.2.7] and the Two-for-One Strong Duality Theorem [8, 3.3.2] give purely finitary sufficient conditions for dualisability. These two theorems are often used to establish natural dualities.

Throughout this text, we will establish dualities using the following deep result, proved independently by L. Zádori [66, 3.5] and R. Willard [14, 4.3]. Fix a finite algebra $\underline{\mathbf{M}}$ and an alter ego $\underset{\sim}{\mathbf{M}} = \langle M; G, H, R, \mathcal{T}\rangle$. We say that $\underset{\sim}{\mathbf{M}}$ is of **finite type** if the set $G \cup H \cup R$ is finite.

1.4.2 Duality Compactness Theorem [66, 14, 8] *Let $\underline{\mathbf{M}}$ be a finite algebra and let $\underset{\sim}{\mathbf{M}}$ be an alter ego of $\underline{\mathbf{M}}$ that is of finite type. If $\underset{\sim}{\mathbf{M}}$ yields a duality on each finite algebra in $\mathbb{ISP}(\underline{\mathbf{M}})$, then $\underset{\sim}{\mathbf{M}}$ yields a duality on $\mathbb{ISP}(\underline{\mathbf{M}})$.*

The Duality Compactness Theorem allows us to give purely combinatorial proofs of dualisability. To show that $\underset{\sim}{\mathbf{M}}$ yields a duality on an algebra $\mathbf{A}$ from $\mathcal{A} := \mathbb{ISP}(\underline{\mathbf{M}})$, we need to show that $e_{\mathbf{A}} : \mathbf{A} \hookrightarrow \mathrm{ED}(\mathbf{A})$ is surjective. So we need to prove that every morphism $\alpha : \mathrm{D}(\mathbf{A}) \to \underset{\sim}{\mathbf{M}}$ is of the form $e_{\mathbf{A}}(a)$, for some $a \in A$. For each $a \in A$, the map $e_{\mathbf{A}}(a) : \mathcal{A}(\mathbf{A}, \underline{\mathbf{M}}) \to M$ is called an **evaluation**. If the algebra $\mathbf{A}$ is finite, then the topology on the dual $\mathrm{D}(\mathbf{A})$ is discrete. Thus $\underset{\sim}{\mathbf{M}}$ yields a duality on a finite algebra $\mathbf{A}$ if and only if every map $\alpha : \mathcal{A}(\mathbf{A}, \underline{\mathbf{M}}) \to M$ that preserves (the structure in) $G \cup H \cup R$ is an evaluation.

1.4.3 Much Ado About Nothing Let $\underline{\mathbf{M}}$ be a finite algebra and let $\mathbf{A}$ belong to the quasi-variety $\mathcal{A} := \mathbb{ISP}(\underline{\mathbf{M}})$. The algebra $\mathbf{A}$ is separated by homomorphisms into $\underline{\mathbf{M}}$, by the $\mathbb{ISP}$ Theorem, 1.1.1. So $\mathcal{A}(\mathbf{A}, \underline{\mathbf{M}})$ must be non-empty, unless $\mathbf{A}$ is a one-element algebra and $\underline{\mathbf{M}}$ has no one-element subalgebras. But, in this case, every map $\alpha : \mathcal{A}(\mathbf{A}, \underline{\mathbf{M}}) \to M$ is vacuously an evaluation. Henceforth, we will skip the case where $\mathcal{A}(\mathbf{A}, \underline{\mathbf{M}}) = \varnothing$.

Since we are not attempting to produce useful dualities, we will be happy to include a lot of structure in the type of an alter ego. Indeed, we shall often be including all the algebraic relations of a particular arity. Let $n \in \omega \backslash \{0\}$ and define R_n to be the set of all n-ary algebraic relations on $\underline{\mathbf{M}}$. (When we need to be explicit, we shall write $R_n(\underline{\mathbf{M}})$ instead of R_n.) The following helpful lemma is implicit in the text by Clark and Davey [8, Chapter 10].

1.4.4 Preservation Lemma *Let $\underline{\mathbf{M}}$ be a finite algebra and let $\mathcal{A} := \mathbb{ISP}(\underline{\mathbf{M}})$. Let $n \in \omega \backslash \{0\}$, let $\mathbf{A} \in \mathcal{A}$ and let $\alpha : \mathcal{A}(\mathbf{A}, \underline{\mathbf{M}}) \to M$. Then α preserves R_n if and only if α agrees with an evaluation on each subset of $\mathcal{A}(\mathbf{A}, \underline{\mathbf{M}})$ with at most n elements.*

Proof Assume that α preserves R_n and let $x_0, \dots, x_{n-1} \in \mathcal{A}(\mathbf{A}, \underline{\mathbf{M}})$. Then the n-ary relation $r := \{\, (x_0(a), \dots, x_{n-1}(a)) \mid a \in A \,\}$ is algebraic over $\underline{\mathbf{M}}$. Therefore α preserves r. As $(x_0, \dots, x_{n-1}) \in r$ in $\mathcal{A}(\mathbf{A}, \underline{\mathbf{M}})$, this implies that $(\alpha(x_0), \dots, \alpha(x_{n-1})) \in r$. Thus there is some $a \in A$ with $\alpha(x_i) = x_i(a)$, for all $i \in \{0, \dots, n-1\}$. So α agrees with $e_{\mathbf{A}}(a)$ on the set $\{x_0, \dots, x_{n-1}\}$.

Now assume that α agrees with an evaluation on each subset of $\mathcal{A}(\mathbf{A}, \underline{\mathbf{M}})$ with at most n elements. Let r be a relation in R_n. To see that α preserves r, choose some $x_0, \dots, x_{n-1} \in \mathcal{A}(\mathbf{A}, \underline{\mathbf{M}})$ with $(x_0, \dots, x_{n-1}) \in r$. There is some $a \in A$ such that α agrees with $e_{\mathbf{A}}(a)$ on the subset $\{x_0, \ldots, x_{n-1}\}$ of $\mathcal{A}(\mathbf{A}, \underline{\mathbf{M}})$. So $(\alpha(x_0), \ldots, \alpha(x_{n-1})) = (x_0(a), \ldots, x_{n-1}(a)) \in r$. Thus α preserves R_n. ∎

Even though we use complicated alter egos to create dualities, the alter egos we use are at least of finite type. An algebra is **finitely dualisable** if it is dualised by an alter ego of finite type. At present, it is not known if there is a dualisable algebra that is not finitely dualisable. This is the *Finite Type Problem* [8].

To obtain our general results on dualisability, we will use the most 'powerful' type of alter ego. Define R_ω to be the set of all finitary algebraic relations on $\underline{\mathbf{M}}$. Then the algebra $\underline{\mathbf{M}}$ is dualisable if and only if $\underset{\sim}{\mathbf{M}}_\omega = \langle M; R_\omega, \mathcal{T} \rangle$ yields a duality on $\mathcal{A}$. (To see this, note that a map $\alpha : \mathcal{A}(\mathbf{A}, \underline{\mathbf{M}}) \to M$ preserves an algebraic partial operation h on $\underline{\mathbf{M}}$ provided α preserves the algebraic relation graph(h).) We call $\underset{\sim}{\mathbf{M}}_\omega$ the **brute-force alter ego of** $\underline{\mathbf{M}}$. For each $\mathbf{A} \in \mathcal{A}$, we say that $\alpha : \mathcal{A}(\mathbf{A}, \underline{\mathbf{M}}) \to M$ is a **brute-force morphism** if $\alpha : \mathrm{D}(\mathbf{A}) \to \underset{\sim}{\mathbf{M}}_\omega$ is a morphism.

The next lemma gives a simple characterisation of brute-force morphisms. Before stating the lemma, we present some more definitions. Consider an algebra $\mathbf{A} \in \mathcal{A}$ and a map $\alpha : \mathcal{A}(\mathbf{A}, \underline{\mathbf{M}}) \to M$. For each $a \in A$, we say that α **is given by evaluation at** a if $\alpha = e_{\mathbf{A}}(a)$. For $Y \subseteq \mathcal{A}(\mathbf{A}, \underline{\mathbf{M}})$ and $a \in A$, we say that α **is given by evaluation at** a **on** Y if $\alpha{\restriction}_Y = e_{\mathbf{A}}(a){\restriction}_Y$. The map α is said to be **locally an evaluation** if it agrees with an evaluation on every finite subset of $\mathcal{A}(\mathbf{A}, \underline{\mathbf{M}})$. Finally, a subset S of A is called a **support** for the map $\alpha : \mathcal{A}(\mathbf{A}, \underline{\mathbf{M}}) \to M$ if, for all $x, y \in \mathcal{A}(\mathbf{A}, \underline{\mathbf{M}})$ with $x{\restriction}_S = y{\restriction}_S$, we have $\alpha(x) = \alpha(y)$.

The following result can be proved using the Preservation Lemma, 1.4.4, and a straightforward topological argument [8, B.6].

1.4.5 Brute Force Lemma *Let* $\underline{\mathbf{M}}$ *be a finite algebra. Define* $\mathcal{A} := \mathbb{ISP}(\underline{\mathbf{M}})$ *and let* $\mathbf{A} \in \mathcal{A}$. *Then a map* $\alpha : \mathcal{A}(\mathbf{A}, \underline{\mathbf{M}}) \to M$ *is a brute-force morphism if and only if* α *has a finite support and is locally an evaluation.*

The easiest way to prove that an algebra is non-dualisable is to use the ghost-element method. This method, which was first used implicity by Davey and Werner [29], has been applied extensively [8, 23, 14, 59]. The following theorem provides a basic description of the ghost-element method.

Let $\mathbf{A}$ be a subalgebra of $\underline{\mathbf{M}}^S$, for some set S. For each $s \in S$, define $\rho_s : \mathbf{A} \to \underline{\mathbf{M}}$ to be the natural projection homomorphism given by $\rho_s := \pi_s{\restriction}_A$. Now consider a map $\alpha : \mathcal{A}(\mathbf{A}, \underline{\mathbf{M}}) \to M$, and define the element g_α of M^S by $g_\alpha(s) := \alpha(\rho_s)$. It is easy to check that, if α is an evaluation, then α must be given by evaluation at g_α. So, if $g_\alpha \notin A$, then α is not an evaluation. In the case that α is a brute-force morphism and $g_\alpha \notin A$, we say that g_α is a **ghost element of A**.

1.4.6 Ghost Element Theorem [8, 10.5.1] *Let* $\underline{\mathbf{M}}$ *be a finite algebra and define* $\mathcal{A} := \mathbb{ISP}(\underline{\mathbf{M}})$. *Assume that there is a subalgebra* $\mathbf{A}$ *of* $\underline{\mathbf{M}}^S$, *for some set* S, *and a brute-force morphism* $\alpha : \mathcal{A}(\mathbf{A}, \underline{\mathbf{M}}) \to M$ *such that* $g_\alpha \notin A$. *Then* $\underline{\mathbf{M}}$ *is not dualisable.*

1.5 Full and strong dualisability

The definition of a duality is (intentionally) biased towards algebras. The more symmetric notion is that of a full duality. Most of the time, we will simply be studying dualities. But we shall consider full dualities in Chapters 4 and 7.

Again, let $\underline{\mathbf{M}}$ be a finite algebra and define $\mathcal{A} := \mathbb{ISP}(\underline{\mathbf{M}})$. Choose an alter ego $\underset{\sim}{\mathbf{M}}$ of $\underline{\mathbf{M}}$ and define $\mathcal{X} := \mathbb{IS}_c\mathbb{P}^+(\underset{\sim}{\mathbf{M}})$. We have seen that, for each $\mathbf{A} \in \mathcal{A}$, there is a natural embedding $e_\mathbf{A} : \mathbf{A} \hookrightarrow \mathrm{ED}(\mathbf{A})$ given by $e_\mathbf{A}(a)(x) := x(a)$, for all $a \in A$ and $x \in \mathcal{A}(\mathbf{A}, \underline{\mathbf{M}})$. Similarly, for each $\mathbf{X} \in \mathcal{X}$, there is a natural embedding

$$\varepsilon_\mathbf{X} : \mathbf{X} \hookrightarrow \mathrm{DE}(\mathbf{X}),$$

given by evaluation: $\varepsilon_\mathbf{X}(x)(\alpha) := \alpha(x)$, for all $x \in X$ and $\alpha \in \mathcal{X}(\mathbf{X}, \underset{\sim}{\mathbf{M}})$. We say that $\underset{\sim}{\mathbf{M}}$ **yields a full duality on** $\mathcal{A}$ **(based on** $\underline{\mathbf{M}}$**)** if the maps $e_\mathbf{A}$ and $\varepsilon_\mathbf{X}$ are isomorphisms, for all $\mathbf{A} \in \mathcal{A}$ and $\mathbf{X} \in \mathcal{X}$. In this case, the categories $\mathcal{A}$ and $\mathcal{X}$ are dually equivalent. The algebra $\underline{\mathbf{M}}$ is called **fully dualisable** if there is an alter ego $\underset{\sim}{\mathbf{M}}$ of $\underline{\mathbf{M}}$ that yields a full duality on $\mathcal{A}$.

As we shall see in Chapter 4, full dualisability can be rather complicated. There is a simpler, stronger notion that is often used instead. Before we introduce this notion, we need to give a few definitions.

For a pair of maps $x, y : A \to B$, we use $\mathrm{eq}(x, y)$ to denote the **equaliser of** x **and** y, which is given by

$$\mathrm{eq}(x, y) := \{\, a \in A \mid x(a) = y(a) \,\}.$$

Now let S be a non-empty set and let $F_{\underline{\mathbf{M}}}(S)$ denote the set of all S-ary term functions of $\underline{\mathbf{M}}$. A subset X of M^S is **term closed** if, for each $y \in M^S \backslash X$, there are term functions $t_1, t_2 \in F_{\underline{\mathbf{M}}}(S)$ that agree on X but differ at y. It follows that X is a term-closed subset of M^S if and only if

$$X = \bigcap \{\, \mathrm{eq}(t_1, t_2) \mid t_1, t_2 \in F_{\underline{\mathbf{M}}}(S) \text{ and } t_1{\restriction}_X = t_2{\restriction}_X \,\}.$$

This definition is related to full dualities by the following theorem.

1.5.1 Full Duality Theorem [6, 8] *Let $\underline{\mathbf{M}}$ be a finite algebra and let $\underset{\sim}{\mathbf{M}}$ be an alter ego of $\underline{\mathbf{M}}$. Then $\underset{\sim}{\mathbf{M}}$ yields a full duality on $\mathbb{ISP}(\underline{\mathbf{M}})$ if and only if*

(i) *$\underset{\sim}{\mathbf{M}}$ yields a duality on $\mathbb{ISP}(\underline{\mathbf{M}})$, and*

(ii) *each closed substructure of each non-zero power of $\underset{\sim}{\mathbf{M}}$ is isomorphic to a term-closed substructure of a non-zero power of $\underset{\sim}{\mathbf{M}}$.*

We now say that $\underset{\sim}{\mathbf{M}}$ **yields a strong duality on $\mathcal{A}$ (based on $\underline{\mathbf{M}}$)** if $\underset{\sim}{\mathbf{M}}$ yields a duality on $\mathcal{A}$ and each closed substructure of each non-zero power of $\underset{\sim}{\mathbf{M}}$ *is* term closed. So each strong duality is also a full duality. At present, it is not known whether every full duality is also strong. This is called the *Full versus Strong Problem* [8]. If there is an alter ego of $\underline{\mathbf{M}}$ that yields a strong duality on $\mathcal{A}$, then the algebra $\underline{\mathbf{M}}$ is called **strongly dualisable**. J. Hyndman [39] has shown that, as with dualisability, the property of strong dualisability really applies to a quasi-variety rather than to the algebra chosen to generate it. It is not known if the corresponding result is true for full dualisability.

1.5.2 Strong Independence Theorem [39] *Let $\underline{\mathbf{M}}$ and $\underline{\mathbf{N}}$ be finite algebras such that $\mathbb{ISP}(\underline{\mathbf{M}}) = \mathbb{ISP}(\underline{\mathbf{N}})$. If $\underline{\mathbf{M}}$ is strongly dualisable, then $\underline{\mathbf{N}}$ is strongly dualisable.*

The structure $\underset{\sim}{\mathbf{M}}$ is **injective** in $\mathcal{X}$ if, for each non-empty set S and each closed substructure $\mathbf{X}$ of $\underset{\sim}{\mathbf{M}}^S$, every morphism $\alpha : \mathbf{X} \to \underset{\sim}{\mathbf{M}}$ extends to a morphism $\beta : \underset{\sim}{\mathbf{M}}^S \to \underset{\sim}{\mathbf{M}}$. By the First Strong Duality Theorem [6, 8], the alter ego $\underset{\sim}{\mathbf{M}}$ yields a strong duality on $\mathcal{A}$ if and only if $\underset{\sim}{\mathbf{M}}$ yields a full duality on $\mathcal{A}$ and $\underset{\sim}{\mathbf{M}}$ is injective in $\mathcal{X}$.

There are close connections between the injectivity of $\underset{\sim}{\mathbf{M}}$ in $\mathcal{X}$ and the injectivity of $\underline{\mathbf{M}}$ in $\mathcal{A}$ [8, 3.2.10]. The strong dualisability of an algebra seems to be related to how close the algebra is to being injective. We shall pursue this further in the appendix where we show, amongst many other things, that every dualisable algebra that is injective in the quasi-variety it generates is also strongly dualisable.

In Chapters 4 and 7, we shall use two general methods for showing that a dualisable algebra is strongly dualisable. The first method is due to Clark, Idziak, Sabourin, Szabó and Willard.

1.5.3 Theorem [14, 4.8] *Let $\underline{\mathbf{M}}$ be a finite algebra and let $\underset{\sim}{\mathbf{M}}$ be an alter ego of $\underline{\mathbf{M}}$. Then $\underset{\sim}{\mathbf{M}}$ yields a strong duality on $\mathbb{ISP}(\underline{\mathbf{M}})$ if and only if*

(i) *$\underset{\sim}{\mathbf{M}}$ yields a duality on $\mathbb{ISP}(\underline{\mathbf{M}})$, and*

(ii) *for every algebra $\mathbf{A}$ in $\mathbb{ISP}(\underline{\mathbf{M}})$ and every proper closed substructure $\mathbf{X}$ of $\mathrm{D}(\mathbf{A})$, the maps in X do not separate the elements of A.*

The second method was introduced by Lampe, McNulty and Willard [45]. It is based on work of R. Willard [65]. For each subset Y of $\mathcal{A}(\underline{\mathbf{M}}^n, \underline{\mathbf{M}})$, where $n \in \omega$, define the natural product homomorphism

$$\sqcap Y : \underline{\mathbf{M}}^n \to \underline{\mathbf{M}}^Y \quad \text{by} \quad \sqcap Y(a)(y) := y(a).$$

We say that $\underline{\mathbf{M}}$ has **enough algebraic operations** if there is a map $f : \omega \to \omega$ for which the following condition holds:

> for all $n \in \omega\backslash\{0\}$, all algebras $\mathbf{B} \leqslant \mathbf{A} \leqslant \underline{\mathbf{M}}^n$ and all homomorphisms $h : \mathbf{A} \to \underline{\mathbf{M}}$, there exists a subset Y of $\mathcal{A}(\underline{\mathbf{M}}^n, \underline{\mathbf{M}})$, with $|Y| \leqslant f(|B|)$, and a homomorphism $h' : \sqcap Y(\mathbf{A}) \to \underline{\mathbf{M}}$ such that $h' \circ \sqcap Y{\restriction}_B = h{\restriction}_B$.

$$\begin{array}{ccccc}
\mathbf{B} & \hookrightarrow & \mathbf{A} & \hookrightarrow & \underline{\mathbf{M}}^n \\
{\scriptstyle h\restriction_B}\downarrow & & \downarrow\!\!\!\downarrow{\scriptstyle \sqcap Y\restriction_A} & & \\
\underline{\mathbf{M}} & \underset{h'}{\leftarrow\,-\,-\,-} & \sqcap Y(\mathbf{A}) & &
\end{array}$$

Although the definition of enough algebraic operations appears technical, it often provides a relatively easy way to lift dualisability up to strong dualisability.

1.5.4 EAO Theorem [45, 4.3] *Let $\underline{\mathbf{M}}$ be a finite algebra. If $\underline{\mathbf{M}}$ is dualisable and has enough algebraic operations, then $\underline{\mathbf{M}}$ is strongly dualisable.*

We shall prove Theorems 1.5.3 and 1.5.4 in the appendix, where we present a new necessary and sufficient condition for a finite algebra to be strongly dualisable. This provides us with an alternative approach to R. Willard's [65] important but complex concept of the rank of a finite algebra.

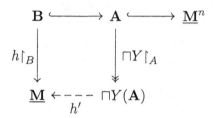

2

Binary homomorphisms and natural dualities

Certain binary homomorphisms of a finite algebra can guarantee its dualisability. Here, we study binary homomorphisms that are lattice, flat-semilattice or group operations. We develop some general tools that we use to prove the dualisability of a large number of unary algebras. For example, we show that any set of endomorphisms of a finite lattice forms the operations of a dualisable unary algebra.

One major focus of natural duality theory has been to produce general dualisability theorems—theorems that can be used to help find a duality based on a given finite algebra. One such theorem, the NU Duality Theorem [8, 2.3.4], stands out as being not only very powerful but also very easy to apply. The NU Duality Theorem was proved, by Davey and Werner [29], while natural duality theory was first being developed. It states that a finite algebra is dualisable if it has a near-unanimity term, and has been used to produce a multitude of useful dualities. For example, every finite lattice-based algebra has a ternary near-unanimity term and is therefore dualisable.

There are many other powerful general theorems within duality theory. However, most of these theorems require the user to do a bit of work. One of the virtues of the NU Duality Theorem is that the theorem supplies its own dualising alter ego. Most other general results in duality theory require users first to guess a potential dualising alter ego for their algebra. For example, the Second Duality Theorem, the Two-for-One Strong Duality Theorem and the Duality Compactness Theorem are all of this type [8]. Using these theorems, and others like them, can involve a fair amount of ingenuity.

The challenge inherent in proving dualisability is illustrated by the relative scarcity of known examples of dualisable algebras that do not have a near-unanimity term. It has long been known that the two-element semilattice is dualisable [37], as is every finite cyclic group [55, 29] and every finite one-dimensional vector space [29]. Applying results of the second author [17], it follows that all finite semilattices, abelian groups and vector spaces are dualisable. The dualisable commutative rings with identity have been characterised by Clark, Idziak, Sabourin, Szabó and Willard [14]. Major progress has been made on the Dualisability Problem for non-abelian groups [27, 59, 60], though it remains unsolved. Beyond these results, knowledge of dualisability remains somewhat scattered, and there is still much to be done.

In this chapter, we develop some results that can be used to establish dualisability quickly and easily. Given a finite algebra $\underline{\mathbf{M}}$, we shall be interested in the homomorphisms from $\underline{\mathbf{M}}^2$ to $\underline{\mathbf{M}}$, which we will be calling the **binary homomorphisms of $\underline{\mathbf{M}}$**. We will find conditions on the binary homomorphisms of $\underline{\mathbf{M}}$ that guarantee that $\underline{\mathbf{M}}$ is dualisable. Our results will be particularly suited to proving that various unary algebras are dualisable. For example, we shall prove that the endomorphisms of a finite cyclic group are the operations of a dualisable unary algebra.

We begin this chapter with a result that exemplifies our approach: a finite algebra is dualisable provided it has a pair of binary homomorphisms that are lattice operations. It follows from this result that every finite unary algebra can be embedded into a dualisable unary algebra.

In this chapter, we use well-behaved binary homomorphisms to show that various algebras are dualisable. Some of the limits of this approach will be exposed in the next chapter. There, we shall find a three-element unary algebra that is non-dualisable even though it has both a binary homomorphism that is a semilattice operation and a ternary homomorphism that is a majority operation. Within this area of investigation, there are many natural questions that are currently unanswered. For instance: 'Is there a non-dualisable algebra with a binary homomorphism that is a group operation?'

This chapter is based on a paper written by the authors and D. M. Clark [12], which contains a number of results and examples not included here.

2.1 Lattice operations

Consider a finite algebra $\underline{\mathbf{M}}$, and assume there is a set G of binary homomorphisms of $\underline{\mathbf{M}}$ such that the algebra $\mathbf{M}_0 := \langle M; G \rangle$ is very well behaved. Every binary homomorphism in G is an algebraic operation on $\underline{\mathbf{M}}$. So we may regard $\mathbf{M}_0$ as the zeroth approximation to a dualising structure for $\underline{\mathbf{M}}$. Our aim will be to try to find a finite family R of algebraic relations on $\underline{\mathbf{M}}$ such that the structure $\underset{\sim}{\mathbf{M}} = \langle M; G, R, \mathcal{T} \rangle$ yields a duality on $\mathbb{ISP}(\underline{\mathbf{M}})$.

In this section, we give a beautiful illustration of this approach. We show that a finite algebra must be dualisable if it has a pair of binary homomorphisms that are lattice operations.

2.1.1 Theorem *Let $\underline{\mathbf{M}}$ be a finite algebra that has binary homomorphisms $\vee$ and $\wedge$ such that $\langle M; \vee, \wedge \rangle$ is a lattice. Then $\underset{\sim}{\mathbf{M}} := \langle M; \vee, \wedge, R_{2|M|}, \mathcal{T} \rangle$ yields a duality on $\mathbb{ISP}(\underline{\mathbf{M}})$.*

Proof We shall use the Duality Compactness Theorem, 1.4.2. Let $\mathbf{A}$ be a finite algebra in $\mathcal{A} := \mathbb{ISP}(\underline{\mathbf{M}})$ and let $\alpha : \mathrm{D}(\mathbf{A}) \to \underset{\sim}{\mathbf{M}}$ be a morphism. We want to show that α is an evaluation. Let $\{m_1, \ldots, m_k\}$ be the image of α, where $k \leqslant |M|$. The dual $\mathrm{D}(\mathbf{A})$ is a substructure of $\underset{\sim}{\mathbf{M}}^A$. Therefore $\mathrm{D}(\mathbf{A})$ is finite and has a lattice reduct. So, for each $i \in \{1, \ldots, k\}$, we can define

$$x_i := \bigwedge \alpha^{-1}(m_i) \quad \text{and} \quad y_i := \bigvee \alpha^{-1}(m_i)$$

in $\mathrm{D}(\mathbf{A})$. By the Preservation Lemma, 1.4.4, we know that α is given by evaluation at some $a \in A$ on the set $\{x_1, y_1, \ldots, x_k, y_k\}$.

Now consider any $z \in \mathcal{A}(\mathbf{A}, \underline{\mathbf{M}})$. There exists some $j \in \{1, \ldots, k\}$ with $\alpha(z) = m_j$. We have $x_j \leqslant z \leqslant y_j$ in $\mathrm{D}(\mathbf{A})$. As α preserves $\vee$ and $\wedge$, we also have $\alpha(x_j) = m_j = \alpha(y_j)$. This gives us

$$m_j = \alpha(x_j) = x_j(a) \leqslant z(a) \leqslant y_j(a) = \alpha(y_j) = m_j$$

in $\underset{\sim}{\mathbf{M}}$. So $\alpha(z) = m_j = z(a)$, and therefore α is an evaluation. Thus $\underset{\sim}{\mathbf{M}}$ yields a duality on $\mathbb{ISP}(\underline{\mathbf{M}})$, by the Duality Compactness Theorem. ∎

The next result follows immediately from the previous theorem. Note that a pair of lattice operations $\vee, \wedge : M^2 \to M$ are binary homomorphisms of a unary algebra $\underline{\mathbf{M}} = \langle M; F \rangle$ if and only if F is a set of endomorphisms of the lattice $\mathbf{M}_0 = \langle M; \vee, \wedge \rangle$.

We use $\mathrm{End}(\mathbf{A})$ to denote the set of all endomorphisms of an algebra $\mathbf{A}$.

2.1.2 Lattice Endomorphism Theorem *Lattice endomorphisms yield dualisable unary algebras. More precisely, if $\underline{\mathbf{M}} = \langle M; F \rangle$ is a finite unary algebra with $F \subseteq \mathrm{End}(\mathbf{M}_0)$, for some lattice $\mathbf{M}_0 = \langle M; \vee, \wedge \rangle$, then the structure $\underset{\sim}{\mathbf{M}} := \langle M; \vee, \wedge, R_{2|M|}, \mathcal{T} \rangle$ yields a duality on $\mathbb{ISP}(\underline{\mathbf{M}})$.*

By using the Second Duality Theorem [8, 2.2.7], rather than the Duality Compactness Theorem, it is possible to prove that the alter ego of the previous theorem is injective in the topological quasi-variety that it generates [12].

The previous theorem is quite powerful when applied to three-element unary algebras. It turns out that there are exactly 699 unary clones on the set $\{0, 1, 2\}$. Of them, precisely 221 consist of order-preserving maps for some total order on $\{0, 1, 2\}$ and therefore determine dualisable unary algebras. In the next chapter, we shall completely characterise the dualisable three-element unary algebras.

The Lattice Endomorphism Theorem above can also be used to solve the Inherent Non-dualisability Problem. We shall say that a finite algebra $\underline{\mathbf{M}}$ is **inherently non-dualisable** if every finite algebra that has $\underline{\mathbf{M}}$ as a subalgebra is non-dualisable. While this is not the original definition of inherent non-dualisability [23], the following lemma shows that the two definitions are equivalent.

2.1.3 Lemma *Let $\underline{\mathbf{M}}$ be a finite algebra. Then $\underline{\mathbf{M}}$ is inherently non-dualisable if and only if every finite algebra $\underline{\mathbf{N}}$ with $\underline{\mathbf{M}} \in \mathbb{ISP}(\underline{\mathbf{N}})$ is non-dualisable.*

Proof For the 'only if' direction, assume that $\underline{\mathbf{M}}$ is inherently non-dualisable and let $\underline{\mathbf{N}}$ be a finite algebra such that $\underline{\mathbf{M}} \in \mathbb{ISP}(\underline{\mathbf{N}})$. We need to show that $\underline{\mathbf{N}}$ is non-dualisable. There is an embedding $\varphi : \underline{\mathbf{M}} \hookrightarrow \underline{\mathbf{N}}^k$, for some $k \in \omega \backslash \{0\}$. The algebra $\underline{\mathbf{N}}^k$ must be non-dualisable, as the algebra $\varphi(\underline{\mathbf{M}})$ is inherently non-dualisable. But $\mathbb{ISP}(\underline{\mathbf{N}}^k) = \mathbb{ISP}(\underline{\mathbf{N}})$, and therefore $\underline{\mathbf{N}}$ is non-dualisable, by the Independence Theorem, 1.4.1. The 'if' direction is trivial. ∎

There are many inherently non-dualisable algebras. For example, all non-dualisable two-element algebras and all non-dualisable graph algebras are inherently non-dualisable [23]. The *Inherent Non-dualisability Problem* [8] asks: 'Is there is a finite algebra that is non-dualisable but not inherently non-dualisable?' A simple application of the Lattice Endomorphism Theorem tells us that the answer is 'Yes'. In fact, we can show that there are no inherently non-dualisable unary algebras at all.

2.1.4 Theorem *There are no inherently non-dualisable unary algebras. In other words, every finite unary algebra can be embedded into a dualisable algebra.*

Proof Let $\underline{\mathbf{M}} = \langle M; F^{\underline{\mathbf{M}}} \rangle$ be a finite unary algebra and let $\mathbf{N}_0 = \langle N; \vee, \wedge \rangle$ be the free distributive lattice generated by the set M. The quasi-variety of distributive lattices is locally finite, as it is generated by the two-element lattice. So $\mathbf{N}_0$ is finite. For each operation symbol $u \in F$, the operation $u^{\underline{\mathbf{M}}} : M \to M$ has a unique extension to an endomorphism $u^{\underline{\mathbf{N}}} : N \to N$ of $\mathbf{N}_0$. Thus $\underline{\mathbf{M}}$ is a subalgebra of $\underline{\mathbf{N}} := \langle N; F^{\underline{\mathbf{N}}} \rangle$. The unary algebra $\underline{\mathbf{N}}$ is dualisable, by the Lattice Endomorphism Theorem, 2.1.2. ∎

We have already met examples of non-dualisable unary algebras: a finite unary algebra $\underline{\mathbf{M}}$, with $\{0,1,2\} \subseteq M$, is non-dualisable if it has each map in $\{0,1\}^M$ as a term function [35]. We shall find more non-dualisable unary algebras in the next chapter.

2.2 Operations with strong idempotents

Fix a finite unary algebra $\underline{\mathbf{M}}$ and an algebra $\mathbf{A}$ in $\mathcal{A} := \mathbb{ISP}(\underline{\mathbf{M}})$. An alter ego $\underset{\sim}{\mathbf{M}}$ of $\underline{\mathbf{M}}$ yields a duality on $\mathbf{A}$ provided every morphism $\alpha : \mathrm{D}(\mathbf{A}) \to \underset{\sim}{\mathbf{M}}$ is an evaluation. Throughout the remainder of this chapter, we investigate methods for ensuring that a map $\alpha : \mathcal{A}(\mathbf{A}, \underline{\mathbf{M}}) \to M$ agrees with an evaluation on particular subsets of $\mathcal{A}(\mathbf{A}, \underline{\mathbf{M}})$ that are of the form $\alpha^{-1}(S)$, for some $S \subseteq M$.

First let $s \in M$. We shall consider what it means for α to agree with an evaluation on the set $\alpha^{-1}(s)$. For any $a \in A$, the map α agrees with $e_{\mathbf{A}}(a)$ on $\alpha^{-1}(s)$ if and only if, for all $x \in \alpha^{-1}(s)$, we have $x(a) = \alpha(x) = s$. So α agrees with an evaluation on $\alpha^{-1}(s)$ if and only if the set

$$A_{\alpha,s} := \bigcap \{\, x^{-1}(s) \mid x \in \alpha^{-1}(s) \,\}$$

is non-empty. We are aiming to find a dualising structure for $\underline{\mathbf{M}}$ that is of finite type. So, by the Duality Compactness Theorem, 1.4.2, we can assume that the algebra $\mathbf{A}$ is finite. Now the set $A_{\alpha,s}$ is non-empty provided that

(a) the set $x^{-1}(s)$ is non-empty, for each $x \in \alpha^{-1}(s)$, and

(b) the set $\mathcal{X}_{\alpha,s} := \{\, x^{-1}(s) \mid x \in \alpha^{-1}(s) \,\}$ is closed under pairwise intersection.

The second condition is equivalent to a condition on binary homomorphisms.

2.2.1 Lemma *Let $\underline{\mathbf{M}}$ be a finite algebra and define $\mathcal{A} := \mathbb{ISP}(\underline{\mathbf{M}})$. Let $s \in M$. Then the following are equivalent:*

(i) *there is an alter ego $\underset{\sim}{\mathbf{M}}$ of $\underline{\mathbf{M}}$ such that, for every $\mathbf{A} \in \mathcal{A}$ and every morphism $\alpha : \mathrm{D}(\mathbf{A}) \to \underset{\sim}{\mathbf{M}}$, the set $\mathcal{X}_{\alpha,s} := \{\, x^{-1}(s) \mid x \in \alpha^{-1}(s) \,\}$ is closed under pairwise intersection;*

(ii) *there is a binary homomorphism g of $\underline{\mathbf{M}}$ such that $g^{-1}(s) = \{(s,s)\}$.*

Moreover, if there is a binary homomorphism g of $\underline{\mathbf{M}}$ with $g^{-1}(s) = \{(s,s)\}$, then, for every $\mathbf{A} \in \mathcal{A}$ and every map $\alpha : \mathcal{A}(\mathbf{A}, \underline{\mathbf{M}}) \to M$ that preserves g, the set $\mathcal{X}_{\alpha,s} := \{\, x^{-1}(s) \mid x \in \alpha^{-1}(s) \,\}$ is closed under pairwise intersection.

Proof Assume that (i) holds. Define $\mathbf{A} := \underline{\mathbf{M}}^2$ and consider the morphism $\alpha : \mathrm{D}(\mathbf{A}) \to \underset{\sim}{\mathbf{M}}$ given by $\alpha := e_{\mathbf{A}}\big((s,s)\big)$. The projections π_0 and π_1 belong to $\mathcal{A}(\mathbf{A}, \underline{\mathbf{M}})$. For each $i \in \{0,1\}$, we have $\alpha(\pi_i) = \pi_i(s,s) = s$ and so

$\pi_i^{-1}(s) \in \mathcal{X}_{\alpha,s}$. By (i), there is some $g \in \alpha^{-1}(s)$ such that

$$g^{-1}(s) = \pi_0^{-1}(s) \cap \pi_1^{-1}(s) = \{(s,s)\}.$$

As $\alpha^{-1}(s) \subseteq \mathcal{A}(\underline{\mathbf{M}}^2, \underline{\mathbf{M}})$, it follows that g is a binary homomorphism of $\underline{\mathbf{M}}$. Therefore (ii) holds.

Before we show that (ii) implies (i), we will prove the 'Moreover' part of the lemma. Assume that there is a binary homomorphism g of $\underline{\mathbf{M}}$ such that $g^{-1}(s) = \{(s,s)\}$. Let $\mathbf{A} \in \mathcal{A}$ and let $\alpha : \mathcal{A}(\mathbf{A}, \underline{\mathbf{M}}) \to M$ preserve g. For all $x, y \in \alpha^{-1}(s)$, we have $g(x,y) \in \mathcal{A}(\mathbf{A}, \underline{\mathbf{M}})$ with

$$\alpha(g(x,y)) = g(\alpha(x), \alpha(y)) = g(s,s) = s,$$

and so $x^{-1}(s) \cap y^{-1}(s) = g(x,y)^{-1}(s) \in \mathcal{X}_{\alpha,s}$. Thus $\mathcal{X}_{\alpha,s}$ is closed under intersection.

Now assume that (ii) holds and define the alter ego $\underset{\sim}{\mathbf{M}} = \langle M; g, \mathcal{T} \rangle$ of $\underline{\mathbf{M}}$. Then (i) holds, by the 'Moreover' claim. ∎

We say that an element s of M is a **strong idempotent** of a binary operation $g : M^2 \to M$ if $g^{-1}(s) = \{(s,s)\}$. We were led, via Lemma 2.2.1, to the notion of a strong idempotent because we were trying to find a condition under which α agrees with an evaluation on sets of the form $\alpha^{-1}(s)$, where $s \in M$. The following lemma shows that strong idempotents can be used to ensure that α agrees with an evaluation on even larger subsets of $\mathcal{A}(\mathbf{A}, \underline{\mathbf{M}})$.

2.2.2 Strong Idempotents Lemma *Let $\underline{\mathbf{M}}$ be a finite algebra and define the quasi-variety $\mathcal{A} := \mathbb{ISP}(\underline{\mathbf{M}})$. Let S be a non-empty subset of M, and assume that each $s \in S$ is a strong idempotent of some binary homomorphism g_s of $\underline{\mathbf{M}}$. Now let $\mathbf{A}$ be a finite algebra in $\mathcal{A}$ and assume that $\alpha : \mathcal{A}(\mathbf{A}, \underline{\mathbf{M}}) \to M$ preserves $\{\, g_s \mid s \in S \,\}$ and $R_{|S|}$. Then α agrees with an evaluation on the set $\alpha^{-1}(S)$.*

Proof Assume that $S \cap \alpha\big(\mathcal{A}(\mathbf{A}, \underline{\mathbf{M}})\big) = \{s_1, \dots, s_k\}$, where $k \leqslant |S|$, and let $i \in \{1, \dots, k\}$. By Lemma 2.2.1, the set

$$\mathcal{X}_{\alpha,s_i} := \{\, x^{-1}(s_i) \mid x \in \alpha^{-1}(s_i) \,\}$$

is closed under pairwise intersection. Since the algebra $\mathbf{A}$ is finite, there is some $x_i \in \alpha^{-1}(s_i)$ such that $x_i^{-1}(s_i) = \bigcap \mathcal{X}_{\alpha,s_i}$. By the Preservation Lemma, 1.4.4, the map α is given by evaluation at some $a \in A$ on the set $\{x_1, \dots, x_k\}$. Now choose any $y \in \alpha^{-1}(S)$. Then $\alpha(y) = s_j$, for some $j \in \{1, \dots, k\}$. We have $s_j = \alpha(x_j) = x_j(a)$, and so $a \in x_j^{-1}(s_j) = \bigcap \mathcal{X}_{\alpha,s_j} \subseteq y^{-1}(s_j)$. Thus $\alpha(y) = s_j = y(a)$, whence α is given by evaluation at a on $\alpha^{-1}(S)$. ∎

Using the previous lemma and the Duality Compactness Theorem, 1.4.2, we can now prove that some algebras are dualisable just by looking at their binary homomorphisms.

2.2.3 Strong Idempotents Theorem *Let $\underline{\mathbf{M}}$ be a finite algebra and assume that each element $s \in M$ is a strong idempotent of some binary homomorphism g_s of $\underline{\mathbf{M}}$. Then $\underline{\mathbf{M}}$ is dualised by the structure $\underset{\sim}{\mathbf{M}} := \langle M; G, R_{|M|}, \mathcal{T}\rangle$, where $G := \{\, g_s \mid s \in M \,\}$.*

At the end of this section, we will use the Strong Idempotents Theorem to construct several dualisable unary algebras. Our next result is a generalisation of this theorem. Assume that $\mathbf{A}$ is finite and consider a map $\alpha : \mathcal{A}(\mathbf{A}, \underline{\mathbf{M}}) \to M$. Define the set

$$\mathcal{X}_\alpha := \{\, x^{-1}(\alpha(x)) \mid x \in \mathcal{A}(\mathbf{A}, \underline{\mathbf{M}}) \,\}.$$

Assume that every element of M is a strong idempotent of a binary homomorphism of $\underline{\mathbf{M}}$ and that α preserves all these binary homomorphisms. For each $s \in M$, the set

$$\mathcal{X}_{\alpha,s} := \{\, x^{-1}(s) \mid x \in \alpha^{-1}(s) \,\}$$

is closed under pairwise intersection, by Lemma 2.2.1. So the minimal elements of $\mathcal{X}_\alpha$ belong to $\{\, \bigcap \mathcal{X}_{\alpha,s} \mid s \in M \,\}$. This implies that $\mathcal{X}_\alpha$ has at most $|M|$ minimal elements. In fact, any choice of $\underset{\sim}{\mathbf{M}}$ that imposes a uniform finite upper bound on the number of minimal elements in the set $\mathcal{X}_\alpha$, for each finite algebra $\mathbf{A} \in \mathcal{A}$ and each morphism $\alpha : \mathrm{D}(\mathbf{A}) \to \underset{\sim}{\mathbf{M}}$, will lead to a duality.

2.2.4 Theorem *Let $\underline{\mathbf{M}}$ be a finite algebra and let $\underset{\sim}{\mathbf{M}} = \langle M; G, H, R, \mathcal{T}\rangle$ be an alter ego of $\underline{\mathbf{M}}$ with finite type. Assume there is some $n \in \omega\backslash\{0\}$ such that, for every finite algebra $\mathbf{A}$ in the quasi-variety $\mathcal{A} := \mathbb{ISP}(\underline{\mathbf{M}})$ and every morphism $\alpha : \mathrm{D}(\mathbf{A}) \to \underset{\sim}{\mathbf{M}}$, the set $\mathcal{X}_\alpha := \{\, x^{-1}(\alpha(x)) \mid x \in \mathcal{A}(\mathbf{A}, \underline{\mathbf{M}}) \,\}$ has at most n minimal elements. Then $\underset{\sim}{\mathbf{M}}' := \langle M; G, H, R \cup R_n, \mathcal{T}\rangle$ yields a duality on $\mathcal{A}$.*

Proof Let $\mathbf{A}$ be a finite algebra in $\mathcal{A}$ and let $\alpha : \mathcal{A}(\mathbf{A}, \underline{\mathbf{M}}) \to M$ preserve the structure induced by $\underset{\sim}{\mathbf{M}}'$. Then α preserves the structure induced by $\underset{\sim}{\mathbf{M}}$, and so $\mathcal{X}_\alpha$ has at most n minimal elements. Choose $x_1, \dots, x_n \in \mathcal{A}(\mathbf{A}, \underline{\mathbf{M}})$ such that

$$\{\, x_i^{-1}(\alpha(x_i)) \mid i \in \{1, \dots, n\} \,\}$$

includes all the minimal elements of $\mathcal{X}_\alpha$. Then α is given by evaluation at some $a \in A$ on $\{x_1, \dots, x_n\}$, by the Preservation Lemma, 1.4.4. Now let $y \in \mathcal{A}(\mathbf{A}, \underline{\mathbf{M}})$. There is some $i \in \{1, \dots, n\}$ with $x_i^{-1}(\alpha(x_i)) \subseteq y^{-1}(\alpha(y))$.

Since $a \in x_i^{-1}(\alpha(x_i))$, we must have $\alpha(y) = y(a)$. So α is an evaluation. Thus $\underset{\sim}{\mathbf{M}}'$ dualises $\underline{\mathbf{M}}$, by the Duality Compactness Theorem, 1.4.2. ∎

An interesting problem is that of finding an application of Theorem 2.2.4 that does not already follow from the Strong Idempotents Theorem, 2.2.3.

We now turn to the case that $\underline{\mathbf{M}}$ has elements that are not strong idempotents. Our strategy for dealing with this case is based on the following observation.

2.2.5 Lemma *Let $\underline{\mathbf{M}}$ be a finite algebra and let $\mathbf{A}$ belong to $\mathcal{A} := \mathbb{ISP}(\underline{\mathbf{M}})$. Assume that $\alpha : \mathcal{A}(\mathbf{A}, \underline{\mathbf{M}}) \to M$ preserves the binary homomorphism g of $\underline{\mathbf{M}}$. Let $s, s' \in \alpha(\mathcal{A}(\mathbf{A}, \underline{\mathbf{M}}))$ and assume that α is given by evaluation at a on $\alpha^{-1}(\{s, s'\})$. Then*

$$g(s, \alpha(y)) = s' \quad \textit{implies} \quad g(s, y(a)) = s',$$

for all $y \in \mathcal{A}(\mathbf{A}, \underline{\mathbf{M}})$.

Proof Let $y \in \mathcal{A}(\mathbf{A}, \underline{\mathbf{M}})$ and assume that $g(s, \alpha(y)) = s'$. Choose some $x \in \mathcal{A}(\mathbf{A}, \underline{\mathbf{M}})$ such that $\alpha(x) = s$. Then $g(x, y) \in \mathcal{A}(\mathbf{A}, \underline{\mathbf{M}})$ and

$$\alpha(g(x, y)) = g(\alpha(x), \alpha(y)) = g(s, \alpha(y)) = s'.$$

Since α is given by evaluation at a on $\alpha^{-1}(\{s, s'\})$, we have $x(a) = s$ and $g(x, y)(a) = s'$. So

$$g(s, y(a)) = g(x(a), y(a)) = g(x, y)(a) = s',$$

as required. ∎

Using the previous result as a jumping-off point, we introduce a new separation condition on the algebra $\underline{\mathbf{M}}$. Let G be a set of binary operations on M, let S be a subset of M and let $t \in M$. We say that G **and** S **distinguish** t **within** M if there is $\{\, g_i \mid i \in I \,\} \subseteq G$ and $\{\, s_i, s_i' \mid i \in I \,\} \subseteq S$ such that

$$\mathop{\&}_{i \in I} g_i(s_i, m) = s_i' \iff m = t,$$

for all $m \in M$.

Without placing any restrictions on the sets G and S, it is very easy to distinguish t within M. Let $\pi_0, \pi_1 : M^2 \to M$ be the two projection functions. Then, for example, the sets $\{\pi_1\}$ and $\{t\}$ distinguish t within M. However, we wish to distinguish t within M using a set G of binary homomorphisms of $\underline{\mathbf{M}}$ and a subset S of M such that $t \notin S$. We will then be able to give conditions under which a map $\alpha : \mathcal{A}(\mathbf{A}, \underline{\mathbf{M}}) \to M$ that agrees with an evaluation on the set $\alpha^{-1}(S)$ also agrees with an evaluation on the set $\alpha^{-1}(S \cup \{t\})$.

The names of the following three results stem from the symbols used in the separation condition.

2.2.6 First GST Lemma *Let* $\underline{\mathbf{M}}$ *be a finite algebra and define* $\mathcal{A} := \mathbb{ISP}(\underline{\mathbf{M}})$. *Let* $S \subseteq M$ *and let* $t \in M$. *Let* $\mathbf{A} \in \mathcal{A}$ *and let* $\alpha : \mathcal{A}(\mathbf{A}, \underline{\mathbf{M}}) \to M$. *Assume there is a set* G *of binary homomorphisms of* $\underline{\mathbf{M}}$ *such that* G *and* $S \cap \alpha(\mathcal{A}(\mathbf{A}, \underline{\mathbf{M}}))$ *distinguish* t *within* M. *If* α *preserves* G *and is given by evaluation at* $a \in A$ *on* $\alpha^{-1}(S)$, *then* α *is also given by evaluation at* a *on* $\alpha^{-1}(t)$.

Proof Assume that α preserves G and that α is given by evaluation at a on $\alpha^{-1}(S)$. Since G and $S \cap \alpha(\mathcal{A}(\mathbf{A}, \underline{\mathbf{M}}))$ distinguish t within M, there are sets $\{ g_i \mid i \in I \} \subseteq G$ and $\{ s_i, s_i' \mid i \in I \} \subseteq S \cap \alpha(\mathcal{A}(\mathbf{A}, \underline{\mathbf{M}}))$ such that

$$\mathop{\&}_{i \in I} g_i(s_i, m) = s_i' \iff m = t, \tag{\$}$$

for all $m \in M$. Now let $y \in \alpha^{-1}(t)$. By (\$), we have $g_i(s_i, \alpha(y)) = s_i'$, for all $i \in I$. Using Lemma 2.2.5, we obtain $g_i(s_i, y(a)) = s_i'$, for each $i \in I$. So, by (\$), we can conclude that $\alpha(y) = t = y(a)$. Thus α is given by evaluation at a on $\alpha^{-1}(t)$. ∎

The following weaker version of the First GST Lemma can sometimes be easier to apply. We use $\mathrm{End}(\underline{\mathbf{M}})(t)$ to denote the set of images of t under the collection of endomorphisms of $\underline{\mathbf{M}}$.

2.2.7 Second GST Lemma *Let* $\underline{\mathbf{M}}$ *be a finite algebra and define the quasi-variety* $\mathcal{A} := \mathbb{ISP}(\underline{\mathbf{M}})$. *Let* $S \subseteq M$ *and let* $t \in M$. *Assume there is a set* G *of binary homomorphisms of* $\underline{\mathbf{M}}$ *such that* G *and* $S \cap \mathrm{End}(\underline{\mathbf{M}})(t)$ *distinguish* t *within* M. *Now let* $\mathbf{A} \in \mathcal{A}$ *and assume that* $\alpha : \mathcal{A}(\mathbf{A}, \underline{\mathbf{M}}) \to M$ *preserves* G *and* $\mathrm{End}(\underline{\mathbf{M}})$. *If* α *is given by evaluation at* $a \in A$ *on* $\alpha^{-1}(S)$, *then* α *is also given by evaluation at* a *on* $\alpha^{-1}(t)$.

Proof The result will follow from the First GST Lemma once we have shown that $\mathrm{End}(\underline{\mathbf{M}})(t) \subseteq \alpha(\mathcal{A}(\mathbf{A}, \underline{\mathbf{M}}))$. To do this, let $s \in \mathrm{End}(\underline{\mathbf{M}})(t)$. There exists $e \in \mathrm{End}(\underline{\mathbf{M}})$ such that $e(t) = s$. We can assume that $\alpha^{-1}(t)$ is not empty. So there is $x \in \mathcal{A}(\mathbf{A}, \underline{\mathbf{M}})$ with $\alpha(x) = t$. Now

$$s = e(t) = e(\alpha(x)) = \alpha(e \circ x) \in \alpha(\mathcal{A}(\mathbf{A}, \underline{\mathbf{M}})).$$

Thus, $\mathrm{End}(\underline{\mathbf{M}})(t) \subseteq \alpha(\mathcal{A}(\mathbf{A}, \underline{\mathbf{M}}))$. ∎

Our next theorem localises all assumptions to the algebra $\underline{\mathbf{M}}$, giving a sufficient condition for dualisability that avoids mention of maps of the form $\alpha : \mathcal{A}(\mathbf{A}, \underline{\mathbf{M}}) \to M$.

2.2.8 GST Theorem *Let* $\underline{\mathbf{M}}$ *be a finite algebra, let* S *be a non-empty subset of* M *and let* G *be a set of binary homomorphisms of* $\underline{\mathbf{M}}$*. Assume that each* $s \in S$ *is a strong idempotent of a map in* G*. Assume further that, for each* $t \in M \backslash S$*, the sets* G *and* $S \cap \mathrm{End}(\underline{\mathbf{M}})(t)$ *distinguish* t *within* M*. Then* $\underset{\sim}{\mathbf{M}} := \langle M; \mathrm{End}(\underline{\mathbf{M}}) \cup G, R_{|S|}, \mathcal{T}\rangle$ *yields a duality on* $\mathbb{ISP}(\underline{\mathbf{M}})$*.*

Proof Let $\mathbf{A}$ be a finite algebra in $\mathbb{ISP}(\underline{\mathbf{M}})$, and let $\alpha : \mathrm{D}(\mathbf{A}) \to \underset{\sim}{\mathbf{M}}$ be a morphism. By the Strong Idempotents Lemma, 2.2.2, there is some $a \in A$ such that α is given by evaluation at a on $\alpha^{-1}(S)$. By the Second GST Lemma, 2.2.7, the map α is given by evaluation at a on $\alpha^{-1}(t)$, for each $t \in M \backslash S$. So α is an evaluation. The Duality Compactness Theorem, 1.4.2, tells us that $\underset{\sim}{\mathbf{M}}$ dualises $\underline{\mathbf{M}}$. ∎

We will be using the following theorem in Chapter 5.

2.2.9 Theorem *Let* $\underline{\mathbf{M}}$ *be a finite algebra and let* $0 \in M$*. Assume that* 0 *is the value of a constant term function of* $\underline{\mathbf{M}}$*. Let* S *be a non-empty subset of* M *and let* G *be a set of binary homomorphisms of* $\underline{\mathbf{M}}$*. Assume that each* $s \in S$ *is a strong idempotent of a map in* G*. Assume further that, for all* $k \in M \backslash \{0\}$ *and* $t \in M \backslash S$*, the sets* G *and* $S \cap \mathrm{End}(\underline{\mathbf{M}})(k)$ *distinguish* t *within* M*. Then* $\underset{\sim}{\mathbf{M}} := \langle M; \mathrm{End}(\underline{\mathbf{M}}) \cup G, R_{|S|}, \mathcal{T}\rangle$ *yields a duality on* $\mathbb{ISP}(\underline{\mathbf{M}})$*.*

Proof Let $\mathbf{A}$ be a finite algebra in $\mathcal{A} := \mathbb{ISP}(\underline{\mathbf{M}})$, and let $\alpha : \mathrm{D}(\mathbf{A}) \to \underset{\sim}{\mathbf{M}}$ be a morphism. First assume that the map α is constant with value 0. Since 0 is the value of a constant term function of $\underline{\mathbf{M}}$, there is an element $0^{\mathbf{A}}$ of $\mathbf{A}$ that is the value of the corresponding constant term function of $\mathbf{A}$. So $\alpha(x) = 0 = x(0^{\mathbf{A}})$, for all $x \in \mathcal{A}(\mathbf{A}, \underline{\mathbf{M}})$. Therefore α is given by evaluation at $0^{\mathbf{A}}$.

Now assume that there is some $k \in \alpha\big(\mathcal{A}(\mathbf{A}, \underline{\mathbf{M}})\big)$ with $k \neq 0$. By the Strong Idempotents Lemma, 2.2.2, there exists $a \in A$ such that α is given by evaluation at a on $\alpha^{-1}(S)$. Since $k \in \alpha\big(\mathcal{A}(\mathbf{A}, \underline{\mathbf{M}})\big)$ and α preserves $\mathrm{End}(\underline{\mathbf{M}})$, we have $\mathrm{End}(\underline{\mathbf{M}})(k) \subseteq \alpha\big(\mathcal{A}(\mathbf{A}, \underline{\mathbf{M}})\big)$. So, by the First GST Lemma, 2.2.6, the map α is given by evaluation at a on $\alpha^{-1}(t)$, for each $t \in M \backslash S$. Thus α is an evaluation. It follows, by the Duality Compactness Theorem, 1.4.2, that $\underset{\sim}{\mathbf{M}}$ dualises $\underline{\mathbf{M}}$. ∎

We end this section by presenting some applications of our results. We want to show, using small examples, that the results in this section can be very easy to apply. The dualisability of some of the algebras we consider will also follow from more general results in the next chapter.

In order to apply the Strong Idempotents Theorem, 2.2.3, to an algebra $\underline{\mathbf{M}}$, it is helpful if $\underline{\mathbf{M}}$ has a binary homomorphism that is a flat-semilattice operation.

To see this, let $0 \in M$ and define the operation $\wedge_0 : M^2 \to M$ by

$$k \wedge_0 \ell := \begin{cases} k & \text{if } k = \ell, \\ 0 & \text{otherwise.} \end{cases}$$

Then $\wedge_0$ is the meet operation of a flat semilattice on M with bottom element 0. More importantly, every element of $M \backslash \{0\}$ is a strong idempotent of $\wedge_0$. If $\underline{\mathbf{M}}$ is a unary algebra, then it is very easy to check whether or not $\wedge_0$ is a binary homomorphism of $\underline{\mathbf{M}}$. We say that a map $u : M \to M$ is **one-to-one away from** 0 if, for all $k, \ell \in M$ with $k \neq \ell$ and $u(k) = u(\ell)$, we have $u(k) = 0$.

2.2.10 Lemma *Let $\underline{\mathbf{M}}$ be a unary algebra such that $0 \in M$. Then the flat-semilattice operation $\wedge_0 : M^2 \to M$ is a binary homomorphism of $\underline{\mathbf{M}}$ if and only if every non-constant operation of $\underline{\mathbf{M}}$ preserves* 0 *and is one-to-one away from* 0.

Proof The map $\wedge_0 : M^2 \to M$ is a binary homomorphism of $\underline{\mathbf{M}}$ if and only if every operation of $\underline{\mathbf{M}}$ is an endomorphism of $\langle M; \wedge_0 \rangle$. It is easy to check that a non-constant operation $u : M \to M$ is an endomorphism of the flat semilattice $\langle M; \wedge_0 \rangle$ if and only if u preserves 0 and is one-to-one away from 0. ∎

In fact, for a small unary algebra $\underline{\mathbf{M}}$, it is easy to check whether or not any given map $g : M^2 \to M$ is a binary homomorphism of $\underline{\mathbf{M}}$. Assume that $M = \{0, \dots, n\}$, for some $n \in \omega$. Recall that we denote a unary operation $u : M \to M$ by the string $u(0) \cdots u(n)$. A binary operation $g : M^2 \to M$ will be denoted by the matrix

$$\begin{pmatrix} g(0,n) & \cdots & g(n,n) \\ \vdots & \iddots & \vdots \\ g(0,0) & \cdots & g(n,0) \end{pmatrix}.$$

In the following examples, we make claims that binary operations on a set M are binary homomorphisms of some unary algebra $\underline{\mathbf{M}}$. To show that a flat-semilattice operation on M is algebraic over $\underline{\mathbf{M}}$, we can invoke the previous lemma. We show that other binary operations are algebraic by checking them against diagrams of $\underline{\mathbf{M}}^2$ and $\underline{\mathbf{M}}$, as in Figures 2.1 and 2.2.

2.2.11 Diagram Checking Using the diagram of $\underline{\mathbf{M}}$, it is a simple matter to verify that the given diagram of $\underline{\mathbf{M}}^2$ is correct. First, check that the diagonal of $\underline{\mathbf{M}}^2$ is a copy of $\underline{\mathbf{M}}$. It then remains to check that each operation on the diagram of $\underline{\mathbf{M}}^2$ preserves rows and columns. A **row** of M^2 is a set $\{ (k, \ell) \mid k \in M \}$, for some $\ell \in M$. A **column** of M^2 is a set $\{ (k, \ell) \mid \ell \in M \}$, for some $k \in M$.

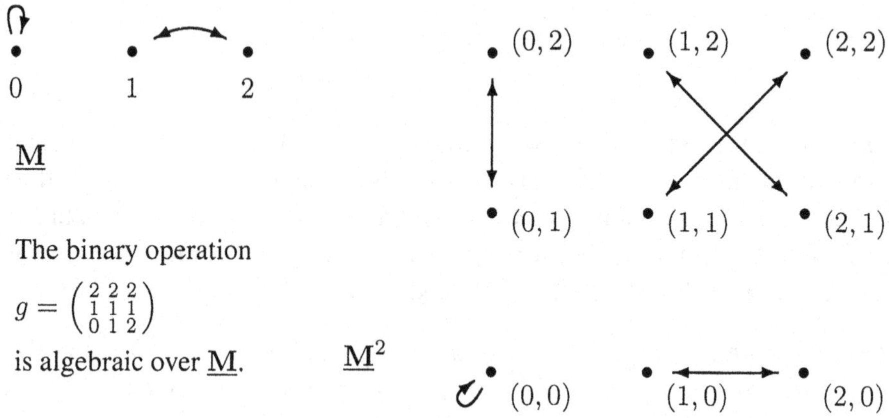

Figure 2.1 $\underline{\mathbf{M}} = \langle\{0,1,2\}; 021\rangle$

For each operation u and each row R of M^2, the elements of $u(R)$ should all belong to the same row of M^2. Similarly, for each operation u and each column C of M^2, the elements of $u(C)$ should all belong to the same column of M^2.

2.2.12 Example *The unary algebras* $\langle\{0,1,2\}; 021\rangle$ *and* $\langle\{0,1,2\}; 021, 000\rangle$ *are dualisable.*

Proof Let $\underline{\mathbf{M}} = \langle\{0,1,2\}; 021\rangle$ and define the binary operations

$$\wedge_0 := \begin{pmatrix} 0 & 0 & 2 \\ 0 & 1 & 0 \\ 0 & 0 & 0 \end{pmatrix} \quad \text{and} \quad g := \begin{pmatrix} 2 & 2 & 2 \\ 1 & 1 & 1 \\ 0 & 1 & 2 \end{pmatrix}$$

on M. The flat-semilattice operation $\wedge_0$ is algebraic over $\underline{\mathbf{M}}$, by Lemma 2.2.10. Using Figure 2.1, it is easy to check that g is a binary homomorphism of $\underline{\mathbf{M}}$. The elements 1 and 2 are strong idempotents of $\wedge_0$, and 0 is a strong idempotent of g. By the Strong Idempotents Theorem, 2.2.3, the algebra $\underline{\mathbf{M}}$ is dualised by the structure $\langle\{0,1,2\}; \wedge_0, g, R_3, \mathcal{T}\rangle$. (So it follows that $\underline{\mathbf{M}}$ is also dualised by the structure $\langle\{0,1,2\}; R_3, \mathcal{T}\rangle$.) Since both the binary operations $\wedge_0$ and g preserve 0, the Strong Idempotents Theorem also tells us that the algebra $\langle\{0,1,2\}; 021, 000\rangle$ is dualisable. ∎

Example 2.2.12 illustrates an easy way to extend results obtained using the Strong Idempotents Theorem, 2.2.3. Assume that every element of M is a strong idempotent of a binary homomorphism of $\underline{\mathbf{M}}$. Then we can add extra fundamental operations to the algebra $\underline{\mathbf{M}}$, provided they preserve these binary homomorphisms, and it will remain dualisable.

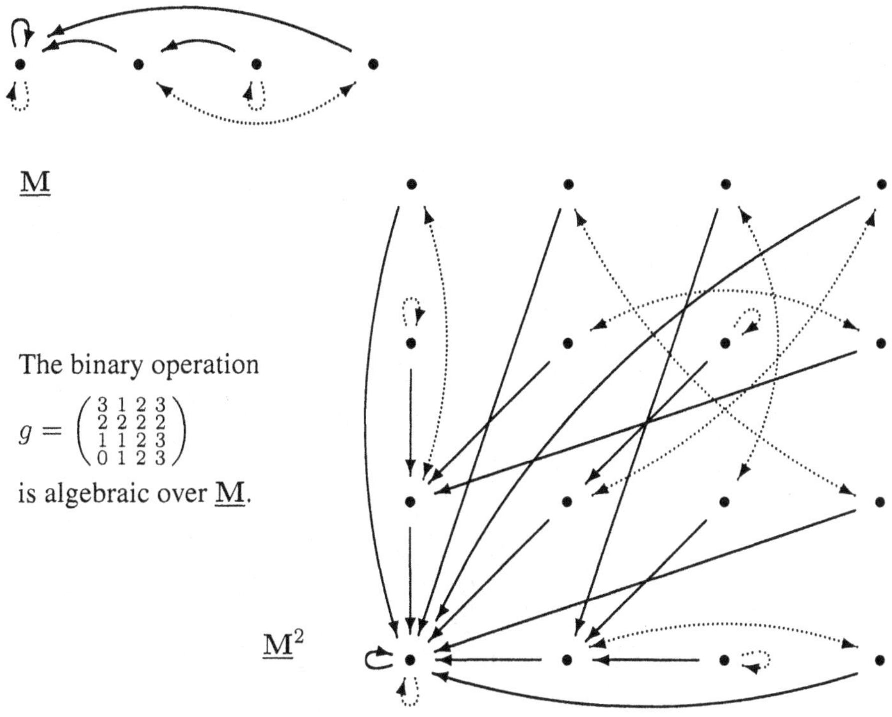
M
The binary operation
$g = \begin{pmatrix} 3&1&2&3\\ 2&2&2&2\\ 1&1&2&3\\ 0&1&2&3 \end{pmatrix}$
is algebraic over M.
M²

Figure 2.2 $\underline{\mathbf{M}} = \langle \{0,1,2,3\}; 0010, 0321 \rangle$

2.2.13 Example *The unary algebra* $\langle \{0,1,2,3\}; 0010, 0321 \rangle$ *is dualisable.*

Proof Define the binary operations

$$\wedge_0 := \begin{pmatrix} 0&0&0&3\\ 0&0&2&0\\ 0&1&0&0\\ 0&0&0&0 \end{pmatrix} \quad \text{and} \quad g := \begin{pmatrix} 3&1&2&3\\ 2&2&2&2\\ 1&1&2&3\\ 0&1&2&3 \end{pmatrix}$$

on $\{0,1,2,3\}$. Using Lemma 2.2.10 and Figure 2.2, it is easy to check that $\wedge_0$ and g are algebraic over $\underline{\mathbf{M}} = \langle \{0,1,2,3\}; 0010, 0321 \rangle$. Every element of M is a strong idempotent of one of these two maps. So $\underline{\mathbf{M}}$ is dualisable, by the Strong Idempotents Theorem, 2.2.3. ∎

Our next example illustrates the GST Theorem, 2.2.8, at work.

2.2.14 Example *The unary algebra* $\langle \{0,1,2,3\}; 0010, 0011 \rangle$ *is dualisable.*

Proof The binary operations g_0, g_{12} and g_3, given in Figure 2.3, are algebraic over $\underline{\mathbf{M}} = \langle \{0,1,2,3\}; 0010, 0011 \rangle$. We will apply the GST Theorem, 2.2.8, with $S := \{0,1,2\}$ and $G := \{g_0, g_{12}, g_3\}$.

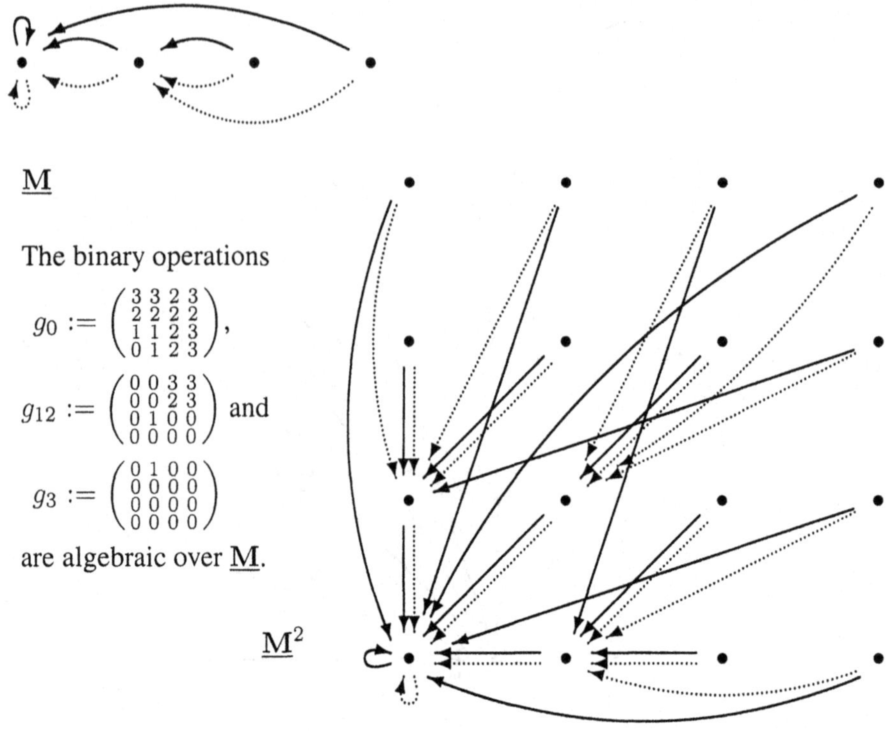
M
The binary operations
$g_0 := \begin{pmatrix} 3&3&2&3\\ 2&2&2&2\\ 1&1&2&3\\ 0&1&2&3 \end{pmatrix}$,
$g_{12} := \begin{pmatrix} 0&0&3&3\\ 0&0&2&3\\ 0&1&0&0\\ 0&0&0&0 \end{pmatrix}$ and
$g_3 := \begin{pmatrix} 0&1&0&0\\ 0&0&0&0\\ 0&0&0&0\\ 0&0&0&0 \end{pmatrix}$
are algebraic over M.
$\mathbf{M}^2$

Figure 2.3 $\underline{\mathbf{M}} = \langle\{0, 1, 2, 3\}; 0010, 0011\rangle$

The element $0 \in M$ is a strong idempotent of g_0, and both 1 and 2 are strong idempotents of g_{12}. (It is possible to show that 3 is not a strong idempotent of any binary homomorphism of $\underline{\mathbf{M}}$.) The only element of $M\backslash S$ is 3. We have

$$g_3(1, m) = 1 \iff m = 3,$$

for all $m \in M$. So $\{g_3\}$ and $\{1\}$ distinguish 3 within M. Since

$$0010 \circ 0011 = 0000 = 0011 \circ 0010,$$

the operation 0011 is an endomorphism of $\underline{\mathbf{M}}$. As $1 = 0011(3)$, this implies that $1 \in S \cap \mathrm{End}(\underline{\mathbf{M}})(3)$. It follows, by the GST Theorem, that $\underline{\mathbf{M}}$ is dualised by the structure

$$\langle\{0, 1, 2, 3\}; \mathrm{End}(\underline{\mathbf{M}}) \cup \{g_0, g_{12}, g_3\}, R_3, \mathcal{T}\rangle,$$

and therefore also by $\langle\{0, 1, 2, 3\}; R_3, \mathcal{T}\rangle$. ∎

A more complicated dualising structure for the algebra of the previous example can be obtained using the Lattice Endomorphisms Theorem, 2.1.2, by regarding the operations as endomorphisms of the lattice on $\{0, 1, 2, 3\}$ with $0 < 1 < 3 < 2$.

Just as a flat-semilattice operation is helpful when applying the Strong Idempotents Theorem, there is an operation that can be helpful when using the GST results. Let $0 \in M$ and define the near-projection operation $*_0 : M^2 \to M$ by

$$k *_0 \ell := \begin{cases} 0 & \text{if } k = \ell, \\ k & \text{otherwise.} \end{cases}$$

Then we have

$$\mathop{\&}_{s \in M \backslash \{0\}} \quad s *_0 m = s \iff m = 0,$$

for all $m \in M$. So the sets $\{*_0\}$ and $M \backslash \{0\}$ distinguish 0 within M.

2.2.15 Lemma *Let $\underline{\mathbf{M}}$ be a unary algebra such that $0 \in M$. Then the near-projection operation $*_0 : M^2 \to M$ is a binary homomorphism of $\underline{\mathbf{M}}$ if and only if every operation of $\underline{\mathbf{M}}$ preserves 0 and is one-to-one away from 0.*

Proof Let $u : M \to M$ be a unary operation. We must show that u preserves $*_0$ if and only if u preserves 0 and is one-to-one away from 0. First assume that u preserves $*_0$. Then

$$u(0) = u(0 *_0 0) = u(0) *_0 u(0) = 0,$$

and therefore u preserves 0. Let $k, \ell \in M$ with $k \neq \ell$ and $u(k) = u(\ell)$. We have

$$u(k) = u(k *_0 \ell) = u(k) *_0 u(\ell) = 0.$$

So u is one-to-one away from 0.

Now assume that u preserves 0 and is one-to-one away from 0. For all $k, \ell \in M$ with $k \neq \ell$, we have

$$u(k *_0 k) = u(0) = 0 = u(k) *_0 u(k)$$

and

$$u(k *_0 \ell) = u(k) = u(k) *_0 u(\ell).$$

So u preserves $*_0$. ∎

The following example uses both $*_0$ and $\wedge_0$ to show that endomorphisms of a prime-order cyclic group form the operations of a dualisable unary algebra. In Section 2.4, we will show that the endomorphisms of a finite cyclic group yield a dualisable unary algebra.

2.2.16 Example *Let p be prime and let F be a set of endomorphisms of the cyclic group $\mathbf{Z}_p$ of order p. Assume that F contains the constant endomorphism of $\mathbf{Z}_p$. Then the unary algebra $\underline{\mathbf{M}} := \langle \mathbb{Z}_p; F\rangle$ is dualised by the structure $\underset{\sim}{\mathbf{M}} := \langle \mathbb{Z}_p; \mathrm{End}(\underline{\mathbf{M}}), \wedge_0, *_0, R_{p-1}, \mathcal{T}\rangle$.*

Proof Every endomorphism of $\mathbf{Z}_p$ preserves 0, and every non-constant endomorphism of $\mathbf{Z}_p$ is a permutation. So both $\wedge_0$ and $*_0$ are binary homomorphisms of $\underline{\mathbf{M}}$, by Lemmas 2.2.10 and 2.2.15. We will use Theorem 2.2.9 with $S = \mathbb{Z}_p\backslash\{0\}$ and $G = \{\wedge_0, *_0\}$. Each $s \in S = \mathbb{Z}_p\backslash\{0\}$ is a strong idempotent of $\wedge_0$. Every endomorphism of the cyclic group $\mathbf{Z}_p$ is also an endomorphism of the unary algebra $\underline{\mathbf{M}}$. So, for all $k \in \mathbb{Z}_p\backslash\{0\}$, we have $\mathrm{End}(\underline{\mathbf{M}})(k) = \mathbb{Z}_p$. The sets $\{*_0\}$ and $\mathbb{Z}_p\backslash\{0\}$ distinguish 0 within $\mathbb{Z}_p$. It follows by Theorem 2.2.9 that $\underset{\sim}{\mathbf{M}}$ dualises $\underline{\mathbf{M}}$. ∎

A much stronger result is proved in the next chapter: a finite unary algebra must be dualisable if each of its operations is constant or a permutation, 3.1.8.

In the next two sections, we develop some extra tools that help us to find more applications of the GST Lemmas.

2.3 Dualisable term retracts

Let $\underline{\mathbf{D}}$ and $\underline{\mathbf{M}}$ be algebras of the same type. If there exists a pair of homomorphisms $\gamma : \underline{\mathbf{M}} \to \underline{\mathbf{D}}$ and $\beta : \underline{\mathbf{D}} \to \underline{\mathbf{M}}$ with $\gamma \circ \beta = \mathrm{id}_D$, then we say that γ is a **retraction**, that β is a **coretraction** and that $\underline{\mathbf{D}}$ is a **retract** of $\underline{\mathbf{M}}$. Note that a retraction is necessarily surjective and a coretraction is necessarily an embedding.

Now assume that $\underline{\mathbf{D}}$ is a subalgebra of $\underline{\mathbf{M}}$ and that there is a retraction $\gamma : \underline{\mathbf{M}} \twoheadrightarrow \underline{\mathbf{D}}$ that is also a unary term function of $\underline{\mathbf{M}}$. We say that γ is a **term retraction** and that $\underline{\mathbf{D}}$ is a **term retract** of $\underline{\mathbf{M}}$. The corresponding coretraction $\beta : \underline{\mathbf{D}} \hookrightarrow \underline{\mathbf{M}}$ need not be the inclusion map.

We will see in Chapter 5 that a term retract of a dualisable algebra is also dualisable. But, in general, a finite algebra with a dualisable term retract does not have to be dualisable; see Example 5.2.5. Here, we shall show that knowing that the algebra $\underline{\mathbf{M}}$ has a dualisable term retract can sometimes help us to prove that $\underline{\mathbf{M}}$ is dualisable. We shall see that, if the algebra $\underline{\mathbf{D}}$ is dualisable, then we can guarantee that a map $\alpha : \mathcal{A}(\mathbf{A}, \underline{\mathbf{M}}) \to M$ agrees with an evaluation on $\alpha^{-1}(D)$ provided it preserves the right structure. We begin by showing that we can assume that a term retraction $\gamma : \underline{\mathbf{M}} \twoheadrightarrow \underline{\mathbf{D}}$ fixes the elements of D.

2.3.1 Lemma *Let $\underline{\mathbf{M}}$ be a finite algebra and let $\gamma : \underline{\mathbf{M}} \twoheadrightarrow \underline{\mathbf{D}}$ be a term retraction. Then there is a term retraction $\gamma' : \underline{\mathbf{M}} \twoheadrightarrow \underline{\mathbf{D}}$ such that $\gamma'{\restriction_D} = \mathrm{id}_D$.*

Proof As $\gamma : \underline{\mathbf{M}} \twoheadrightarrow \underline{\mathbf{D}}$ is a retraction, there is a coretraction $\beta : \underline{\mathbf{D}} \hookrightarrow \underline{\mathbf{M}}$ with $\gamma \circ \beta = \mathrm{id}_D$. As γ is a term function of $\underline{\mathbf{M}}$ and $\beta(\underline{\mathbf{D}})$ is a subalgebra of $\underline{\mathbf{M}}$, we have $\gamma(\beta(D)) \subseteq \beta(D)$. Since $\gamma \circ \beta = \mathrm{id}_D$, the map $\gamma{\restriction}_{\beta(D)} : \beta(D) \to D$ is a bijection. As $\beta(D)$ is finite, it follows that $D = \gamma(\beta(D)) = \beta(D)$. We have shown that $\gamma{\restriction}_D : D \to D$ is a permutation. So there must be some $n \in \omega\backslash\{0\}$ such that $\gamma^n{\restriction}_D = \mathrm{id}_D$. Since $\gamma^n(M) = \gamma^{n-1}(D) = D$, the homomorphism $\gamma^n : \underline{\mathbf{M}} \twoheadrightarrow \underline{\mathbf{D}}$ is a term retraction. ∎

We want to be able to transform algebraic structure on $\underline{\mathbf{D}}$ into algebraic structure on $\underline{\mathbf{M}}$. Every algebraic relation on $\underline{\mathbf{D}}$ is also an algebraic relation on $\underline{\mathbf{M}}$, and every algebraic partial operation on $\underline{\mathbf{D}}$ is also an algebraic partial operation on $\underline{\mathbf{M}}$. There is a straightforward method for transforming algebraic operations on $\underline{\mathbf{D}}$ into algebraic operations on $\underline{\mathbf{M}}$. For each $n \in \omega$, let $\gamma^{[n]}$ denote the natural product homomorphism $\gamma \times \cdots \times \gamma : \underline{\mathbf{M}}^n \to \underline{\mathbf{D}}^n$. Now, for each algebraic operation $g : \underline{\mathbf{D}}^n \to \underline{\mathbf{D}}$, where $n \in \omega$, define the algebraic operation $g_\gamma : \underline{\mathbf{M}}^n \to \underline{\mathbf{M}}$ by $g_\gamma := g \circ \gamma^{[n]}$. For any set G of algebraic operations on $\underline{\mathbf{D}}$, define $G_\gamma := \{\, g_\gamma \mid g \in G \,\}$.

2.3.2 Term Retract Lemma *Let $\underline{\mathbf{M}}$ be a finite algebra and let $\gamma : \underline{\mathbf{M}} \twoheadrightarrow \underline{\mathbf{D}}$ be a term retraction such that $\gamma{\restriction}_D = \mathrm{id}_D$. Assume that $\underline{\mathbf{D}}$ is dualised by the structure $\underset{\sim}{\mathbf{D}} = \langle D; G, H, R, \mathcal{T}\rangle$ and define $\underset{\sim}{\mathbf{M}} := \langle M; \{\gamma\} \cup G_\gamma, H, R, \mathcal{T}\rangle$. Let $\mathbf{A} \in \mathbb{ISP}(\underline{\mathbf{M}})$ and let $\alpha : \mathrm{D}(\mathbf{A}) \to \underset{\sim}{\mathbf{M}}$ be a morphism. Then α agrees with an evaluation on $\alpha^{-1}(D)$.*

Proof We can assume that $\mathbf{A} \leqslant \underline{\mathbf{M}}^S$, for some set S. The pointwise extension $\gamma^{[S]} : \underline{\mathbf{M}}^S \to \underline{\mathbf{D}}^S$ of γ is both a homomorphism and a term function of $\underline{\mathbf{M}}^S$. So $\gamma^{[S]}{\restriction}_A : \mathbf{A} \to \mathbf{A}$ is a homomorphism and also a term function of $\mathbf{A}$. We shall simply write this map as $\gamma : \mathbf{A} \to \mathbf{A}$.

The algebra $\gamma(\mathbf{A})$ is a subalgebra of $\underline{\mathbf{D}}^S$, and so $\gamma(\mathbf{A})$ belongs to the quasi-variety $\mathcal{A} := \mathbb{ISP}(\underline{\mathbf{D}})$. We want to define the map

$$\beta : \mathcal{A}(\gamma(\mathbf{A}), \underline{\mathbf{D}}) \to D \quad \text{by} \quad \beta(y) := \alpha(y \circ \gamma).$$

To see that this is allowed, let $y \in \mathcal{A}(\gamma(\mathbf{A}), \underline{\mathbf{D}})$. Then $y \circ \gamma : \mathbf{A} \to \underline{\mathbf{M}}$ is a homomorphism in $\mathrm{D}(\mathbf{A})$, with

$$\alpha(y \circ \gamma) = \alpha(\gamma \circ y \circ \gamma) = \gamma(\alpha(y \circ \gamma)),$$

since $\gamma{\restriction}_D = \mathrm{id}_D$ and α preserves γ. So $\alpha(y \circ \gamma) \in D$, whence β is well defined.

We now want to show that the map $\beta : \mathcal{A}(\gamma(\mathbf{A}), \underline{\mathbf{D}}) \to D$ preserves the structure induced by $\underset{\sim}{\mathbf{D}}$. We will prove that β preserves each operation in G.

Let $n \in \omega$ and choose an n-ary operation $g \in G$. Then, for all $y_0, \dots, y_{n-1}$ in $\mathcal{A}(\gamma(\mathbf{A}), \underline{\mathbf{D}})$, we have

$$\begin{aligned}
\beta(g(y_0, \dots, y_{n-1})) &= \alpha(g(y_0, \dots, y_{n-1}) \circ \gamma) \\
&= \alpha(g \circ \gamma^{[n]}(y_0 \circ \gamma, \dots, y_{n-1} \circ \gamma)) && \text{as } \gamma\restriction_D = \mathrm{id}_D \\
&= \alpha(g_\gamma(y_0 \circ \gamma, \dots, y_{n-1} \circ \gamma)) \\
&= g_\gamma(\alpha(y_0 \circ \gamma), \dots, \alpha(y_{n-1} \circ \gamma)) && \text{as } \alpha \text{ preserves } g_\gamma \\
&= g \circ \gamma^{[n]}(\beta(y_0), \dots, \beta(y_{n-1})) \\
&= g(\beta(y_0), \dots, \beta(y_{n-1})) && \text{as } \gamma\restriction_D = \mathrm{id}_D.
\end{aligned}$$

Therefore β preserves G. It is also easy to show that β preserves H and R, and that β is continuous. So $\beta : \mathcal{A}(\gamma(\mathbf{A}), \underline{\mathbf{D}}) \to D$ preserves the structure induced by $\underset{\sim}{\mathrm{D}}$. Since $\underset{\sim}{\mathrm{D}}$ dualises $\underline{\mathbf{D}}$, the map β must be given by evaluation at some $b \in \gamma(A) \subseteq A$. We will show that α is given by evaluation at b on $\alpha^{-1}(D)$.

Let $x \in \alpha^{-1}(D)$. For all $a \in A$, we have $x(\gamma(a)) = \gamma(x(a)) \in D$, as γ is a term function. So $x\restriction_{\gamma(A)} \in \mathcal{A}(\gamma(\mathbf{A}), \underline{\mathbf{D}})$. This gives us

$$\begin{aligned}
\alpha(x) &= \gamma(\alpha(x)) && \text{as } \alpha(x) \in D \text{ and } \gamma\restriction_D = \mathrm{id}_D \\
&= \alpha(\gamma \circ x) && \text{as } \alpha \text{ preserves } \gamma \\
&= \alpha(x \circ \gamma) && \text{as } \gamma \text{ is a term function} \\
&= \beta(x\restriction_{\gamma(A)}) && \text{as } x\restriction_{\gamma(A)} \in \mathcal{A}(\gamma(\mathbf{A}), \underline{\mathbf{D}}) \\
&= x(b) && \text{as } \beta \text{ is given by evaluation at } b.
\end{aligned}$$

Thus α is given by evaluation at b on $\alpha^{-1}(D)$. ∎

We can use the Term Retract Lemma to obtain new dualities when we team it with the First GST Lemma, 2.2.6.

2.3.3 Term Retract Theorem *Let $\underline{\mathbf{M}}$ be a finite algebra and let $\gamma : \underline{\mathbf{M}} \twoheadrightarrow \underline{\mathbf{D}}$ be a term retraction such that $\gamma\restriction_D = \mathrm{id}_D$. Assume that $\underset{\sim}{\mathrm{D}} = \langle D; G, H, R, \mathcal{T}\rangle$ is a dualising structure for $\underline{\mathbf{D}}$ of finite type. Let S be a non-empty subset of M, let G' be a set of binary homomorphisms of $\underline{\mathbf{M}}$ and define*

$$\underset{\sim}{\mathrm{M}} := \big\langle M; \mathrm{End}(\underline{\mathbf{M}}) \cup G' \cup G_\gamma, H, R \cup R_{|S|}(\underline{\mathbf{M}}), \mathcal{T}\big\rangle.$$

Assume that each $s \in S$ is a strong idempotent of a map in G'. Assume further that, for all $k \in M \backslash D$ and $t \in M \backslash S$, the sets G' and $S \cap \mathrm{End}(\underline{\mathbf{M}})(k)$ distinguish t within M. Then $\underset{\sim}{\mathrm{M}}$ yields a duality on $\mathbb{ISP}(\underline{\mathbf{M}})$.

Proof Let $\mathbf{A}$ be a finite algebra in $\mathcal{A} := \mathbb{ISP}(\underline{\mathbf{M}})$ and let $\alpha : \mathrm{D}(\mathbf{A}) \to \underset{\sim}{\mathrm{M}}$ be a morphism. By the Term Retract Lemma, if $\alpha\big(\mathcal{A}(\mathbf{A}, \underline{\mathbf{M}})\big) \subseteq D$, then α is an evaluation. So we can assume that there exists $k \in \alpha\big(\mathcal{A}(\mathbf{A}, \underline{\mathbf{M}})\big)\backslash D$. By the Strong Idempotents Lemma, 2.2.2, the map α is given by evaluation at some $a \in A$ on the set $\alpha^{-1}(S)$. Since $k \in \alpha\big(\mathcal{A}(\mathbf{A}, \underline{\mathbf{M}})\big)$ and α preserves $\mathrm{End}(\underline{\mathbf{M}})$, we have $\mathrm{End}(\underline{\mathbf{M}})(k) \subseteq \alpha\big(\mathcal{A}(\mathbf{A}, \underline{\mathbf{M}})\big)$. Thus, by the First GST Lemma, 2.2.6, the map α is also given by evaluation at a on $\alpha^{-1}(t)$, for each $t \in M\backslash S$. Hence α is an evaluation. It follows that $\underset{\sim}{\mathbf{M}}$ yields a duality on $\mathcal{A}$, by the Duality Compactness Theorem, 1.4.2. ∎

This result is easily modified to cover the case where there is a finite number of term retractions onto dualisable subalgebras.

2.3.4 Multi Term Retract Theorem *Let $n \in \omega\backslash\{0\}$ and let $\underline{\mathbf{M}}$ be a finite algebra. Assume that, for each $i \in \{1, \dots, n\}$, the map $\gamma_i : \underline{\mathbf{M}} \twoheadrightarrow \underline{\mathbf{D}}_i$ is a term retraction with $\gamma_i{\restriction}_{D_i} = \mathrm{id}_{D_i}$ and that $\underset{\sim}{\mathbf{D}}_i = \langle D_i; G^{(i)}, H^{(i)}, R^{(i)}, \mathcal{T}\rangle$ is a dualising structure for $\underline{\mathbf{D}}_i$ of finite type. Let S be a non-empty subset of M, let G' be a set of binary homomorphisms of $\underline{\mathbf{M}}$ and define*

$$\underset{\sim}{\mathbf{M}} := \Big\langle M; \mathrm{End}(\underline{\mathbf{M}}) \cup G' \cup \bigcup_{i=1}^{n} G^{(i)}_{\gamma_i}, \bigcup_{i=1}^{n} H^{(i)}, \bigcup_{i=1}^{n} R^{(i)} \cup R_{|S|}(\underline{\mathbf{M}}), \mathcal{T}\Big\rangle.$$

Assume that each element $s \in S$ is a strong idempotent of a map in G'. Assume further that, for every transversal T of $\{\, M\backslash D_i \mid i \in \{1, \dots, n\}\,\}$ and every $t \in M\backslash S$, the sets G' and $S \cap \mathrm{End}(\underline{\mathbf{M}})(T)$ distinguish t within M. Then $\underset{\sim}{\mathbf{M}}$ yields a duality on $\mathbb{ISP}(\underline{\mathbf{M}})$.

The following example illustrates the Term Retract Theorem.

2.3.5 Example *The algebras $\langle\{0,1,2\}; 121\rangle$ and $\langle\{0,1,2\}; 121, 111, 222\rangle$ are dualisable.*

Proof Let $\underline{\mathbf{M}}$ be either $\langle\{0,1,2\}; 121\rangle$ or $\langle\{0,1,2\}; 121, 111, 222\rangle$, and let $\underline{\mathbf{D}}$ be the subalgebra of $\underline{\mathbf{M}}$ on the set $D := \{1,2\}$. Then $212 : \underline{\mathbf{M}} \twoheadrightarrow \underline{\mathbf{D}}$ is a term retraction, as $212 = 121 \circ 121$. The algebra $\underline{\mathbf{D}}$ is either a two-element set with involution or a two-element doubly pointed set with involution. In either case, there is a finite set G of algebraic operations on $\underline{\mathbf{D}}$ such that $\underset{\sim}{\mathbf{D}} = \langle\{1,2\}; G, \mathcal{T}\rangle$ yields a duality on $\mathbb{ISP}(\underline{\mathbf{D}})$. This can be shown using the general results in the text by Clark and Davey [8, Table 10.2].

The diagrams in Figure 2.4 make it straightforward to check that

$$g_0 := \begin{pmatrix} 2\,2\,2 \\ 1\,1\,1 \\ 0\,2\,2 \end{pmatrix}, \quad g_1 := \begin{pmatrix} 2\,1\,2 \\ 0\,1\,2 \\ 2\,1\,0 \end{pmatrix} \quad \text{and} \quad g_2 := \begin{pmatrix} 0\,2\,2 \\ 1\,1\,1 \\ 2\,0\,2 \end{pmatrix}$$

are all algebraic over $\underline{\mathbf{M}}$. We will use the Term Retract Theorem, 2.3.3, with $S := \{0\}$ and $G' := \{g_0, g_1, g_2\}$. The element 0 is a strong idempotent of g_0.

$\underline{\mathbf{M}}$

The binary operations

$$\begin{pmatrix} 2&2&2\\ 1&1&1\\ 0&2&2 \end{pmatrix}, \begin{pmatrix} 2&1&2\\ 0&1&2\\ 2&1&0 \end{pmatrix} \text{ and } \begin{pmatrix} 0&2&2\\ 1&1&1\\ 2&0&2 \end{pmatrix}$$

are algebraic over $\underline{\mathbf{M}}$.

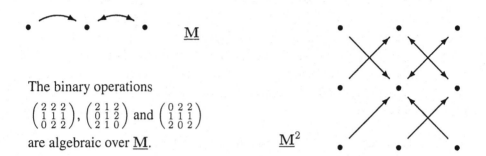

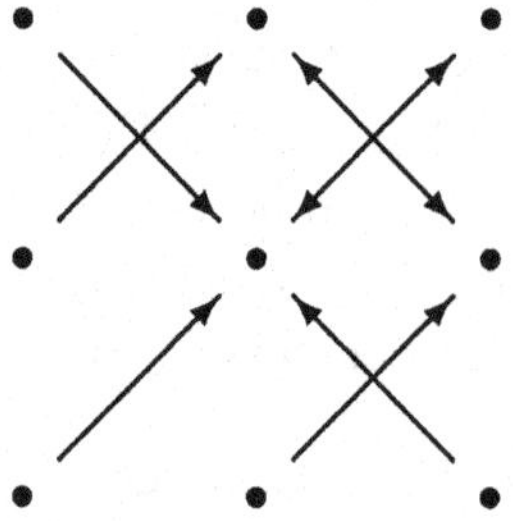

$\underline{\mathbf{M}}^2$

Figure 2.4 $\underline{\mathbf{M}} = \langle\{0,1,2\}; 121\rangle$

The only element of $M \backslash D$ is 0. We have

$$g_1(0,m) = 0 \iff m = 1 \quad \text{and} \quad g_2(0,m) = 0 \iff m = 2,$$

for all $m \in M$. So $\{g_1, g_2\}$ and $\{0\}$ distinguish both 1 and 2 within M. Thus $\underline{\mathbf{M}}$ is dualised by $\underset{\sim}{\mathbf{M}} = \langle\{0,1,2\}; \mathrm{End}(\underline{\mathbf{M}}) \cup G' \cup G_{212}, R_1(\underline{\mathbf{M}}), \mathcal{T}\rangle$. ∎

We can build on the previous example to find another dualisable algebra.

2.3.6 Example *The unary algebra* $\langle\{0,1,2,3\}; 1212, 0121\rangle$ *is dualisable.*

Proof Define the algebras

$$\underline{\mathbf{M}} := \langle\{0,1,2,3\}; 1212, 0121\rangle \quad \text{and} \quad \underline{\mathbf{D}} := \langle\{0,1,2\}; 121, 012\rangle.$$

As $0121 \circ 1212 = 1212 = 1212 \circ 0121$, the map 0121 is term retraction from $\underline{\mathbf{M}}$ onto $\underline{\mathbf{D}}$. By the previous example, the algebra $\underline{\mathbf{D}}$ is dualised by an alter ego of finite type. Using Figure 2.5, it is easy to see that the binary operations

$$g_{03} := \begin{pmatrix} 1&1&1&3\\ 2&2&2&2\\ 1&1&1&1\\ 0&2&2&2 \end{pmatrix}, \quad g_1 := \begin{pmatrix} 1&1&1&1\\ 2&2&2&2\\ 1&1&1&3\\ 2&2&2&2 \end{pmatrix} \quad \text{and} \quad g_2 := \begin{pmatrix} 2&2&2&2\\ 1&1&1&3\\ 2&2&2&2\\ 1&1&1&1 \end{pmatrix}$$

are algebraic over $\underline{\mathbf{M}}$. Both the elements 0 and 3 are strong idempotents of g_{03}. We have

$$g_1(3,m) = 3 \iff m = 1 \quad \text{and} \quad g_2(3,m) = 3 \iff m = 2,$$

for all $m \in M$. So $\{g_1, g_2\}$ and $\{3\}$ distinguish both 1 and 2 within M. Thus $\underline{\mathbf{M}}$ is dualisable, by the Term Retract Theorem, 2.3.3. ∎

As an illustration of the Multi Term Retract Theorem, 2.3.4, we give an alternative dualisability proof for the algebra $\langle\{0,1,2\}; 010, 002\rangle$ of lattice endomorphisms.

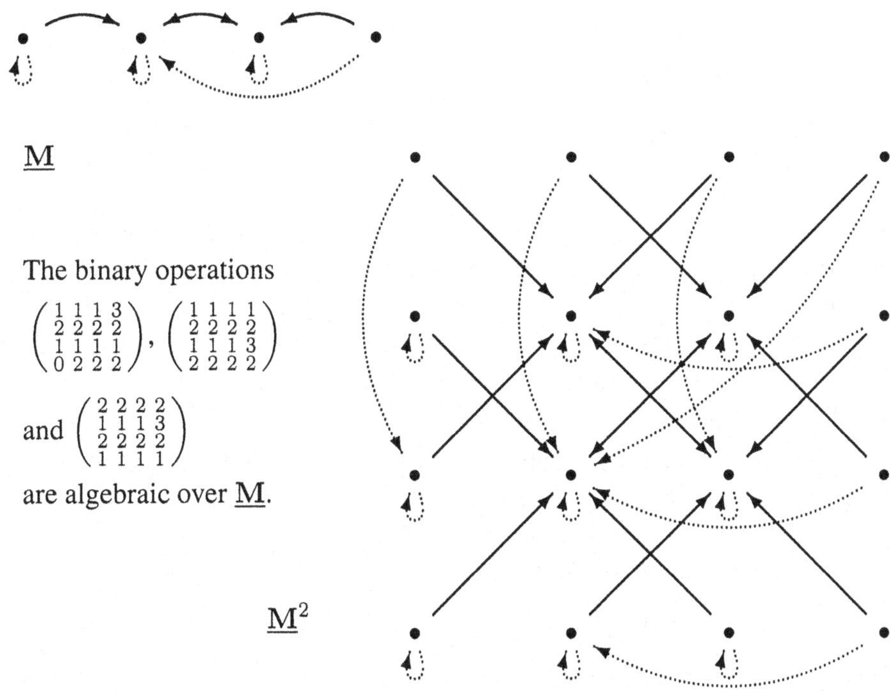
M
The binary operations
$\begin{pmatrix} 1 & 1 & 1 & 3 \\ 2 & 2 & 2 & 2 \\ 1 & 1 & 1 & 1 \\ 0 & 2 & 2 & 2 \end{pmatrix}$, $\begin{pmatrix} 1 & 1 & 1 & 1 \\ 2 & 2 & 2 & 2 \\ 1 & 1 & 1 & 3 \\ 2 & 2 & 2 & 2 \end{pmatrix}$
and $\begin{pmatrix} 2 & 2 & 2 & 2 \\ 1 & 1 & 1 & 3 \\ 2 & 2 & 2 & 2 \\ 1 & 1 & 1 & 1 \end{pmatrix}$
are algebraic over M.
M²

Figure 2.5 $\underline{\mathbf{M}} = \langle\{0,1,2,3\}; 1212, 0121\rangle$

2.3.7 Example *The unary algebra* $\langle\{0,1,2\}; 010, 002\rangle$ *is dualisable.*

Proof Define $\underline{\mathbf{M}} = \langle\{0,1,2\}; 010, 002\rangle$. Then 010 is a term retraction from $\underline{\mathbf{M}}$ onto the subalgebra $\underline{\mathbf{D}}_1$ of $\underline{\mathbf{M}}$ with $D_1 := \{0,1\}$, and 002 is a term retraction from $\underline{\mathbf{M}}$ onto the subalgebra $\underline{\mathbf{D}}_2$ of $\underline{\mathbf{M}}$ with $D_2 := \{0,2\}$. Since $\underline{\mathbf{D}}_1$ and $\underline{\mathbf{D}}_2$ are both pointed sets, they are finitely dualisable [8, Table 10.2]. (In fact, their dualisability also follows from either the Lattice Endomorphism Theorem, 2.1.2, or the Strong Idempotents Theorem, 2.2.3.)

Now consider the binary operations

$$\wedge_0 := \begin{pmatrix} 0 & 0 & 2 \\ 0 & 1 & 0 \\ 0 & 0 & 0 \end{pmatrix} \quad \text{and} \quad *_0 := \begin{pmatrix} 0 & 1 & 0 \\ 0 & 0 & 2 \\ 0 & 1 & 2 \end{pmatrix}$$

on M. These operations are binary homomorphisms of $\underline{\mathbf{M}}$, by Lemmas 2.2.10 and 2.2.15. The elements in $S := \{1,2\}$ are strong idempotents of $\wedge_0$. The only transversal of the set $\{M\backslash D_1, M\backslash D_2\}$ is $T := \{1,2\}$. The only element of $M\backslash S$ is 0. For each $m \in M$, we have

$$1 *_0 m = 1 \;\&\; 2 *_0 m = 2 \iff m = 0.$$

So $\{*_0\}$ and $\{1, 2\}$ distinguish 0 within M. It now follows by the Multi Term Retract Theorem, 2.3.4, that $\underline{\mathbf{M}}$ is dualisable. ∎

2.4 Group operations

A binary homomorphism that is also a group operation can play a central role in producing a duality.

2.4.1 Group Lemma *Let $\underline{\mathbf{M}}$ be a finite algebra that has a binary homomorphism $*$ such that $\langle M; *\rangle$ is a group. Let $\mathbf{A}$ be an algebra in $\mathcal{A} := \mathbb{ISP}(\underline{\mathbf{M}})$ and assume that $\alpha : \mathcal{A}(\mathbf{A}, \underline{\mathbf{M}}) \to M$ preserves $*$. Let S be a non-empty subset of $\alpha(\mathcal{A}(\mathbf{A}, \underline{\mathbf{M}}))$ and assume that α is given by evaluation at $a \in A$ on $\alpha^{-1}(S)$. Then α is given by evaluation at a on $\alpha^{-1}(T)$, where $\langle T; *\rangle$ is the subgroup of $\langle M; *\rangle$ generated by S.*

Proof Let $\mathbf{M}_0$ denote the group $\langle M; *\rangle$. As $*$ is algebraic over $\underline{\mathbf{M}}$, the set $\mathcal{A}(\mathbf{A}, \underline{\mathbf{M}})$ forms a subgroup of $(\mathbf{M}_0)^A$. So, since $\alpha : \mathcal{A}(\mathbf{A}, \underline{\mathbf{M}}) \to M$ preserves $*$, the set $\alpha(\mathcal{A}(\mathbf{A}, \underline{\mathbf{M}}))$ forms a subgroup of $\mathbf{M}_0$.

Now define the subset S_a of M by

$$S_a := \{\, s \in \alpha(\mathcal{A}(\mathbf{A}, \underline{\mathbf{M}})) \mid \alpha \text{ is given by evaluation at } a \text{ on } \alpha^{-1}(s) \,\}.$$

We want to prove that S_a also forms a subgroup of $\mathbf{M}_0$. Since $S \subseteq S_a$, the set S_a is non-empty.

Let $' : M \to M$ be the inverse operation of the group $\mathbf{M}_0$. Since $\mathbf{M}_0$ is finite, the operation $'$ can be derived from the group operation $*$. We begin by showing that S_a is closed under the operation $'$. Let $r \in S_a \subseteq \alpha(\mathcal{A}(\mathbf{A}, \underline{\mathbf{M}}))$. Then we must have $r' \in \alpha(\mathcal{A}(\mathbf{A}, \underline{\mathbf{M}}))$. To see that $r' \in S_a$, let $x \in \alpha^{-1}(r')$. Then $x' \in \mathcal{A}(\mathbf{A}, \underline{\mathbf{M}})$ with $\alpha(x') = \alpha(x)' = r'' = r$. As $r \in S_a$, this implies that $\alpha(x') = x'(a)$ and therefore $\alpha(x) = \alpha(x')' = x'(a)' = x(a)$. Therefore $r' \in S_a$, whence S_a is closed under $'$.

Now we can show that S_a is closed under the operation $*$, by using the First GST Lemma, 2.2.6. Let $r, s \in S_a$. Then $r, s \in \alpha(\mathcal{A}(\mathbf{A}, \underline{\mathbf{M}}))$, and therefore $r * s \in \alpha(\mathcal{A}(\mathbf{A}, \underline{\mathbf{M}}))$. For all $m \in M$, we have

$$r' * m = s \iff m = r * s.$$

So $\{*\}$ and S_a distinguish $r * s$ within M, since $r', s \in S_a$. As α is given by evaluation at a on $\alpha^{-1}(S_a)$, the First GST Lemma tells us that α is given by evaluation at a on $\alpha^{-1}(r * s)$. Thus $r * s \in S_a$.

We have shown that S_a forms a subgroup of $\mathbf{M}_0$. Since $S \subseteq S_a$ and T is the underlying set of the subgroup of $\mathbf{M}_0$ generated by S, it follows that $T \subseteq S_a$. Hence α is given by evaluation at a on the set $\alpha^{-1}(T)$. ∎

We will now use the Group Lemma and the Term Retract Lemma to find a large class of dualisable unary algebras. If we begin with a finite group $\mathbf{M}_0 = \langle M; * \rangle$, we can ask whether the unary algebra $\underline{\mathbf{M}} = \langle M; \mathrm{End}(\mathbf{M}_0) \rangle$ is dualisable. In particular, we want to know whether or not we can augment the structure $\underset{\sim}{\mathbf{M}}_0 = \langle M; *, \mathcal{T} \rangle$ so that it will yield a duality on $\mathbb{ISP}(\underline{\mathbf{M}})$. The general problem of determining for which finite groups $\mathbf{M}_0$ the unary algebra $\underline{\mathbf{M}} = \langle M; \mathrm{End}(\mathbf{M}_0) \rangle$ is dualisable remains open. We shall show that $\underline{\mathbf{M}}$ is dualisable when the group $\mathbf{M}_0$ is cyclic.

2.4.2 Lemma *Let p be prime, let $r \in \omega \backslash \{0\}$ and consider the cyclic group $\mathbf{Z}_{p^r}$ of order p^r. There is a map $g : (\mathbb{Z}_{p^r})^2 \to \mathbb{Z}_{p^r}$ such that*

(i) *g preserves every endomorphism of $\mathbf{Z}_{p^r}$, and*

(ii) *the element p^{r-1} of $\mathbb{Z}_{p^r}$ is a strong idempotent of g.*

Proof For each integer n, define the endomorphism

$$\lambda_n : \mathbf{Z}_{p^r} \to \mathbf{Z}_{p^r} \quad \text{by} \quad \lambda_n(k) = nk \pmod{p^r}.$$

Then $\mathrm{End}(\mathbf{Z}_{p^r}) = \{\, \lambda_n \mid n \in \mathbb{Z}_{p^r} \,\}$. So the endomorphisms of $\mathbf{Z}_{p^r}$ are generated, as a monoid, by the set $\{\lambda_p\} \cup \{\, \lambda_n \mid n \in \mathbb{Z}_{p^r} \text{ with } p \nmid n \,\}$. Now define $g : (\mathbb{Z}_{p^r})^2 \to \mathbb{Z}_{p^r}$ by

$$g(k, \ell) = \begin{cases} k & \text{if } \lambda_n(k) = \lambda_n(\ell) = p^{r-1}, \text{ for some integer } n, \\ 0 & \text{otherwise,} \end{cases}$$

for all $k, \ell \in \mathbb{Z}_{p^r}$. We will prove that g satisfies conditions (i) and (ii).

Claim 1 The map g preserves every endomorphism of $\mathbf{Z}_{p^r}$.

Let $m \in \{p\} \cup \{\, n \in \mathbb{Z}_{p^r} \mid p \nmid n \,\}$. To see that g preserves the endomorphism λ_m of $\mathbf{Z}_{p^r}$, let $k, \ell \in \mathbb{Z}_{p^r}$. First assume that $g(\lambda_m(k), \lambda_m(\ell)) \neq 0$. There is some integer a with

$$\lambda_a(\lambda_m(k)) = \lambda_a(\lambda_m(\ell)) = p^{r-1}.$$

So $\lambda_{am}(k) = \lambda_{am}(\ell) = p^{r-1}$, and therefore

$$\lambda_m(g(k, \ell)) = \lambda_m(k) = g(\lambda_m(k), \lambda_m(\ell)).$$

Since $\lambda_m(0) = 0$, we can now assume that $g(\lambda_m(k), \lambda_m(\ell)) = 0$ and that $g(k, \ell) \neq 0$. So there is some integer a such that

$$\lambda_a(k) = \lambda_a(\ell) = p^{r-1},$$

but there is no integer b such that

$$\lambda_b(\lambda_m(k)) = \lambda_b(\lambda_m(\ell)) = p^{r-1}.$$

Suppose that $p \nmid m$. Then $\gcd(m, p^r) = 1$, and so there are integers c and d for which $cm + dp^r = 1$. But this gives us

$$\lambda_{ca}(\lambda_m(k)) = \lambda_a(k) = p^{r-1} \quad \text{and} \quad \lambda_{ca}(\lambda_m(\ell)) = \lambda_a(\ell) = p^{r-1},$$

which is a contradiction.

We have shown that $m = p$. We have $\lambda_a(k) = \lambda_a(\ell) = p^{r-1}$, but there is no integer b such that $\lambda_{bp}(k) = \lambda_{bp}(\ell) = p^{r-1}$. This implies that $p \nmid a$, and therefore $p^{r-1} \mid k$. So

$$\lambda_p(g(k, \ell)) = \lambda_p(k) = 0 = g(\lambda_p(k), \lambda_p(\ell)),$$

whence g preserves λ_m. Thus g preserves all the endomorphisms of $\mathbf{Z}_{p^r}$.

Claim 2 The element p^{r-1} of $\mathbb{Z}_{p^r}$ is a strong idempotent of g.

Clearly, we have $g(p^{r-1}, p^{r-1}) = p^{r-1}$. Now assume that $k, \ell \in \mathbb{Z}_{p^r}$ with $g(k, \ell) = p^{r-1}$. Then $k = p^{r-1}$ and there is some integer n such that

$$\lambda_n(p^{r-1}) = \lambda_n(k) = \lambda_n(\ell) = p^{r-1}.$$

We must have $p \nmid n$, and therefore λ_n is a permutation. So $\ell = k = p^{r-1}$. Thus p^{r-1} is a strong idempotent of g. ∎

2.4.3 Theorem *The endomorphisms of a finite cyclic group are the operations of a dualisable unary algebra. More precisely, for each $n \in \omega\backslash\{0\}$, the unary algebra $\underline{\mathbf{M}} := \langle \mathbb{Z}_n; \mathrm{End}(\mathbf{Z}_n)\rangle$, of endomorphisms of the cyclic group $\mathbf{Z}_n$, is dualised by $\underset{\sim}{\mathbf{M}} := \langle \mathbb{Z}_n; +, g, R_1, \mathcal{T}\rangle$, for some binary homomorphism g of $\underline{\mathbf{M}}$.*

Proof We will argue by induction on the number of prime factors of the order of the cyclic group. The induction commences by observing that the endomorphisms of $\mathbf{Z}_1$ yield a dualisable unary algebra.

Let $n \in \omega$, let $p_0, \dots, p_n$ be distinct primes and let $r_0, \dots, r_n \in \omega\backslash\{0\}$. Define the group $\mathbf{M}_0 := \mathbf{Z}_{p_0^{r_0}} \times \cdots \times \mathbf{Z}_{p_n^{r_n}}$ and define the unary algebra $\underline{\mathbf{M}} := \langle M; \mathrm{End}(\mathbf{M}_0)\rangle$, where $M := \mathbb{Z}_{p_0^{r_0}} \times \cdots \times \mathbb{Z}_{p_n^{r_n}}$.

By Lemma 2.4.2, for all $i \in \{0, \dots, n\}$, there is a map $g_i : (\mathbb{Z}_{p_i^{r_i}})^2 \to \mathbb{Z}_{p_i^{r_i}}$ that preserves each of the endomorphisms of the group $\mathbf{Z}_{p_i^{r_i}}$ and has the element $p_i^{r_i - 1}$ as a strong idempotent. Define the map

$$g : M^2 \to M \quad \text{by} \quad g(k, \ell)(i) = g_i(k(i), \ell(i)).$$

As the group $\mathbf{M}_0$ is cyclic, every endomorphism e of $\mathbf{M}_0$ is a product, of the form $e = e_0 \times \cdots \times e_n$, of endomorphisms of the factor groups $\mathbf{Z}_{p_0^{r_0}}, \dots, \mathbf{Z}_{p_n^{r_n}}$. It follows that g preserves all the endomorphisms of $\mathbf{M}_0$. So g is a binary homomorphism of $\underline{\mathbf{M}}$. We will show that $\underline{\mathbf{M}}$ is dualised by $\underset{\sim}{\mathbf{M}} := \langle M; +, g, R_1, \mathcal{T}\rangle$.

Let $\mathbf{A}$ be a finite algebra in $\mathcal{A} := \mathbb{ISP}(\underline{\mathbf{M}})$ and let $\alpha : \mathrm{D}(\mathbf{A}) \to \underset{\sim}{\mathbf{M}}$ be a morphism. We will prove that α is an evaluation. It will then follow, by the Duality Compactness Theorem, 1.4.2, that $\underset{\sim}{\mathbf{M}}$ dualises $\underline{\mathbf{M}}$.

Since the group $\mathbf{M}_0$ is cyclic, every endomorphism of $\mathbf{M}_0$ is also an endomorphism of $\underline{\mathbf{M}}$. (In fact, $\mathrm{End}(\underline{\mathbf{M}}) = \mathrm{End}(\mathbf{M}_0)$.) Since α preserves the group operation $+$, the map α also preserves every endomorphism of the cyclic group $\mathbf{M}_0$.

As $\mathbf{M}_0$ is cyclic, there is some $m \in M$ such that m generates $\alpha\big(\mathcal{A}(\mathbf{A}, \underline{\mathbf{M}})\big)$ as a subgroup of $\mathbf{M}_0$. We split our proof that α is an evaluation into two cases.

Case 1: $m(j) = 0$, for some $j \in \{0, \dots, n\}$. Define the endomorphism γ of $\mathbf{M}_0$ by

$$\gamma(k)(i) = \begin{cases} 0 & \text{if } i = j, \\ k(i) & \text{otherwise,} \end{cases}$$

for every $k \in M$ and $i \in \{0, \dots, n\}$. Then $\gamma : M \to M$ is an operation of $\underline{\mathbf{M}}$ and $\alpha\big(\mathcal{A}(\mathbf{A}, \underline{\mathbf{M}})\big) \subseteq \gamma(M)$. It is easy to check that $\gamma : \underline{\mathbf{M}} \twoheadrightarrow \gamma(\underline{\mathbf{M}})$ is a term retraction of $\underline{\mathbf{M}}$ with $\gamma{\restriction}_{\gamma(M)} = \mathrm{id}_{\gamma(M)}$. The operations of the unary algebra $\gamma(\underline{\mathbf{M}})$ are the endomorphisms of the cyclic group $\gamma(\mathbf{M}_0)$. The order of the group $\gamma(\mathbf{M}_0)$ has one less prime factor than the order of $\mathbf{M}_0$. So, by the inductive assumption, the unary algebra $\gamma(\underline{\mathbf{M}})$ is dualised by

$$\big\langle \gamma(M); +{\restriction}_{\gamma(M)}, g{\restriction}_{\gamma(M)}, R_1{\restriction}_{\gamma(M)}, \mathcal{T} \big\rangle.$$

The map $\alpha : \mathcal{A}(\mathbf{A}, \underline{\mathbf{M}}) \to M$ preserves the endomorphism γ of $\mathbf{M}_0$, and so it follows that α preserves the structure induced by

$$\big\langle M; \gamma, +_\gamma, g_\gamma, R_1{\restriction}_{\gamma(M)}, \mathcal{T} \big\rangle.$$

As $\alpha\big(\mathcal{A}(\mathbf{A}, \underline{\mathbf{M}})\big) \subseteq \gamma(M)$, the Term Retract Lemma, 2.3.2, tells us that α is an evaluation.

Case 2: $m(i) \neq 0$, for all $i \in \{0, \dots, n\}$. By the Group Lemma, 2.4.1, it suffices to show that α agrees with an evaluation on $\alpha^{-1}(m)$. Define the element s of M by $s(i) = p_i^{r_i - 1}$. Since $p_i^{r_i-1}$ is a strong idempotent of g_i, for each $i \in \{0, \dots, n\}$, it follows that s is a strong idempotent of g. Since α preserves g and R_1, we can use the Strong Idempotents Lemma, 2.2.2, to find some $a \in A$ such that α is given by evaluation at a on $\alpha^{-1}(s)$.

As $m \in \alpha\big(\mathcal{A}(\mathbf{A}, \underline{\mathbf{M}})\big)$, there is some $z \in \mathcal{A}(\mathbf{A}, \underline{\mathbf{M}})$ such that $\alpha(z) = m$. Let $i \in \{0, \dots, n\}$. Then $m(i) \neq 0$, and therefore $\gcd(m(i), p_i^{r_i}) \mid p_i^{r_i - 1}$. So there are integers k and ℓ such that $k\,m(i) + \ell p_i^{r_i} = p_i^{r_i - 1}$. It follows that there is an endomorphism v_i of $\mathbf{Z}_{p_i^{r_i}}$ with $v_i(m(i)) = p_i^{r_i-1}$. Now define the endomorphism $v := v_0 \times \cdots \times v_n$ of $\mathbf{M}_0$. Then $v(m) = s$, and so

$$\alpha(v \circ z) = v(\alpha(z)) = v(m) = s.$$

As α is given by evaluation at a on $\alpha^{-1}(s)$, this gives $s = \alpha(v \circ z) = v \circ z(a)$.

Again, let $i \in \{0, \ldots, n\}$. Since $v \circ z(a) = s = v(m)$, we have

$$v_i(z(a)(i)) = p_i^{r_i-1} = v_i(m(i)).$$

So $\gcd(z(a)(i), p_i^{r_i}) = \gcd(m(i), p_i^{r_i})$, and therefore $\gcd(z(a)(i), p_i^{r_i}) \mid m(i)$. Hence there is an endomorphism w_i of $\mathbf{Z}_{p_i^{r_i}}$ for which $w_i(z(a)(i)) = m(i)$. Now define $w := w_0 \times \cdots \times w_n$. This gives us

$$z(w(a))(i) = w(z(a))(i) = w_i(z(a)(i)) = m(i),$$

so that $z(w(a)) = m$.

The map α is given by evaluation at a on $\alpha^{-1}(s)$. This implies that, for all $x \in \alpha^{-1}(s)$, we have

$$x(w(a)) = w(x(a)) = w(s) = w(v \circ z(a)) = v(z(w(a))) = v(m) = s.$$

Thus α is also given by evaluation at $w(a)$ on $\alpha^{-1}(s)$. We will finish the proof by showing that α is given by evaluation at $w(a)$ on $\alpha^{-1}(m)$.

Let $y \in \alpha^{-1}(m)$. Then

$$\alpha(v \circ z - z + y) = v(\alpha(z)) - \alpha(z) + \alpha(y) = v(m) - m + m = s.$$

Since α is given by evaluation at $w(a)$ on $\alpha^{-1}(s)$, we have

$$\begin{aligned} s &= (v \circ z - z + y)(w(a)) = v \circ z(w(a)) - z(w(a)) + y(w(a)) \\ &= v(m) - m + y(w(a)) = s - m + y(w(a)), \end{aligned}$$

which implies that $y(w(a)) = m$. Thus α is given by evaluation at $w(a)$ on $\alpha^{-1}(m)$. By the Group Lemma, it follows that α is an evaluation. ∎

3

The complexity of dualisability: three-element unary algebras

We solve the Dualisability Problem restricted to the class of three-element unary algebras. The intricacy of the solution demonstrates the difficulty of the general Dualisability Problem. Our solution also provides a source of examples and counterexamples for us to use later in the text.

The most fundamental problem in the theory of natural dualities is the Dualisability Problem: deciding exactly which finite algebras are dualisable. In this chapter, we demonstrate the difficulty of this problem. We do this by solving the Dualisability Problem restricted to a class of apparently very simple algebras: three-element unary algebras. We will find that the complexity of dualisability is evident even within this humble class.

A complete solution to the Dualisability Problem for two-element algebras is given in the last chapter of the text by Clark and Davey [8]. The characterisation follows comparatively easily from the general results developed throughout that text. Unfortunately, these general results do not apply so readily to three-element unary algebras, and we will be establishing dualities from scratch, using knowledge of the structure of the algebras in the quasi-varieties.

Unary algebras are actually rather complicated from the point of view of duality theory. Amongst the two-element algebras, it is the highly structured algebras that have simple dualising structures, and the simpler algebras that have complicated dualising structures. For example, the two-element implicative lattice $\langle \{0,1\}; \vee, \wedge, \rightarrow \rangle$ is dualised by the discrete pointed set $\langle \{0,1\}; 1, \mathcal{T} \rangle$, and the two-element pointed set $\langle \{0,1\}; 1 \rangle$ is dualised by the discrete implicative lattice $\langle \{0,1\}; \vee, \wedge, \rightarrow, \mathcal{T} \rangle$. The dualising structures we obtain for three-element unary algebras will be quite unwieldy. But, rather than aiming to create useful dualities, we are aiming to shed light on the general Dualisability Problem.

The dualisability of an algebra seems to be related to finiteness properties of the quasi-variety and, perhaps unexpectedly, of the variety that it generates. The exact nature of this relationship remains mysterious. In the positive direction, every known dualisable algebra has a finitely based equational theory. However, a dualisable algebra need not have a finitely based quasi-equational theory. The three-element unary algebras $\underline{\mathbf{M}}_{\mathrm{V}}$ and $\underline{\mathbf{M}}_{\mathrm{L}}$, defined in Figure 4.1, are dualisable, by the Lattice Endomorphism Theorem, 2.1.2. But neither has a finite basis for its quasi-equational theory [4].

In the other direction, it seems as though finite algebras that generate badly 'non-finite' varieties must be non-dualisable. A variety is said to be residually large if, up to isomorphism, it contains a proper class of subdirectly irreducible algebras. We conjecture that a finite algebra that generates a residually large variety is non-dualisable.

In this chapter, some of our proofs of dualisability make direct use of very strong finiteness conditions on the generated quasi-varieties. However, we shall also find dualisable and non-dualisable algebras where the difference between the quasi-varieties they generate appears to be slight.

The dualisability of a three-element unary algebra is strongly related to the number of different patterns that its unary term functions have. To make this more precise, consider a unary algebra $\underline{\mathbf{M}}$. A **kernel of $\underline{\mathbf{M}}$** is an equivalence relation on M of the form $\ker(u)$, for some unary term function u of $\underline{\mathbf{M}}$ that is neither a constant map nor a permutation. We shall say that $\underline{\mathbf{M}}$ is an **n-kernel unary algebra** if n is the number of different kernels of $\underline{\mathbf{M}}$.

There are only three non-trivial partitions of a three-element set. So the class of three-element unary algebras divides into zero-, one-, two- and three-kernel algebras. The main result of this chapter is the following characterisation of dualisability for three-element unary algebras.

3.0.1 Theorem *Let $\underline{\mathbf{M}}$ be a three-element unary algebra.*

(i) *If $\underline{\mathbf{M}}$ is a zero-kernel or one-kernel algebra, then $\underline{\mathbf{M}}$ is dualisable.*

(ii) *Assume that $\underline{\mathbf{M}}$ is a two-kernel algebra, on the set $\{0, 1, 2\}$, with kernels $\{01|2\}$ and $\{02|1\}$. Then $\underline{\mathbf{M}}$ is dualisable if and only if all the following conditions hold:*

   (a) *if ppq and pqp are term functions of $\underline{\mathbf{M}}$, for some $p, q \in \{0, 1, 2\}$ with $p \neq q$, then* 010 *and* 002 *are also term functions of $\underline{\mathbf{M}}$;*

   (b) *if* 010, 001 *and* 110 *are term functions of $\underline{\mathbf{M}}$, then so is* 222;

   (c) *if* 002, 020 *and* 202 *are term functions of $\underline{\mathbf{M}}$, then so is* 111.

(iii) *If $\underline{\mathbf{M}}$ is a three-kernel algebra, then $\underline{\mathbf{M}}$ is not dualisable.*

operations	*kernels*	*dualisable?*
$F_0 = \{012, 000, 001\}$	1	yes
$F_1 = \{012, 000, 001, 010\}$	2	no
$F_2 = \{012, 000, 001, 010, 002\}$	2	yes
$F_3 = \{012, 000, 001, 010, 002, 110, 111\}$	2	no
$F_4 = \{012, 000, 001, 010, 002, 110, 111, 222\}$	2	yes
$F_5 = \{012, 000, 001, 010, 002, 110, 111, 222, 011\}$	3	no

Table 3.1 Six three-element unary algebras

Given any two-kernel three-element unary algebra $\underline{\mathbf{M}}$, there is a straightforward method for constructing an isomorphic copy of $\underline{\mathbf{M}}$, on the set $\{0, 1, 2\}$, that has kernels $\{01|2\}$ and $\{02|1\}$. (See Lemma 3.3.1 and the discussion on conjugation that precedes it.) So the previous result really does completely characterise dualisability for three-element unary algebras.

Broadly speaking, our characterisation says that three-element unary algebras with few unary term functions are dualisable, while those with many unary term functions are non-dualisable. However, the dualisable and non-dualisable two-kernel algebras are tightly entangled. Indeed, there is a chain $F_0 \subseteq F_1 \subseteq \cdots \subseteq F_5$ of sets of unary operations on $\{0, 1, 2\}$ such that the corresponding algebras

$$\langle\{0,1,2\}; F_0\rangle, \quad \langle\{0,1,2\}; F_1\rangle, \quad \ldots, \quad \langle\{0,1,2\}; F_5\rangle$$

are alternately dualisable and non-dualisable. The definitions of the operation sets $F_0, \ldots, F_5$ are given in Table 3.1. For each $i \in \{0, \ldots, 5\}$, it is easy to check that every unary term function of the algebra $\langle\{0,1,2\}; F_i\rangle$ belongs to F_i, and then to use Theorem 3.0.1 to determine whether or not this algebra is dualisable.

A three-element unary algebra with universe $\{0, 1, 2\}$ is determined, up to term equivalence, by its monoid of unary term functions. It turns out that there are exactly 699 such monoids on $\{0, 1, 2\}$. These 699 monoids determine 160 non-isomorphic unary algebras. In Tables 3.2 and 3.3, we indicate how the 699 monoids on $\{0, 1, 2\}$ and the 160 non-isomorphic unary algebras that they determine are distributed amongst the different kernel types.

A surprising number of very different arguments seems to be required to solve the Dualisability Problem for three-element unary algebras. We shall examine the four kernel types section by section. We use both general and particular

kernels	0	1	2	3	*total*
dualisable	24	198	210		432
non-dualisable			126	141	267

Table 3.2 The 699 monoids on $\{0,1,2\}$

kernels	0	1	2	3	*total*
dualisable	12	44	43		99
non-dualisable			24	37	61

Table 3.3 The 160 non-isomorphic three-element unary algebras

results to complete our characterisation. As a bonus, we end by using some general results from this and the previous chapter to prove that every finite unary algebra with only one fundamental operation is dualisable.

The first four sections of this chapter are based on a paper written by the authors and D. M. Clark [13]. The result in the final section was found by the first author [51].

3.1 Dividing unary algebras into petals

The zero-kernel unary algebras are exactly those whose fundamental operations are all constants or permutations. In this section, we prove that every finite zero-kernel unary algebra (of any size) is dualisable. To do this, we show that finite zero-kernel algebras generate extremely simple quasi-varieties. Each of these quasi-varieties can be created, via coproducts, from a finite number of building-block algebras. The following example gives an easy illustration of this phenomenon.

3.1.1 Example Consider the three-element unary algebra

$$\underline{\mathbf{M}} := \langle \{0,1,2\}; 021, 000 \rangle,$$

shown in Figure 3.1. The quasi-variety $\mathcal{A} := \mathbb{ISP}(\underline{\mathbf{M}})$ is determined by the quasi-equations

$$021(021(x)) \approx x,\ 000(x) \approx 000(y) \text{ and } 021(x) \approx x \implies x \approx 000(x).$$

Figure 3.1 gives an example of a typical algebra $\mathbf{A}$ from $\mathcal{A}$. It is easy to check that every algebra in $\mathcal{A}$ is a coproduct of copies of $\underline{\mathbf{M}}$.

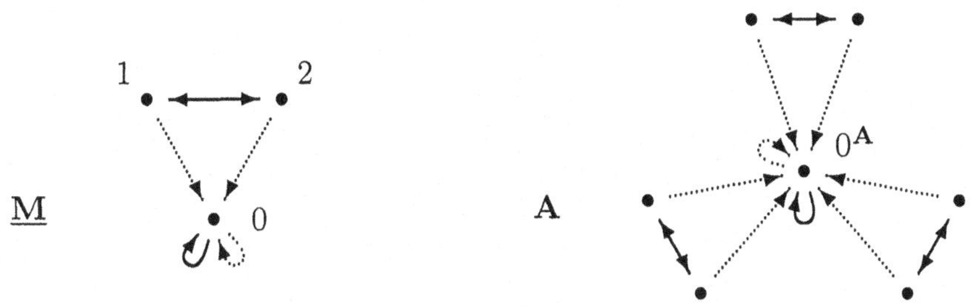

Figure 3.1 A zero-kernel unary algebra

Let $\underline{\mathbf{M}} = \langle M; F \rangle$ be an arbitrary unary algebra and let $\mathbf{A}$ belong to the quasi-variety $\mathcal{A} := \mathbb{ISP}(\underline{\mathbf{M}})$. Recall that we associate a directed graph $G(\mathbf{A})$ with the algebra $\mathbf{A}$, where

$$G(\mathbf{A}) = \langle A; E_{\mathbf{A}} \rangle \quad \text{and} \quad E_{\mathbf{A}} := \big\{\, (a, u(a)) \;\big|\; a \in A \text{ and } u \in F \,\big\}.$$

Now define the **centre of A** to be the subuniverse of $\mathbf{A}$ given by

$$C_{\mathbf{A}} := \big\{\, m^{\mathbf{A}} \;\big|\; m \text{ is the value of a constant term function of } \underline{\mathbf{M}} \,\big\}.$$

Let $G^*(\mathbf{A})$ denote the induced subgraph of $G(\mathbf{A})$ with vertex set $A \backslash C_{\mathbf{A}}$. Then a subalgebra $\mathbf{P}$ of $\mathbf{A}$ is called a **petal of A** if $P \backslash C_{\mathbf{A}}$ is non-empty and is the vertex set of a connected component of the graph $G^*(\mathbf{A})$. An algebra $\mathbf{P}$ is a **petal of $\mathcal{A}$** if $\mathbf{P}$ is a petal of some $\mathbf{A} \in \mathcal{A}$.

Note that the centre and the petals of a unary algebra $\mathbf{A}$ depend on which quasi-variety $\mathbb{ISP}(\underline{\mathbf{M}})$ we choose as its home. Whenever we are talking about centres and petals, we shall make sure that the chosen quasi-variety is clear.

The following example demonstrates the origin of the names 'petal' and 'centre'.

3.1.2 Example Again, consider the algebra $\underline{\mathbf{M}} := \langle \{0, 1, 2\}; 021, 000 \rangle$. The only element of $\underline{\mathbf{M}}$ that is the value of a constant term function is 0. In Figure 3.1, the algebra $\mathbf{A}$, from $\mathbb{ISP}(\underline{\mathbf{M}})$, has been drawn to resemble a flower. This algebra has centre $C_{\mathbf{A}} = \{0^{\mathbf{A}}\}$ and three petals.

3.1.3 Remark Let $\underline{\mathbf{M}}$ be a unary algebra and choose some $\mathbf{A}$ in $\mathbb{ISP}(\underline{\mathbf{M}})$. Then we can define a partition of A by

$$\{\, P \backslash C_{\mathbf{A}} \mid \mathbf{P} \text{ is a petal of } \mathbf{A} \,\} \cup \big(\{C_{\mathbf{A}}\} \backslash \{\varnothing\}\big).$$

This partition reflects the structure of $\mathbf{A}$, since $C_{\mathbf{A}}$ is a subuniverse of $\mathbf{A}$ and, for each petal $\mathbf{P}$ of $\mathbf{A}$, the set $P = (P \backslash C_{\mathbf{A}}) \cup C_{\mathbf{A}}$ is a subuniverse of $\mathbf{A}$. The

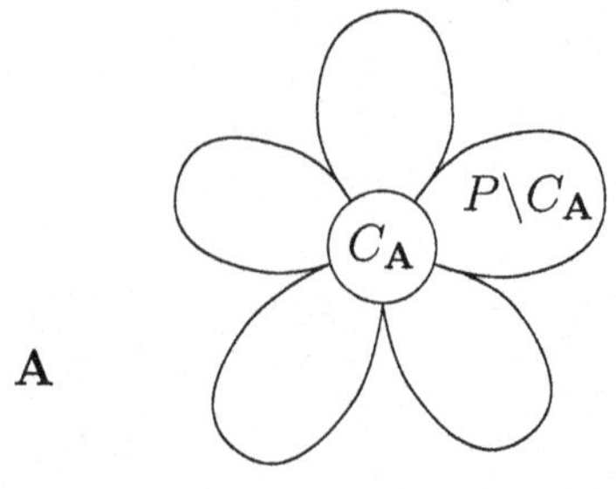

Figure 3.2 Petals

diagram in Figure 3.2 gives a general picture of how a unary algebra is divided into petals.

The next lemma reveals the importance of petals.

3.1.4 Lemma *Let* $\underline{\mathbf{M}}$ *be a unary algebra and let* $\mathbf{A}$ *belong to* $\mathcal{A} := \mathbb{ISP}(\underline{\mathbf{M}})$. *Then* $\mathbf{A}$ *is the coproduct (in* $\mathcal{A}$*) of its petals, unless* $\mathbf{A}$ *is trivial and* $\underline{\mathbf{M}}$ *has at least two constant unary term functions.*

Proof Assume that $\mathbf{A}$ is non-trivial or that $\underline{\mathbf{M}}$ has at most one constant unary term function.

Case 1: $A = C_{\mathbf{A}}$. The algebra $\mathbf{A}$ has no petals. Choose an algebra $\mathbf{B}$ from $\mathcal{A}$. Then there is a unique homomorphism $x : \mathbf{A} \to \mathbf{B}$, given by $x(m^{\mathbf{A}}) = m^{\mathbf{B}}$, for each $m \in M$ that is the value of a constant term function of $\underline{\mathbf{M}}$. To check that x is well defined, assume that $m_1, m_2 \in M$, with $m_1 \neq m_2$, such that both m_1 and m_2 are values of constant term functions of $\underline{\mathbf{M}}$. By our initial assumption, the algebra $\mathbf{A}$ must be non-trivial. Since $\mathbf{A} \in \mathbb{ISP}(\underline{\mathbf{M}})$, there is a homomorphism from $\mathbf{A}$ into $\underline{\mathbf{M}}$, which implies that $m_1^{\mathbf{A}} \neq m_2^{\mathbf{A}}$. Therefore x is well defined, whence $\mathbf{A}$ is the empty-indexed coproduct in $\mathcal{A}$.

Case 2: $A \neq C_{\mathbf{A}}$. Let $\{\, \mathbf{P}_i \mid i \in I \,\}$ be the set of petals of $\mathbf{A}$. Then $C_{\mathbf{A}} \subseteq P_i$, for all $i \in I$, and $\{\, P_i \backslash C_{\mathbf{A}} \mid i \in I \,\}$ is a partition of $A \backslash C_{\mathbf{A}}$. Let $\mathbf{B} \in \mathcal{A}$ and, for each $i \in I$, let $x_i : \mathbf{P}_i \to \mathbf{B}$ be a homomorphism. We must have $x_i(m^{\mathbf{A}}) = m^{\mathbf{B}}$, for all $i \in I$ and $m^{\mathbf{A}} \in C_{\mathbf{A}}$. So the homomorphism

$$x : \mathbf{A} \to \underline{\mathbf{M}}, \quad \text{given by } x := \bigcup \{\, x_i \mid i \in I \,\},$$

is the unique extension of the family $\{\, x_i \mid i \in I \,\}$. It follows that $\mathbf{A}$ is the coproduct of $\{\, \mathbf{P}_i \mid i \in I \,\}$ in $\mathcal{A}$. ∎

Within classes of non-unary algebras, most algebras cannot be written as a coproduct of simpler algebras. Similarly, in many quasi-varieties of unary algebras, the petals are just as complicated as the algebras in general. This is the case in the quasi-variety generated by the three-element unary algebra $\underline{\mathbf{Q}}$ defined in 1.2.3. The petals of $\mathbb{ISP}(\underline{\mathbf{Q}})$ are those algebras $\mathbf{A}$ for which the graph that is obtained from $\Gamma(\mathbf{A})$ by removing the vertex $0^{\mathbf{A}}$ is connected; see Figure 1.4 and the proof of Lemma 1.2.5. Connected graphs are really just as complicated as graphs in general. So focusing on the petals of $\mathbb{ISP}(\underline{\mathbf{Q}})$ is not very helpful.

Nevertheless, there are quasi-varieties of unary algebras whose petals are particularly well behaved: for example, the quasi-variety $\mathbb{ISP}(\underline{\mathbf{R}})$ defined in 1.2.1. The algebra $\underline{\mathbf{R}}$ has no constant term functions. So the petals of an algebra in $\mathbb{ISP}(\underline{\mathbf{R}})$ are simply its connected components. In Lemma 1.2.2, we showed that $\mathbb{ISP}(\underline{\mathbf{R}})$ has only two kinds of non-trivial petals. So the petals of $\mathbb{ISP}(\underline{\mathbf{R}})$ are less complicated than the general algebras in $\mathbb{ISP}(\underline{\mathbf{R}})$.

If the petals of a quasi-variety $\mathcal{A}$ are simple, it makes sense to establish a duality for $\mathcal{A}$ in two steps. First find a structure that yields a duality on the petals of $\mathcal{A}$, and then enrich this structure so that it yields a duality on all of $\mathcal{A}$. The following two lemmas show how the second step can be done.

3.1.5 Lemma *Let $\underline{\mathbf{M}}$ be a finite unary algebra and define $\mathcal{A} := \mathbb{ISP}(\underline{\mathbf{M}})$. Let $\mathbf{A}$ be a finite algebra in $\mathcal{A}$, with $A \neq C_{\mathbf{A}}$, and assume that $\alpha : \mathcal{A}(\mathbf{A}, \underline{\mathbf{M}}) \to M$ preserves R_4. Then there is a petal $\mathbf{P}$ of $\mathbf{A}$ such that P is a support for α.*

Proof Let $\mathbf{P}_1, \dots, \mathbf{P}_n$ be the petals of $\mathbf{A}$, where $n \in \omega\backslash\{0\}$. (Since $A \neq C_{\mathbf{A}}$, the algebra $\mathbf{A}$ has at least one petal.) If the map α is constant, then every subset of A is a support for α. So we can assume that there are

$$w, x \in \mathcal{A}(\mathbf{A}, \underline{\mathbf{M}}) \quad \text{with} \quad \alpha(w) \neq \alpha(x).$$

We shall construct a sequence $w = w_0, \dots, w_n = x$ of homomorphisms in $\mathcal{A}(\mathbf{A}, \underline{\mathbf{M}})$ such that w_i and w_{i+1} agree on $A\backslash P_{i+1}$, for all $i \in \{0, \dots, n-1\}$. First define $w_0 := w$. For each $i \in \{0, \dots, n-1\}$, we can define

$$w_{i+1} : \mathbf{A} \to \underline{\mathbf{M}} \quad \text{by} \quad w_{i+1} := w_i{\restriction}_{A\backslash P_{i+1}} \cup x{\restriction}_{P_{i+1}},$$

since $\mathbf{P}_{i+1}$ is a cofactor of $\mathbf{A}$. As $A = P_1 \cup \dots \cup P_n$, we have $w_n = x$.

Since $\alpha(w_0) = \alpha(w) \neq \alpha(x) = \alpha(w_n)$, there is some $j \in \{0, \dots, n-1\}$ with $\alpha(w_j) \neq \alpha(w_{j+1})$. To see that P_{j+1} is a support for α, let

$$y, z \in \mathcal{A}(\mathbf{A}, \underline{\mathbf{M}}) \quad \text{with} \quad y{\restriction}_{P_{j+1}} = z{\restriction}_{P_{j+1}}.$$

The map α preserves R_4. So, by the Preservation Lemma, 1.4.4, there is some $a \in A$ such that α is given by evaluation at a on $\{w_j, w_{j+1}, y, z\}$. The maps w_j

and w_{j+1} can only differ on P_{j+1}. As $w_j(a) = \alpha(w_j) \neq \alpha(w_{j+1}) = w_{j+1}(a)$, it follows that $a \in P_{j+1}$. Thus $\alpha(y) = y(a) = z(a) = \alpha(z)$, whence P_{j+1} is a support for α. ∎

3.1.6 Petal Duality Lemma *Let $\underline{\mathbf{M}}$ be a finite unary algebra and let $n \in \omega$ with $n \geqslant 2$. Assume that R_n yields a duality on every finite petal of $\mathbb{ISP}(\underline{\mathbf{M}})$. Then $\underset{\sim}{\mathbf{M}} := \langle M; R_{n+2}, \mathcal{T}\rangle$ yields a duality on $\mathbb{ISP}(\underline{\mathbf{M}})$.*

Proof Let $\mathbf{A}$ be a finite algebra in $\mathcal{A} := \mathbb{ISP}(\underline{\mathbf{M}})$ and let $\alpha : \mathrm{D}(\mathbf{A}) \to \underset{\sim}{\mathbf{M}}$ be a morphism. We shall prove that α is an evaluation. It will then follow, by the Duality Compactness Theorem, 1.4.2, that $\underset{\sim}{\mathbf{M}}$ yields a duality on $\mathcal{A}$.

We can assume that $\mathcal{A}(\mathbf{A}, \underline{\mathbf{M}})$ is not empty. So choose some $z \in \mathcal{A}(\mathbf{A}, \underline{\mathbf{M}})$. For each petal $\mathbf{P}$ of $\mathbf{A}$ and each $w \in \mathcal{A}(\mathbf{P}, \underline{\mathbf{M}})$, we can define $\overline{w} \in \mathcal{A}(\mathbf{A}, \underline{\mathbf{M}})$ by $\overline{w} := w \cup z{\restriction}_{A\backslash P}$, as $\mathbf{P}$ is a cofactor of $\mathbf{A}$. For each petal $\mathbf{P}$ of $\mathbf{A}$, define

$$\alpha_{\mathbf{P}} : \mathcal{A}(\mathbf{P}, \underline{\mathbf{M}}) \to M \quad \text{by} \quad \alpha_{\mathbf{P}}(w) := \alpha(\overline{w}).$$

We split our proof that α is an evaluation into three cases.

Case 1: $A = C_{\mathbf{A}}$. There is only one homomorphism from $\mathbf{A}$ to $\underline{\mathbf{M}}$, and therefore $|\mathcal{A}(\mathbf{A}, \underline{\mathbf{M}})| = 1$. Since α preserves R_{n+2}, it follows by the Preservation Lemma, 1.4.4, that α is an evaluation.

Case 2: $A \neq C_{\mathbf{A}}$ and α is not constant. There exist $x_1, x_2 \in \mathcal{A}(\mathbf{A}, \underline{\mathbf{M}})$ with $\alpha(x_1) \neq \alpha(x_2)$. Since $n + 2 \geqslant 4$, we know from Lemma 3.1.5 that there is a petal $\mathbf{P}$ of $\mathbf{A}$ such that P is a support for α. We have

$$y_1 := x_1{\restriction}_P \cup z{\restriction}_{A\backslash P} \quad \text{and} \quad y_2 := x_2{\restriction}_P \cup z{\restriction}_{A\backslash P}$$

in $\mathcal{A}(\mathbf{A}, \underline{\mathbf{M}})$ with $\alpha(y_1) = \alpha(x_1) \neq \alpha(x_2) = \alpha(y_2)$.

We want to show that $\alpha_{\mathbf{P}} : \mathcal{A}(\mathbf{P}, \underline{\mathbf{M}}) \to M$ preserves R_n. By the Preservation Lemma, it is enough to show that $\alpha_{\mathbf{P}}$ agrees with an evaluation on every subset of $\mathcal{A}(\mathbf{P}, \underline{\mathbf{M}})$ with at most n elements. Let $w_1, \dots, w_n \in \mathcal{A}(\mathbf{P}, \underline{\mathbf{M}})$. Since α preserves R_{n+2}, the map α is given by evaluation at some $a \in A$ on $\{y_1, y_2, \overline{w_1}, \dots, \overline{w_n}\}$, by the Preservation Lemma. We must have $a \in P$, since

$$y_1(a) = \alpha(y_1) \neq \alpha(y_2) = y_2(a) \quad \text{and} \quad y_1{\restriction}_{A\backslash P} = y_2{\restriction}_{A\backslash P}.$$

Now, for all $i \in \{1, \dots, n\}$, we have $\alpha_{\mathbf{P}}(w_i) = \alpha(\overline{w_i}) = \overline{w_i}(a) = w_i(a)$. So $\alpha_{\mathbf{P}}$ is given by evaluation at a on $\{w_1, \dots, w_n\}$, whence $\alpha_{\mathbf{P}}$ preserves R_n.

As R_n yields a duality on the finite petal $\mathbf{P}$, there exists $b \in P$ such that $\alpha_{\mathbf{P}}$ is given by evaluation at b. Since P is a support for α, it follows that

$$\alpha(x) = \alpha(\overline{x{\restriction}_P}) = \alpha_{\mathbf{P}}(x{\restriction}_P) = x{\restriction}_P(b) = x(b),$$

for all $x \in \mathcal{A}(\mathbf{A}, \underline{\mathbf{M}})$. So α is an evaluation.

Case 3: $A \neq C_{\mathbf{A}}$ and α is constant. Suppose, by way of contradiction, that the map $\alpha_{\mathbf{P}}$ is not an evaluation, for every petal $\mathbf{P}$ of $\mathbf{A}$. Now let $\mathbf{P}$ be any petal of $\mathbf{A}$. Since R_n yields a duality on $\mathbf{P}$ and $\alpha_{\mathbf{P}}$ is not an evaluation, the map $\alpha_{\mathbf{P}}$ does not preserve R_n. So, by the Preservation Lemma, there must be homomorphisms $w_{\mathbf{P}1}, \dots, w_{\mathbf{P}n} \in \mathcal{A}(\mathbf{P}, \underline{\mathbf{M}})$ such that $\alpha_{\mathbf{P}}$ does not agree with an evaluation on the set $\{w_{\mathbf{P}1}, \dots, w_{\mathbf{P}n}\}$. By Lemma 3.1.4, the algebra $\mathbf{A}$ is the coproduct of its petals. This implies that, for each $i \in \{1, \dots, n\}$, we can define the homomorphism $w_i \in \mathcal{A}(\mathbf{A}, \underline{\mathbf{M}})$ by

$$w_i := \bigcup \{\, w_{\mathbf{P}i} \mid \mathbf{P} \text{ is a petal of } \mathbf{A} \,\}.$$

Since α preserves R_{n+2}, the map α is given by evaluation at some $a \in A$ on $\{w_1, \dots, w_n\}$. Let $\mathbf{Q}$ be a petal of $\mathbf{A}$ containing a. We are assuming that α is constant. So, for each $i \in \{1, \dots, n\}$, we have

$$\alpha_{\mathbf{Q}}(w_{\mathbf{Q}i}) = \alpha(\overline{w_{\mathbf{Q}i}}) = \alpha(w_i) = w_i(a) = w_{\mathbf{Q}i}(a).$$

Therefore $\alpha_{\mathbf{Q}}$ is given by evaluation at a on the set $\{w_{\mathbf{Q}1}, \dots, w_{\mathbf{Q}n}\}$, which is a contradiction.

We have shown that there is a petal $\mathbf{P}$ of $\mathbf{A}$ for which $\alpha_{\mathbf{P}}$ is an evaluation. Let $b \in P$ such that $\alpha_{\mathbf{P}}$ is given by evaluation at b. Then, since α is constant, we have

$$\alpha(x) = \alpha(\overline{x{\restriction}_P}) = \alpha_{\mathbf{P}}(x{\restriction}_P) = x{\restriction}_P(b) = x(b),$$

for every $x \in \mathcal{A}(\mathbf{A}, \underline{\mathbf{M}})$. Thus α is an evaluation. ∎

The previous lemma can be used to show very easily that every finite zero-kernel unary algebra is dualisable. In order to apply the lemma, we will first show that the quasi-variety generated by a finite zero-kernel algebra is extremely simple.

3.1.7 Lemma *Let $\underline{\mathbf{M}}$ be a finite unary algebra and define $\mathcal{A} := \mathbb{ISP}(\underline{\mathbf{M}})$.*

(i) *If $\underline{\mathbf{M}}$ is a zero-kernel algebra, then, up to isomorphism, the quasi-variety $\mathcal{A}$ has finitely many petals, all of which are finite.*

(ii) *If $\underline{\mathbf{M}}$ is not a zero-kernel algebra, then $\mathcal{A}$ has both arbitrarily large finite petals and infinite petals.*

Proof Assume that $\underline{\mathbf{M}}$ is a zero-kernel algebra. Let F_{p} be the set of all unary term functions of $\underline{\mathbf{M}}$ that are permutations, and let F_{c} be the set of all constant unary term functions of $\underline{\mathbf{M}}$. Then every unary term function of $\underline{\mathbf{M}}$ belongs to $F_{\mathrm{p}} \cup F_{\mathrm{c}}$. For each $u \in F_{\mathrm{p}}$, the map $u^{-1} : M \to M$ is a term function of $\underline{\mathbf{M}}$, as M is finite. Now let $\mathbf{P}$ be a petal of $\mathcal{A}$. Since the graph $G^*(\mathbf{P})$ is connected, it follows that $P \backslash C_{\mathbf{P}}$ is an orbit of F_{p}. Therefore

$$|P| = |C_{\mathbf{P}}| + |P \backslash C_{\mathbf{P}}| \leqslant |F_{\mathrm{c}}| + |F_{\mathrm{p}}|.$$

The quasi-variety $\mathcal{A}$ is locally finite. So, up to isomorphism, there are only finitely many petals of $\mathcal{A}$, all of which are finite.

Now assume that $\underline{\mathbf{M}}$ is not a zero-kernel algebra. There is a unary term function u of $\underline{\mathbf{M}}$ that is neither a constant map nor a permutation. So there are three distinct elements a, b and c of M such that $u(a) = u(b) \neq u(c)$. Let S be a non-empty set and define the algebra $\mathbf{A} := \underline{\mathbf{M}}^S \times \underline{\mathbf{M}}$ in $\mathcal{A}$. Now define the subset $X := \{a, b\}^S \times \{c\}$ of A. Then $u(x) = u(y) \notin C_{\mathbf{A}}$, for all $x, y \in X$. It follows that X is a subset of the vertices of a connected component of the graph $G^*(\mathbf{A})$. So X is contained in a petal of $\mathbf{A}$. Thus $\mathcal{A}$ has both arbitrarily large finite petals and infinite petals. ∎

3.1.8 Theorem *Every finite zero-kernel unary algebra is dualisable.*

Proof Let $\underline{\mathbf{M}}$ be a finite zero-kernel unary algebra and define $\mathcal{A} := \mathbb{ISP}(\underline{\mathbf{M}})$. Using Lemma 3.1.7, we know that, up to isomorphism, there are only finitely many petals of $\mathcal{A}$, all of which are finite. So there is some $n \in \omega \backslash \{0, 1\}$ such that $|\mathcal{A}(\mathbf{P}, \underline{\mathbf{M}})| \leqslant n$, for each petal $\mathbf{P}$ of $\mathcal{A}$. By the Preservation Lemma, 1.4.4, it follows that R_n yields a duality on every petal of $\mathcal{A}$. Thus $\underline{\mathbf{M}}$ is dualised by $\underset{\sim}{\mathbf{M}} = \langle M; R_{n+2}, \mathcal{T} \rangle$, by the Petal Duality Lemma, 3.1.6. ∎

We can use the previous theorem to conclude that all the two-element unary algebras are dualisable, since every unary operation on a two-element set is either a constant map or a permutation.

3.2 One-kernel unary algebras

One-kernel algebras are almost as simple as zero-kernel algebras. In this section, we show that the quasi-variety generated by a finite one-kernel unary algebra $\underline{\mathbf{M}}$ satisfies a very strong finiteness condition. There is a finite set $\mathcal{B}$ of petals of $\mathbb{ISP}(\underline{\mathbf{M}})$ such that every finite petal of $\mathbb{ISP}(\underline{\mathbf{M}})$ is 'nearly isomorphic' to a petal from $\mathcal{B}$. We will then use this finiteness condition to help prove that every finite one-kernel algebra is dualisable.

3.2.1 Example We have already carefully studied the quasi-variety generated by a particular one-kernel algebra. Define $\underline{\mathbf{R}} := \langle \{0, 1, 2\}; 121, 010 \rangle$, as in Definition 1.2.1. The set of unary term functions of $\underline{\mathbf{R}}$ is

$$\{012, 121, 010, 212, 101\}.$$

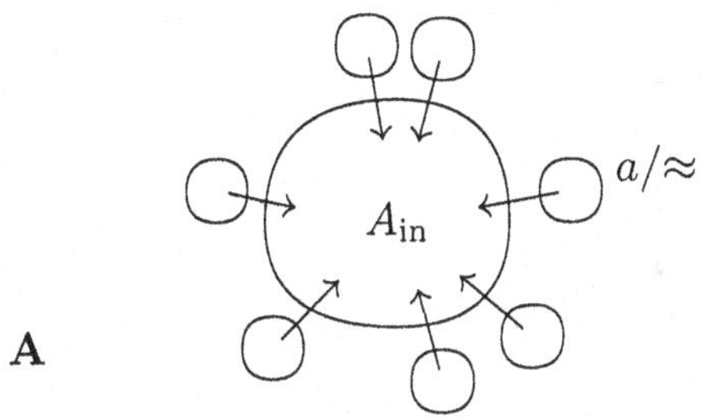

Figure 3.3 Outer and inner elements of a unary algebra

So $\underset{\sim}{\mathbf{R}}$ is a one-kernel unary algebra, with kernel $\{1|02\}$. The algebra $\underset{\sim}{\mathbf{R}}$ does not have any constant term functions, and therefore the petals of an algebra in $\mathbb{ISP}(\underset{\sim}{\mathbf{R}})$ are simply its connected components. In Lemma 1.2.2, we exhibited all the petals of $\mathbb{ISP}(\underset{\sim}{\mathbf{R}})$. We showed that every non-trivial petal of $\mathbb{ISP}(\underset{\sim}{\mathbf{R}})$ is isomorphic to $\mathbf{Tr}_I$, for some set I, or to $\mathbf{Sq}_{IJ}$, for some sets I and J. So the non-trivial petals of $\mathbb{ISP}(\underset{\sim}{\mathbf{R}})$ come in two different kinds. We shall now develop a way to describe the similarity between the petals of the same kind.

Let $\mathbf{A}$ be an algebra. For each subset B of A, we define $\mathrm{sg}_{\mathbf{A}}(B)$ to be the subuniverse of $\mathbf{A}$ generated by B. For each $a \in A$, we shall write $\mathrm{sg}_{\mathbf{A}}(a)$ rather than $\mathrm{sg}_{\mathbf{A}}(\{a\})$.

Now assume that $\mathbf{A}$ is a finite unary algebra. We say that $a \in A$ is an **outer element of A** if $\mathrm{sg}_{\mathbf{A}}(a)$ is a maximal one-generated subuniverse of $\mathbf{A}$. The members of A that are not outer elements of $\mathbf{A}$ are called **inner elements of A**. Define A_{out} to be the set of all outer elements of $\mathbf{A}$, and A_{in} to be the set of all inner elements of $\mathbf{A}$. Then A_{out} is a generating set for $\mathbf{A}$, and A_{in} is a subuniverse of $\mathbf{A}$.

3.2.2 Remark The previous two definitions give us a new way to partition unary algebras. Let $\mathbf{A}$ be a finite unary algebra. There is a natural equivalence relation $\approx$ on A_{out}, given by $a \approx b \iff \mathrm{sg}_{\mathbf{A}}(a) = \mathrm{sg}_{\mathbf{A}}(b)$. So

$$\{\, a/{\approx} \mid a \in A_{\text{out}} \,\} \cup \big(\{A_{\text{in}}\}\backslash\{\varnothing\}\big)$$

is a partition of A. In Figure 3.3, this partition is shown for some particular algebra $\mathbf{A}$. The figure also shows how this partition interacts with the structure of $\mathbf{A}$: we know that A_{in} is a subuniverse of $\mathbf{A}$ and, for each $a \in A_{\text{out}}$, we have $\mathrm{sg}_{\mathbf{A}}(a) \subseteq (a/{\approx}) \cup A_{\text{in}}$.

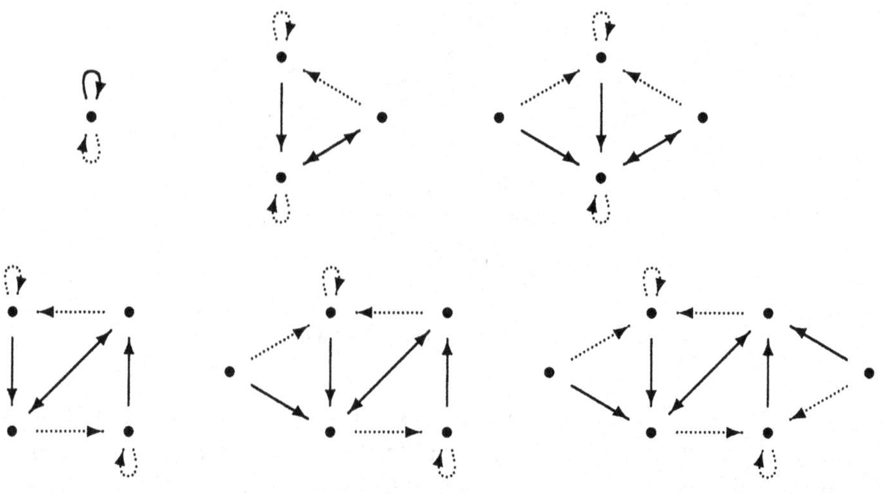

Figure 3.4 A gentle basis for $\mathbb{ISP}(\underline{\mathbf{R}})$

A surjection $\varphi : \mathbf{A} \twoheadrightarrow \mathbf{B}$ is called **gentle** if φ is one-to-one on $A_{\text{in}} \cup \text{sg}_{\mathbf{A}}(a)$, for all $a \in A_{\text{out}}$. Now let $\underline{\mathbf{M}}$ be a finite unary algebra and let $\mathcal{B}$ be a set of finite petals of $\mathcal{A} := \mathbb{ISP}(\underline{\mathbf{M}})$. We say that $\mathcal{B}$ is a **gentle basis for $\mathcal{A}$** if, for every finite petal $\mathbf{P}$ of $\mathcal{A}$, there is a gentle surjection $\varphi : \mathbf{P} \twoheadrightarrow \mathbf{B}$, for some $\mathbf{B} \in \mathcal{B}$.

3.2.3 Example Again, consider the unary algebra $\underline{\mathbf{R}} = \langle\{0, 1, 2\}; 121, 010\rangle$ first defined in 1.2.1. We know that every non-trivial petal of $\mathbb{ISP}(\underline{\mathbf{R}})$ is isomorphic to $\mathbf{Tr}_I$, for some set I, or to $\mathbf{Sq}_{IJ}$, for some sets I and J. It is now easy to check that every finite petal of $\mathbb{ISP}(\underline{\mathbf{R}})$ has a gentle surjection onto one of the six algebras shown in Figure 3.4. For example, for each non-empty set I, there is a gentle surjection from the petal $\mathbf{Tr}_I$ onto the petal $\mathbf{Tr}_{\{0\}}$, which is shown in the top-right of the figure. So the petals in Figure 3.4 form a gentle basis for $\mathbb{ISP}(\underline{\mathbf{R}})$.

Gentle surjections are so called because they do not destroy too much of the structure of the algebra they act on. Loosely speaking, gentle surjections can only collapse repeated structure on the outside of the algebra. This is the case, for example, for the gentle surjection from $\mathbf{Tr}_I$ onto $\mathbf{Tr}_{\{0\}}$, for each non-empty set I.

3.2.4 Lemma *Let $\mathbf{A}$ be a finite unary algebra and let $\varphi : \mathbf{A} \twoheadrightarrow \mathbf{B}$ be a gentle surjection. Then φ is a retraction and, for each subalgebra $\mathbf{C}$ of $\mathbf{A}$ such that $\varphi{\restriction}_C$ is one-to-one, there is a coretraction $\psi : \mathbf{B} \hookrightarrow \mathbf{A}$ for φ with $C \subseteq \psi(B)$.*

Proof Let $\mathbf{C}$ be a subalgebra of $\mathbf{A}$ such that $\varphi{\restriction}_C$ is one-to-one. (For example, we could choose $\mathbf{sg_A}(a)$, for any $a \in A$.) Define the equivalence relation $\equiv_\varphi$ on A_{out} by

$$a \equiv_\varphi b \iff \varphi(\mathrm{sg}_{\mathbf{A}}(a)) = \varphi(\mathrm{sg}_{\mathbf{A}}(b)).$$

We seek a transversal $\mathcal{T}$ of the blocks of $\equiv_\varphi$ for which $C \cap A_{\text{out}} \subseteq \mathrm{sg}_{\mathbf{A}}(\mathcal{T})$. To see that such transversals exist, let $a, b \in C \cap A_{\text{out}}$ with $a \equiv_\varphi b$. Then

$$\varphi(a) \in \varphi(\mathrm{sg}_{\mathbf{A}}(a)) = \varphi(\mathrm{sg}_{\mathbf{A}}(b)) \text{ and } \varphi(b) \in \varphi(\mathrm{sg}_{\mathbf{A}}(b)) = \varphi(\mathrm{sg}_{\mathbf{A}}(a)).$$

Since $\varphi{\restriction}_C$ is one-to-one, this implies that $a \in \mathrm{sg}_{\mathbf{A}}(b)$ and $b \in \mathrm{sg}_{\mathbf{A}}(a)$, whence $\mathrm{sg}_{\mathbf{A}}(a) = \mathrm{sg}_{\mathbf{A}}(b)$. It follows that there is a transversal $\mathcal{T}$ of $A_{\text{out}}/{\equiv_\varphi}$ such that $C \cap A_{\text{out}} \subseteq \mathrm{sg}_{\mathbf{A}}(\mathcal{T})$. This gives us $C \subseteq A_{\text{in}} \cup \mathrm{sg}_{\mathbf{A}}(\mathcal{T})$.

As the homomorphism φ is gentle, it is one-to-one on $A_{\text{in}} \cup \mathrm{sg}_{\mathbf{A}}(a)$, for each $a \in \mathcal{T}$. We wish to show that $\psi : \mathbf{B} \to \mathbf{A}$, given by

$$\psi := \bigcup \{ \left(\varphi{\restriction}_{A_{\text{in}} \cup \mathrm{sg}_{\mathbf{A}}(a)}\right)^{-1} \mid a \in \mathcal{T} \},$$

is a well-defined coretraction for φ with $C \subseteq \psi(B)$.

Claim 1 The relation ψ is a well-defined homomorphism.

The range of ψ is certainly contained in A. The domain of ψ is B, since

$$B = \varphi(A) = \varphi(A_{\text{in}} \cup \mathrm{sg}_{\mathbf{A}}(\mathcal{T})) = \bigcup \{ \varphi(A_{\text{in}} \cup \mathrm{sg}_{\mathbf{A}}(a)) \mid a \in \mathcal{T} \}.$$

Now let $a, b \in \mathcal{T}$ with $a \neq b$. It remains to check that the homomorphisms

$$\left(\varphi{\restriction}_{A_{\text{in}} \cup \mathrm{sg}_{\mathbf{A}}(a)}\right)^{-1} \text{ and } \left(\varphi{\restriction}_{A_{\text{in}} \cup \mathrm{sg}_{\mathbf{A}}(b)}\right)^{-1}$$

agree on

$$\varphi(A_{\text{in}} \cup \mathrm{sg}_{\mathbf{A}}(a)) \cap \varphi(A_{\text{in}} \cup \mathrm{sg}_{\mathbf{A}}(b)).$$

So let $c_a \in A_{\text{in}} \cup \mathrm{sg}_{\mathbf{A}}(a)$ and $c_b \in A_{\text{in}} \cup \mathrm{sg}_{\mathbf{A}}(b)$ such that $\varphi(c_a) = \varphi(c_b)$. We shall prove that $c_a = c_b$. Then we will have

$$\left(\varphi{\restriction}_{A_{\text{in}} \cup \mathrm{sg}_{\mathbf{A}}(a)}\right)^{-1}(\varphi(c_a)) = c_a = c_b = \left(\varphi{\restriction}_{A_{\text{in}} \cup \mathrm{sg}_{\mathbf{A}}(b)}\right)^{-1}(\varphi(c_b)),$$

and it will follow that ψ is a well-defined homomorphism.

Suppose, by way of contradiction, that both c_a and c_b are outer elements of $\mathbf{A}$. Then $\mathrm{sg}_{\mathbf{A}}(c_a) = \mathrm{sg}_{\mathbf{A}}(a)$ and $\mathrm{sg}_{\mathbf{A}}(c_b) = \mathrm{sg}_{\mathbf{A}}(b)$. But this gives us

$$\begin{aligned}\varphi(\mathrm{sg}_{\mathbf{A}}(a)) &= \varphi(\mathrm{sg}_{\mathbf{A}}(c_a)) = \mathrm{sg}_{\mathbf{B}}(\varphi(c_a)) \\ &= \mathrm{sg}_{\mathbf{B}}(\varphi(c_b)) = \varphi(\mathrm{sg}_{\mathbf{A}}(c_b)) = \varphi(\mathrm{sg}_{\mathbf{A}}(b)),\end{aligned}$$

and therefore $a \equiv_\varphi b$. Since $a, b \in \mathcal{T}$ with $a \neq b$, this is a contradiction.

We can now assume that $c_a \in A_{\text{in}}$, whence $c_a, c_b \in A_{\text{in}} \cup \text{sg}_{\mathbf{A}}(b)$. As φ is one-to-one on $A_{\text{in}} \cup \text{sg}_{\mathbf{A}}(b)$ and $\varphi(c_a) = \varphi(c_b)$, we get $c_a = c_b$, as required.

Claim 2 The homomorphism ψ is a coretraction for φ with $C \subseteq \psi(B)$.

Let $b \in B$. Then there is some $a \in \mathcal{T}$ with $b \in \varphi\big(A_{\text{in}} \cup \text{sg}_{\mathbf{A}}(a)\big)$. We have

$$\varphi \circ \psi(b) = \varphi \circ \big(\varphi{\restriction}_{A_{\text{in}} \cup \text{sg}_{\mathbf{A}}(a)}\big)^{-1}(b) = b.$$

So ψ is a coretraction for φ, with $C \subseteq A_{\text{in}} \cup \text{sg}_{\mathbf{A}}(\mathcal{T}) = \psi(B)$. ∎

We can now clarify the statements in the introduction to this section. Consider a gentle surjection $\varphi : \mathbf{A} \twoheadrightarrow \mathbf{B}$. For each $a \in A$, the map φ is one-to-one on $\text{sg}_{\mathbf{A}}(a)$. So, for every $a \in A$, there is a coretraction $\psi_a : \mathbf{B} \hookrightarrow \mathbf{A}$ for φ with $a \in \psi_a(B)$, by Lemma 3.2.4. Therefore $\mathbf{A}$ can be covered with the images of coretractions for φ. This tells us that $\mathbf{B}$ retains most of the structure of $\mathbf{A}$. So gentle surjections really are 'nearly isomorphisms', and having a finite gentle basis is a very strong finiteness condition on a quasi-variety. We will prove that the quasi-variety generated by a finite one-kernel algebra must have a finite gentle basis.

3.2.5 Lemma *Let $\underline{\mathbf{M}}$ be a one-kernel unary algebra with kernel θ. Then every unary term function of $\underline{\mathbf{M}}$ preserves θ.*

Proof There must be a unary term function u of $\underline{\mathbf{M}}$ with $\ker(u) = \theta$. The map u is neither constant nor a permutation. Now let v be any unary term function of $\underline{\mathbf{M}}$. If v is constant or $\ker(v) = \theta$, then v preserves θ. So we can assume that v is a permutation. Let $(a, b) \in \theta$. The term function $u \circ v$ of $\underline{\mathbf{M}}$ is neither constant nor a permutation. Therefore $\ker(u \circ v) = \theta$. This implies that $u \circ v(a) = u \circ v(b)$, and so $(v(a), v(b)) \in \theta$. Thus v preserves θ. ∎

3.2.6 Lemma *Let $\underline{\mathbf{M}}$ be a finite one-kernel unary algebra and let $n := |M|$. Then each finite petal of $\mathbb{ISP}(\underline{\mathbf{M}})$ has at most $n! + n$ inner elements.*

Proof Let θ be the kernel of $\underline{\mathbf{M}}$ and let $\mathbf{P}$ be a finite petal of $\mathbb{ISP}(\underline{\mathbf{M}})$. Every subalgebra of $\mathbf{P}$ contains the centre $C_{\mathbf{P}}$. Since $P \backslash C_{\mathbf{P}}$ must be non-empty, we have $C_{\mathbf{P}} \subseteq P_{\text{in}}$.

We can assume that $\mathbf{P} \leqslant \underline{\mathbf{M}}^S$, for some non-empty set S. Each element $a \in P$ determines two partitions of S:

$$\mathcal{P}(a) := \{\, a^{-1}(m) \mid m \in M \,\} \backslash \{\varnothing\}$$

and

$$\mathcal{P}_\theta(a) := \{\, a^{-1}(m/\theta) \mid m \in M \,\} \backslash \{\varnothing\}.$$

We shall prove that there is a partition $\mathcal{Q}$ of S, with at most n blocks, such that $\mathcal{P}(a) = \mathcal{Q}$, for all $a \in P_{\text{in}} \backslash C_{\mathbf{P}}$. This will tell us that, for each $a \in P_{\text{in}} \backslash C_{\mathbf{P}}$, the element a takes a different value from M on each block of $\mathcal{Q}$. As $|M| = n$, it will follow that $|P_{\text{in}} \backslash C_{\mathbf{P}}| \leqslant n!$, and therefore

$$|P_{\text{in}}| = |C_{\mathbf{P}}| + |P_{\text{in}} \backslash C_{\mathbf{P}}| \leqslant n + n!,$$

as required.

Now choose some $b \in P_{\text{out}}$ and define the partition $\mathcal{Q} := \mathcal{P}_\theta(b)$ of S. The number of blocks of $\mathcal{Q}$ is at most $|M/\theta| \leqslant |M| = n$. Next define the subset

$$P_{\mathcal{Q}} := \{\, a \in P_{\text{in}} \backslash C_{\mathbf{P}} \mid \mathcal{P}(a) = \mathcal{Q} \,\} \cup \{\, a \in P_{\text{out}} \mid \mathcal{P}_\theta(a) = \mathcal{Q} \,\}$$

of $P \backslash C_{\mathbf{P}}$. We wish to show that $P_{\mathcal{Q}} = P \backslash C_{\mathbf{P}}$. Since $\mathbf{P}$ is a petal, the graph $G^*(\mathbf{P})$ is connected. So it is enough to show that $P_{\mathcal{Q}}$ forms a connected component of $G^*(\mathbf{P})$. Let $a \in P \backslash C_{\mathbf{P}}$ and let u be a unary term function of $\underline{\mathbf{M}}$ such that $u(a) \notin C_{\mathbf{P}}$. We will show that $a \in P_{\mathcal{Q}}$ if and only if $u(a) \in P_{\mathcal{Q}}$. As $b \in P_{\mathcal{Q}}$, it will then follow that $P_{\mathcal{Q}} = P \backslash C_{\mathbf{P}}$.

Case 1: $u(a) \in P_{\text{out}}$. Since $\operatorname{sg}_{\mathbf{P}}(u(a)) \subseteq \operatorname{sg}_{\mathbf{P}}(a)$ and $u(a)$ is an outer element of $\mathbf{P}$, we must have $\operatorname{sg}_{\mathbf{P}}(u(a)) = \operatorname{sg}_{\mathbf{P}}(a)$ and $a \in P_{\text{out}}$. There is a unary term function v of $\underline{\mathbf{M}}$ with $v \circ u(a) = a$. Both u and v preserve θ, by Lemma 3.2.5. So $\mathcal{P}_\theta(a) = \mathcal{P}_\theta(u(a))$. Thus $a \in P_{\mathcal{Q}}$ if and only if $u(a) \in P_{\mathcal{Q}}$.

Case 2: $a \in P_{\text{out}}$ and $u(a) \in P_{\text{in}}$. We must have $a \notin \operatorname{sg}_{\mathbf{P}}(u(a))$. Therefore the term function u of $\underline{\mathbf{M}}$ is not a permutation. (Otherwise, the finiteness of M implies that u^{-1} is a term function of $\underline{\mathbf{M}}$.) Since $u(a) \notin C_{\mathbf{P}}$, the term function u of $\underline{\mathbf{M}}$ cannot be constant. Therefore $\ker(u) = \theta$, and it follows that $\mathcal{P}(u(a)) = \mathcal{P}_\theta(a)$. So $a \in P_{\mathcal{Q}}$ if and only if $u(a) \in P_{\mathcal{Q}}$.

Case 3: $a \in P_{\text{in}}$. This implies that $u(a) \in P_{\text{in}}$. The algebra $\mathbf{P}$ is generated by its outer elements. So there is $c \in P_{\text{out}}$ and a unary term function v of $\underline{\mathbf{M}}$ such that $v(c) = a$. By the previous case, we have $c \in P_{\mathcal{Q}}$ if and only if $v(c) \in P_{\mathcal{Q}}$. The previous case also tells us that $c \in P_{\mathcal{Q}}$ if and only if $u \circ v(c) \in P_{\mathcal{Q}}$. So $a \in P_{\mathcal{Q}}$ if and only if $u(a) \in P_{\mathcal{Q}}$.

We have now shown that $\mathcal{P}(a) = \mathcal{Q}$, for all $a \in P_{\text{in}} \backslash C_{\mathbf{P}}$. So, as explained earlier, it follows that $|P_{\text{in}}| \leqslant n + n!$. ∎

3.2.7 Lemma *Let $\underline{\mathbf{M}}$ be a finite one-kernel unary algebra. Then $\mathbb{ISP}(\underline{\mathbf{M}})$ has a finite gentle basis.*

Proof Define $n := |M|$. Then the one-generated free algebra in the quasi-variety $\mathcal{A} := \mathbb{ISP}(\underline{\mathbf{M}})$ has at most n^n elements. So there is some $k \in \omega$ such that, up to isomorphism, there are only k one-generated algebras in $\mathcal{A}$. Now let $\mathbf{P}$ be a finite petal of $\mathcal{A}$. We will show that there is a gentle surjection from $\mathbf{P}$ onto a petal $\mathbf{B}$ of $\mathcal{A}$ with $|B| \leqslant n^n k(n! + n)!$. Since $\mathcal{A}$ is locally finite, it will then follow that $\mathcal{A}$ has a finite gentle basis.

Define $\mathcal{S}$ to be the set of all maximal one-generated subuniverses of $\mathbf{P}$. Then $\mathcal{S} = \{\, \mathrm{sg}_{\mathbf{P}}(a) \mid a \in P_{\mathrm{out}} \,\}$. Define the equivalence relation $\equiv$ on $\mathcal{S}$ by $S \equiv T$ if and only if there is an isomorphism from $\mathbf{S}$ to $\mathbf{T}$ that fixes each element of $S \cap P_{\mathrm{in}}$. Choose a transversal $\mathcal{T}$ of the blocks of $\equiv$ and define the subalgebra $\mathbf{B}$ of $\mathbf{P}$ by $B := \bigcup \mathcal{T}$.

Claim 1 There is a gentle surjection from $\mathbf{P}$ onto $\mathbf{B}$.

For each pair $(S, T) \in \equiv$, choose an isomorphism $\eta_{ST} : \mathbf{S} \hookrightarrow\!\!\!\!\rightarrow \mathbf{T}$ that fixes each element of $S \cap P_{\mathrm{in}}$. We want to define a gentle surjection $\varphi : \mathbf{P} \twoheadrightarrow \mathbf{B}$ by

$$\varphi := \bigcup \{\, \eta_{ST} \mid T \in \mathcal{T} \text{ and } S \in T/{\equiv} \,\}.$$

First note that

$$P = \bigcup \mathcal{S} = \bigcup \{\, S \mid T \in \mathcal{T} \text{ and } S \in T/{\equiv} \,\}.$$

Now consider $\eta_{S_1T_1}$ and $\eta_{S_2T_2}$ such that $S_1 \neq S_2$, where $T_i \in \mathcal{T}$ and $S_i \in T_i/{\equiv}$, for $i \in \{1, 2\}$. Since S_1 is a maximal one-generated subuniverse of $\mathbf{P}$, the subuniverse $S_1 \cap S_2$ of $\mathbf{P}$ is properly contained in S_1. So $S_1 \cap S_2 \subseteq P_{\mathrm{in}}$, and therefore $\eta_{S_1T_1}$ and $\eta_{S_2T_2}$ agree on $S_1 \cap S_2$. It follows that φ is a well-defined surjection, with $\varphi{\restriction}_{P_{\mathrm{in}}} = \mathrm{id}_{P_{\mathrm{in}}}$.

To check that φ is gentle, let $a \in P_{\mathrm{out}}$. Then $S := \mathrm{sg}_{\mathbf{P}}(a)$ belongs to $\mathcal{S}$, and so $S \in T/{\equiv}$, for some $T \in \mathcal{T}$. We want to show that φ is one-to-one on $P_{\mathrm{in}} \cup \mathrm{sg}_{\mathbf{P}}(a)$. We have

$$\varphi{\restriction}_{P_{\mathrm{in}} \cup \mathrm{sg}_{\mathbf{P}}(a)} = \varphi{\restriction}_{P_{\mathrm{in}} \cup S} = \mathrm{id}_{P_{\mathrm{in}}} \cup \eta_{ST}.$$

Since η_{ST} is one-to-one, it is enough to show that $\eta_{ST}(S \backslash P_{\mathrm{in}}) \subseteq T \backslash P_{\mathrm{in}}$. The maximality of S and T guarantees that $S_{\mathrm{out}} = S \cap P_{\mathrm{out}}$ and $T_{\mathrm{out}} = T \cap P_{\mathrm{out}}$. As η_{ST} is an isomorphism, we get

$$\eta_{ST}(S \backslash P_{\mathrm{in}}) = \eta_{ST}(S_{\mathrm{out}}) = T_{\mathrm{out}} = T \backslash P_{\mathrm{in}}.$$

Thus φ is one-to-one on $P_{\mathrm{in}} \cup \mathrm{sg}_{\mathbf{P}}(a)$, whence φ is a gentle surjection.

Claim 2 The algebra $\mathbf{B}$ is a petal of $\mathcal{A}$.

By Claim 1, we know there is a gentle surjection $\varphi : \mathbf{P} \twoheadrightarrow \mathbf{B}$. We begin by showing that $\varphi^{-1}(C_{\mathbf{B}}) = C_{\mathbf{P}}$. Clearly, $C_{\mathbf{P}} \subseteq \varphi^{-1}(C_{\mathbf{B}})$. So let $a \in \varphi^{-1}(C_{\mathbf{B}})$. By Lemma 3.2.4, there is a coretraction $\psi : \mathbf{B} \hookrightarrow \mathbf{P}$ for φ with $a \in \psi(B)$. This implies that

$$a = \psi \circ \varphi(a) \in \psi(C_{\mathbf{B}}) \subseteq C_{\mathbf{P}}.$$

Thus $\varphi^{-1}(C_{\mathbf{B}}) = C_{\mathbf{P}}$. Since $\mathbf{P}$ is a petal of $\mathcal{A}$, the graph $G^*(\mathbf{P})$ is connected. As $\varphi : \mathbf{P} \twoheadrightarrow \mathbf{B}$ is a surjective homomorphism such that $\varphi^{-1}(C_{\mathbf{B}}) = C_{\mathbf{P}}$, it follows that the graph $G^*(\mathbf{B})$ is also connected. So $\mathbf{B}$ is a petal of $\mathcal{A}$.

Claim 3 We have $|B| \leqslant n^n k(n! + n)!$.

Each one-generated algebra in $\mathcal{A}$ has at most n^n elements. Therefore we have $|B| = \left|\bigcup \mathcal{T}\right| \leqslant n^n|\mathcal{T}|$. Each $T \in \mathcal{T}$ determines a one-generated algebra $\mathbf{T}$ from $\mathcal{A}$. We know that, up to isomorphism, there are only k one-generated algebras in $\mathcal{A}$. So we want to bound the number of members of $\mathcal{T}$ that can determine isomorphic algebras.

Let $\mathbf{A}$ be any one-generated algebra in $\mathcal{A}$ and assume that $T_1, T_2 \in \mathcal{T}$, with $T_1 \neq T_2$, such that there are isomorphisms $\varphi_1 : \mathbf{A} \hookrightarrow\!\!\!\!\rightarrow \mathbf{T}_1$ and $\varphi_2 : \mathbf{A} \hookrightarrow\!\!\!\!\rightarrow \mathbf{T}_2$. Then $\varphi_2 \circ \varphi_1^{-1} : \mathbf{T}_1 \hookrightarrow\!\!\!\!\rightarrow \mathbf{T}_2$ is an isomorphism. Since $T_1 \not\equiv T_2$, we must have $\varphi_2 \circ \varphi_1^{-1}(a) \neq a$, for some $a \in T_1 \cap P_{\text{in}}$. As

$$\varphi_1(A_{\text{in}}) = (T_1)_{\text{in}} = T_1 \cap P_{\text{in}},$$

this gives us $b := \varphi_1^{-1}(a) \in A_{\text{in}}$ with $\varphi_1(b) \neq \varphi_2(b)$. So the embeddings $\varphi_1{\restriction}_{A_{\text{in}}} : \mathbf{A}_{\text{in}} \hookrightarrow \mathbf{P}_{\text{in}}$ and $\varphi_2{\restriction}_{A_{\text{in}}} : \mathbf{A}_{\text{in}} \hookrightarrow \mathbf{P}_{\text{in}}$ are different. By the previous lemma, we have $|P_{\text{in}}| \leqslant n! + n$. So the number of different embeddings from $\mathbf{A}_{\text{in}}$ into $\mathbf{P}_{\text{in}}$ is at most $|P_{\text{in}}|! \leqslant (n! + n)!$. This implies that there are at most $(n! + n)!$ subuniverses T of $\mathbf{P}$ belonging to $\mathcal{T}$ such that $\mathbf{T}$ is isomorphic to $\mathbf{A}$. Finally, as there are exactly k non-isomorphic one-generated algebras in $\mathcal{A}$, we get $|B| \leqslant n^n|\mathcal{T}| \leqslant n^n k(n! + n)!$. ∎

The next lemma shows that a finite unary algebra must be a zero- or one-kernel algebra in order to generate a quasi-variety with a finite gentle basis.

3.2.8 Lemma *Let $\underline{\mathbf{M}}$ be a finite unary algebra with at least two kernels. Then $\mathbb{ISP}(\underline{\mathbf{M}})$ does not have a finite gentle basis.*

Proof There are unary term functions u and v of $\underline{\mathbf{M}}$, neither of which is a constant map or a permutation, such that $\ker(u) \nsubseteq \ker(v)$. So there are elements $a, b \in M$ with $u(a) = u(b)$ and $v(a) \neq v(b)$. As u is not constant and v is not a permutation, there exist $c, d, e \in M$, with $d \neq e$, such that $u(b) \neq u(c)$ and $v(d) = v(e)$.

Now let $n \in \omega\backslash\{0\}$. We will construct a finite petal $\mathbf{P}$ of $\mathbb{ISP}(\underline{\mathbf{M}})$ with $|P_{\text{in}}| \geqslant 2^n$. Each gentle surjection from $\mathbf{P}$ must be one-to-one on P_{in}. So it will then follow that $\mathbb{ISP}(\underline{\mathbf{M}})$ does not have a finite gentle basis.

Define the subset X of M^{n+3} by

$$X := \{a, b\}^n \times \{c\} \times \{d\} \times \{e\}.$$

Since $u(a) = u(b) \neq u(c)$, we must have $u(x) = u(y) \notin C_{\underline{\mathbf{M}}^{n+3}}$, for all $x, y \in X$. So X is a subset of the vertices of a connected component of the graph $G^*(\underline{\mathbf{M}}^{n+3})$. Consequently, the set X is contained in a petal $\mathbf{P}$ of $\underline{\mathbf{M}}^{n+3}$. For every $x \in X$, we have $x \notin \mathrm{sg}_{\mathbf{P}}(v(x))$, as $v(d) = v(e)$. Therefore $v(x) \in P_{\mathrm{in}}$, for each $x \in X$. Since $v(a) \neq v(b)$, this implies that

$$|P_{\mathrm{in}}| \geqslant \big|\{\, v(x) \mid x \in X \,\}\big| = 2^n.$$

Thus $\mathbb{ISP}(\underline{\mathbf{M}})$ does not have a finite gentle basis. ∎

Lemmas 3.1.7, 3.2.7 and 3.2.8 tell us exactly which finite unary algebras generate a quasi-variety with a finite gentle basis.

3.2.9 Theorem *Let $\underline{\mathbf{M}}$ be a finite unary algebra. Then $\mathbb{ISP}(\underline{\mathbf{M}})$ has a finite gentle basis if and only if $\underline{\mathbf{M}}$ is a zero-kernel or one-kernel algebra.*

We can now finish this section by establishing the dualisability of finite one-kernel unary algebras.

3.2.10 Theorem *Every finite one-kernel unary algebra is dualisable.*

Proof Let $\underline{\mathbf{M}}$ be a finite one-kernel unary algebra. By Theorem 3.2.9, there is a finite gentle basis $\mathcal{B}$ for $\mathcal{A} := \mathbb{ISP}(\underline{\mathbf{M}})$. As $\mathcal{B}$ consists of a finite number of finite algebras, we can choose $n \in \omega$ such that $n \geqslant |\mathcal{A}(\mathbf{B}, \underline{\mathbf{M}})| + 3$, for all $\mathbf{B} \in \mathcal{B}$. We shall prove that R_n yields a duality on every finite petal of $\mathcal{A}$. By the Petal Duality Lemma, 3.1.6, it will then follow that $\underset{\sim}{\mathbf{M}} := \langle M; R_{n+2}, \mathcal{T}\rangle$ dualises $\underline{\mathbf{M}}$.

Let $\mathbf{P}$ be a finite petal of $\mathcal{A}$. There is a gentle surjection $\varphi_0 : \mathbf{P} \twoheadrightarrow \mathbf{B}_0$, for some $\mathbf{B}_0 \in \mathcal{B}$. Since $\varphi_0{\restriction}_{P_{\mathrm{in}}}$ is one-to-one, we can use Lemma 3.2.4 to find a coretraction $\psi_0 : \mathbf{B}_0 \hookrightarrow \mathbf{P}$ for φ_0 such that $P_{\mathrm{in}} \subseteq \psi_0(B_0)$. Define the subalgebra $\mathbf{B} := \psi_0(\mathbf{B}_0)$ of $\mathbf{P}$. Then $\mathbf{B}$ is isomorphic to $\mathbf{B}_0$, and there is a gentle surjection $\varphi : \mathbf{P} \twoheadrightarrow \mathbf{B}$, given by $\varphi := \psi_0 \circ \varphi_0$, with $\varphi{\restriction}_{P_{\mathrm{in}}} = \mathrm{id}_{P_{\mathrm{in}}}$.

Now let $\alpha : \mathcal{A}(\mathbf{P}, \underline{\mathbf{M}}) \to M$ preserve R_n. We will be finished once we have shown that α is an evaluation.

Case 1: $\alpha(x) = \alpha(x \circ \varphi)$, for all $x \in \mathcal{A}(\mathbf{P}, \underline{\mathbf{M}})$. Since $\mathbf{B}$ is isomorphic to an algebra in $\mathcal{B}$, we have $|\mathcal{A}(\mathbf{B}, \underline{\mathbf{M}})| \leqslant n - 3$. As the map α preserves R_n, we can use the Preservation Lemma, 1.4.4, to find some $a \in P$ such that α is given by evaluation at a on the subset $\{\, w \circ \varphi \mid w \in \mathcal{A}(\mathbf{B}, \underline{\mathbf{M}}) \,\}$ of $\mathcal{A}(\mathbf{P}, \underline{\mathbf{M}})$. For all $x \in \mathcal{A}(\mathbf{P}, \underline{\mathbf{M}})$, we must have

$$\alpha(x) = \alpha(x \circ \varphi) = x \circ \varphi(a).$$

So α is given by evaluation at $\varphi(a)$.

Case 2: $\alpha(y) \neq \alpha(y \circ \varphi)$, for some $y \in \mathcal{A}(\mathbf{P}, \underline{\mathbf{M}})$. Since $\mathbf{P}$ is finite, there is some $k \in \omega \backslash \{0\}$ with $P_{\mathrm{out}} = \{a_1, \ldots, a_k\}$. We shall construct a sequence

$y = y_0, \dots, y_k = y \circ \varphi$ of homomorphisms in $\mathcal{A}(\mathbf{P}, \underline{\mathbf{M}})$ such that

$$P_{\text{in}} \subseteq \text{eq}(y_i, y \circ \varphi) \quad \text{and} \quad P \backslash \text{sg}_{\mathbf{P}}(a_{i+1}) \subseteq \text{eq}(y_i, y_{i+1}),$$

for all $i \in \{0, \dots, k-1\}$. As $\varphi{\restriction}_{P_{\text{in}}} = \text{id}_{P_{\text{in}}}$, we can define $y_0 := y$. Now let $i \in \{0, \dots, k-1\}$ and assume that y_i has been defined. As $\text{sg}_{\mathbf{P}}(a_{i+1})$ is a maximal one-generated subuniverse of $\mathbf{P}$, we have

$$\text{sg}_{\mathbf{P}}(a) \cap \text{sg}_{\mathbf{P}}(a_{i+1}) \subseteq P_{\text{in}},$$

for all $a \in P \backslash \text{sg}_{\mathbf{P}}(a_{i+1})$. Since y_i and $y \circ \varphi$ agree on P_{in}, this means that we can define the homomorphism y_{i+1} in $\mathcal{A}(\mathbf{P}, \underline{\mathbf{M}})$ by

$$y_{i+1} := y_i{\restriction}_{P \backslash \text{sg}_{\mathbf{P}}(a_{i+1})} \cup y \circ \varphi{\restriction}_{\text{sg}_{\mathbf{P}}(a_{i+1})}.$$

As P_{out} is a generating set for $\mathbf{P}$, we get $y_k = y \circ \varphi$.

Since $\alpha(y_0) = \alpha(y) \neq \alpha(y \circ \varphi) = \alpha(y_k)$, there is some $j \in \{0, \dots, k-1\}$ such that $\alpha(y_j) \neq \alpha(y_{j+1})$. By the Preservation Lemma, there is some $b \in P$ for which α is given by evaluation at b on

$$\{y_j, y_{j+1}\} \cup \{\, w \circ \varphi \mid w \in \mathcal{A}(\mathbf{B}, \underline{\mathbf{M}}) \,\}.$$

The maps y_j and y_{j+1} agree on $P \backslash \text{sg}_{\mathbf{P}}(a_{j+1})$. Since

$$y_j(b) = \alpha(y_j) \neq \alpha(y_{j+1}) = y_{j+1}(b),$$

it follows that $b \in \text{sg}_{\mathbf{P}}(a_{j+1})$, We will finish the proof for this case by showing that α is given evaluation at b.

Let $x \in \mathcal{A}(\mathbf{P}, \underline{\mathbf{M}})$. Then there is some $c \in \text{sg}_{\mathbf{P}}(a_{j+1})$ such that α is given by evaluation at c on

$$\{x, y_j, y_{j+1}\} \cup \{\, w \circ \varphi \mid w \in \mathcal{A}(\mathbf{B}, \underline{\mathbf{M}}) \,\}.$$

As φ is gentle, the map φ is one-to-one on $\text{sg}_{\mathbf{P}}(a_{j+1})$. So, by Lemma 3.2.4, there is a coretraction $\psi : \mathbf{B} \hookrightarrow \mathbf{P}$ for φ with $\text{sg}_{\mathbf{P}}(a_{j+1}) \subseteq \psi(B)$. We have $\psi \circ \varphi{\restriction}_{\psi(B)} = \text{id}_{\psi(B)}$ and $b, c \in \text{sg}_{\mathbf{P}}(a_{j+1}) \subseteq \psi(B)$. Therefore

$$\alpha(x) = x(c) = x \circ \psi \circ \varphi(c) = \alpha(x \circ \psi \circ \varphi) = x \circ \psi \circ \varphi(b) = x(b),$$

whence α is an evaluation. ∎

3.3 Dualisable two-kernel three-element unary algebras

Were it not for the two-kernel algebras, the characterisation of dualisability for three-element unary algebras would be very simple. All of the zero- and one-kernel algebras are dualisable, and none of the three-kernel algebras is dualisable. It is only amongst the two-kernel algebras that the dualisable and non-dualisable algebras are hard to differentiate. We shall split the class of two-kernel three-element unary algebras into four further types. Two of these types will be exclusively dualisable, and the other two will be exclusively non-dualisable. First, we will show that it is enough to consider the two-kernel algebras with kernels $\{01|2\}$ and $\{02|1\}$.

Isomorphic copies of a unary algebra can be created via **conjugation**. Consider a unary algebra $\underline{\mathbf{M}} = \langle\{0,1,2\}; F\rangle$ and let $v : \{0,1,2\} \to \{0,1,2\}$ be a permutation. For each $u \in F$, we define the unary operation ${}^{v}u$ on $\{0,1,2\}$ by ${}^{v}u := v \circ u \circ v^{-1}$. Now $v : \{0,1,2\} \to \{0,1,2\}$ is an isomorphism from $\underline{\mathbf{M}}$ onto the algebra

$$^{v}\underline{\mathbf{M}} := \langle\{0,1,2\}; {}^{v}F\rangle, \quad \text{where} \quad {}^{v}F := \{\, {}^{v}u \mid u \in F \,\}.$$

Furthermore, every isomorphic copy of $\underline{\mathbf{M}}$ on the set $\{0,1,2\}$ can be obtained via conjugation in this way.

3.3.1 Lemma *Let $\underline{\mathbf{M}}$ be a two-kernel three-element unary algebra. Then there is an isomorphic copy of $\underline{\mathbf{M}}$, on the set $\{0,1,2\}$, that has kernels $\{01|2\}$ and $\{02|1\}$.*

Proof Let $\underline{\mathbf{M}} = \langle\{0,1,2\}; F\rangle$ be a two-kernel unary algebra with kernels θ_1 and θ_2. There is some $m \in M$ for which $|m/\theta_1| = 2 = |m/\theta_2|$. Let $v : M \to M$ be a permutation sending m to 0. Then ${}^{v}\underline{\mathbf{M}} = \langle\{0,1,2\}; {}^{v}F\rangle$ is isomorphic to $\underline{\mathbf{M}}$ and has kernels $\{01|2\}$ and $\{02|1\}$. ∎

We will often be using symmetry to reduce our work load. Given any unary algebra $\underline{\mathbf{M}}$ on the set $\{0,1,2\}$ with kernels $\{01|2\}$ and $\{02|1\}$, the algebra ${}^{021}\underline{\mathbf{M}}$ is isomorphic to $\underline{\mathbf{M}}$ and also has kernels $\{01|2\}$ and $\{02|1\}$.

3.3.2 Lemma *Let $\underline{\mathbf{M}}$ be a two-kernel unary algebra, on the set $\{0,1,2\}$, with kernels $\{01|2\}$ and $\{02|1\}$. Then the unary term functions of $\underline{\mathbf{M}}$ all belong to the set $\{012, 021\} \cup \{\, ppq, pqp \mid p, q \in \{0,1,2\} \,\}$.*

Proof We just need to show that 012 and 021 are the only permutations that can be term functions of $\underline{\mathbf{M}}$. There exist unary term functions u_1 and u_2 of $\underline{\mathbf{M}}$ such that $\ker(u_1) = \{02|1\}$ and $\ker(u_2) = \{01|2\}$. Now let v be a unary term function of $\underline{\mathbf{M}}$ that is a permutation. Then $u_1 \circ v$ is neither a constant map nor a permutation. So $\ker(u_1 \circ v) = \{01|2\}$ or $\ker(u_1 \circ v) = \{02|1\}$. In either case, we have $u_1 \circ v(1) \neq u_1 \circ v(2)$, which implies that $v(1) = 1$ or $v(2) = 1$. Symmetrically, we have $u_2 \circ v(1) \neq u_2 \circ v(2)$, which tells us that $v(1) = 2$ or $v(2) = 2$. It follows that $v = 012$ or $v = 021$. ∎

The following theorem introduces four types of unary algebras with kernels $\{01|2\}$ and $\{02|1\}$. Each of these types is preserved under conjugation by 021. The two unary operations $f_1 := 010$ and $f_2 := 002$ on $\{0, 1, 2\}$ play a very important role in our characterisation.

3.3.3 Theorem *Let $\underline{\mathbf{M}}$ be a two-kernel unary algebra, on the set $\{0, 1, 2\}$, with kernels $\{01|2\}$ and $\{02|1\}$. Let F be the set of unary term functions of $\underline{\mathbf{M}}$. Then at least one of the following is true:*

$(2)_{\mathrm{O}}$ *each map in F preserves the order $\preccurlyeq$ with $2 \preccurlyeq 0 \preccurlyeq 1$;*

$(2)_{\mathrm{P}}$ *$\{ppq, pqp\} \subseteq F$, for some distinct $p, q \in \{0, 1, 2\}$, and $\{f_1, f_2\} \nsubseteq F$;*

$(2)_{\mathrm{M}}$ *either $\{010, 001, 110\} \subseteq F$ and $222 \notin F$, or $\{002, 020, 202\} \subseteq F$ and $111 \notin F$;*

$(2)_{\mathrm{R}}$ *$\{f_1, f_2\} \subseteq F$, and condition $(2)_{\mathrm{M}}$ fails.*

Proof Assume that $\underline{\mathbf{M}}$ is not of type $(2)_{\mathrm{O}}$, type $(2)_{\mathrm{P}}$ nor type $(2)_{\mathrm{M}}$. To prove that $\underline{\mathbf{M}}$ is of type $(2)_{\mathrm{R}}$, it suffices to show that $\{f_1, f_2\} \subseteq F$. As $\underline{\mathbf{M}}$ is not of type $(2)_{\mathrm{P}}$, we can assume that $\{ppq, pqp\} \nsubseteq F$, for all $p, q \in M$ with $p \neq q$.

Since $\underline{\mathbf{M}}$ is not of type $(2)_{\mathrm{O}}$, there is a map in F that does not preserve the order $\preccurlyeq$. Using Lemma 3.3.2, the only maps that can belong to F and that do not preserve $\preccurlyeq$ are

$$021,\ 220,\ 221,\ 001,\ 121,\ 101 \text{ and } 020.$$

Since $\{01|2\}$ and $\{02|1\}$ are kernels of $\underline{\mathbf{M}}$, there exist $p, q, r, s \in M$, with $p \neq q$ and $r \neq s$, for which $ppq \in F$ and $rsr \in F$. Since $\{ppq, pqp\} \nsubseteq F$ and $\{rrs, rsr\} \nsubseteq F$, we must have $pqp \notin F$ and $rrs \notin F$. This implies that $021, 121, 020 \notin F$, as

$$ppq \circ 021 = ppq \circ 121 = ppq \circ 020 = pqp \notin F.$$

Similarly, we have $221, 001 \notin F$, as

$$rsr \circ 221 = rsr \circ 001 = rrs \notin F.$$

Since F contains a map that does not preserve $\preccurlyeq$, it follows that $101 \in F$ or $220 \in F$.

First assume that $101 \in F$. Then $f_1 = 010 = 101 \circ 101 \in F$. This implies that $110 \notin F$ and $001 \notin F$. We have $112 \notin F$ and $221 \notin F$, since

$$101 \circ 112 = 010 \circ 221 = 001 \notin F.$$

We know there is a map in F with kernel $\{01|2\}$. So $002 \in F$ or $220 \in F$. As $f_2 = 002 = 220 \circ 220$, we have $\{f_1, f_2\} \subseteq F$.

Now assume that $220 \in F$. This case is symmetric, under conjugation by 021, to the case $101 \in F$. We have $101 = {}^{021}220 \in {}^{021}F$. By the previous case, it follows that $\{f_1, f_2\} \subseteq {}^{021}F$. So $\{f_1, f_2\} = \{{}^{021}f_2, {}^{021}f_1\} \subseteq F$. ∎

The names of the types in the previous theorem are meant to serve as *aide-mémoires*. Type-$(2)_{\mathrm{O}}$ algebras have Order-preserving operations; type-$(2)_{\mathrm{P}}$ algebras have operations with the Patterns *ppq* and *pqp*, for some $p, q \in M$; type-$(2)_{\mathrm{M}}$ algebras are Missing a constant operation; and the Rest are type-$(2)_{\mathrm{R}}$ algebras.

It follows straight from the Lattice Endomorphism Theorem, 2.1.2, that every algebra of type $(2)_{\mathrm{O}}$ is dualisable. In the next section, we shall show that every algebra of type $(2)_{\mathrm{P}}$ or type $(2)_{\mathrm{M}}$ is non-dualisable. The remainder of this section is devoted to proving that the rest, all the algebras of type $(2)_{\mathrm{R}}$, are dualisable.

Let $\underline{\mathbf{M}}$ be a three-element unary algebra of type $(2)_{\mathrm{R}}$. Within the quasi-variety $\mathcal{A} := \mathbb{ISP}(\underline{\mathbf{M}})$, we can restrict our attention to subalgebras of powers of $\underline{\mathbf{M}}$. Let S be a non-empty set. For each $a \in M^S$, we say that

$$\mathcal{P}(a) := \{a^{-1}(0), a^{-1}(1), a^{-1}(2)\} \backslash \{\varnothing\}$$

is the **partition of** S **determined by** a.

Now let $\mathbf{A}$ be a subalgebra of $\underline{\mathbf{M}}^S$. The structure of $\mathbf{A}$ may be quite complicated. However, we shall show that the homomorphisms in $\mathcal{A}(\mathbf{A}, \underline{\mathbf{M}})$ are all determined by their restrictions to a very simple subalgebra $\mathbf{A}_{\downarrow 2}$ of $\mathbf{A}$. The underlying set of $\mathbf{A}_{\downarrow 2}$ is given by

$$A_{\downarrow 2} := \{\, a \in A \mid |a(S)| \leqslant 2 \,\}.$$

Since $\underline{\mathbf{M}}$ is of type $(2)_{\mathrm{R}}$, we know that both $f_1 = 010$ and $f_2 = 002$ are term functions of $\underline{\mathbf{M}}$. This implies that $A_{\downarrow 2}$ is not empty. Indeed, we have $f_1(A) \cup f_2(A) \subseteq A_{\downarrow 2}$.

We will be making constant use of the well-behaved term functions f_1 and f_2 of $\underline{\mathbf{M}}$. The maps f_1 and f_2 separate the elements of M. Moreover, for each $m \in \{1, 2\}$, the map f_m on M is idempotent, with image $\{0, m\}$ and $f_m^{-1}(m) = \{m\}$.

3.3.4 Lemma *Assume* $\underline{\mathbf{M}}$ *is of type* $(2)_{\mathrm{R}}$. *Let* $\mathbf{A}$ *be a subalgebra of* $\underline{\mathbf{M}}^S$, *for some non-empty set* S, *and let* $\mathbf{P}$ *be a petal of* $\mathbf{A}_{\downarrow 2}$. *Then all non-centre elements of* $\mathbf{P}$ *determine the same partition of* S.

Proof Let $a \in P \backslash C_{\mathbf{A}}$ and let u be a unary term function of $\underline{\mathbf{M}}$ with $u(a) \notin C_{\mathbf{A}}$. Once we have shown that $\mathcal{P}(a) = \mathcal{P}(u(a))$, the result will follow since the graph $G^*(\mathbf{P})$ is connected. Since $a \in A_{\downarrow 2}$, we must have $|a(S)| \leqslant 2$. First assume that $a \in \{0, m\}^S$, for some $m \in \{1, 2\}$. Since f_m fixes 0 and m, we have $u(f_m(a)) = u(a) \notin C_{\mathbf{A}}$. This implies that the term function $u \circ f_m$ of $\underline{\mathbf{M}}$ is not constant. So $u(0) \neq u(m)$, and therefore $\mathcal{P}(a) = \mathcal{P}(u(a))$. Now assume that $a \in \{1, 2\}^S$. As $u(a) \notin C_{\mathbf{A}}$, the term function u of $\underline{\mathbf{M}}$ is not constant. Since $\underline{\mathbf{M}}$ does not have any term functions with kernel $\{0|12\}$, we have $u(1) \neq u(2)$. Thus $\mathcal{P}(a) = \mathcal{P}(u(a))$. ∎

The previous lemma tells us that every petal of $\mathbf{A}_{\downarrow 2}$ is isomorphic to a subalgebra of $\underline{\mathbf{M}}^2$. So, by Lemma 3.1.4, the subalgebra $\mathbf{A}_{\downarrow 2}$ of $\mathbf{A}$ is simply a coproduct of subalgebras of $\underline{\mathbf{M}}^2$.

We will now show that homomorphisms from $\mathbf{A}$ to $\underline{\mathbf{M}}$ are determined by their restrictions to $\mathbf{A}_{\downarrow 2}$. We shall obtain this as a corollary of the following more general result, which will also be used in Chapter 4.

3.3.5 Lemma *Let $\underline{\mathbf{M}}$ be a two-kernel unary algebra, on the set $\{0, 1, 2\}$, such that f_1 and f_2 are term functions of $\underline{\mathbf{M}}$. Let $\mathbf{B} \leqslant \mathbf{A}$ in $\mathbb{ISP}(\underline{\mathbf{M}})$ with $f_1(A) \cup f_2(A) \subseteq B$, and let $x : \mathbf{B} \to \underline{\mathbf{M}}$ be a homomorphism. Then x extends to $\mathbf{A}$ if and only if $x(f_1(a)) = 0$ or $x(f_2(a)) = 0$, for all $a \in A \backslash B$. Furthermore, if x extends to $\mathbf{A}$, then x extends to $\mathbf{A}$ uniquely.*

Proof First let $a \in A \backslash B$ and let $m \in \{1, 2\}$. Then $f_m(a) \in B$ and, since f_m is idempotent, we have

$$x(f_m(a)) = x(f_m \circ f_m(a)) = f_m(x(f_m(a))) \in \{0, m\}.$$

We have shown that

$$x(f_m(a)) \in \{0, m\}, \text{ for each } m \in \{1, 2\} \text{ and } a \in A \backslash B. \qquad (*)$$

Now assume that x extends to a homomorphism $\overline{x} : \mathbf{A} \to \underline{\mathbf{M}}$. For all $a \in A \backslash B$ and $m \in \{1, 2\}$, we have

$$\overline{x}(a) = m \iff f_m(\overline{x}(a)) = m \iff x(f_m(a)) = m.$$

Thus $\overline{x}$ is the unique extension of x to $\mathbf{A}$. It also follows that $x(f_1(a)) \neq 1$ or $x(f_2(a)) \neq 2$, for every $a \in A \backslash B$. So, by $(*)$, we have $x(f_1(a)) = 0$ or $x(f_2(a)) = 0$, for every $a \in A \backslash B$.

Conversely, assume that $x(f_1(a)) = 0$ or $x(f_2(a)) = 0$, for all $a \in A \backslash B$. Using $(*)$, we can define the extension $\overline{x} : A \to M$ of the map x so that, for each $a \in A \backslash B$, we have

$$\overline{x}(a) = \begin{cases} 0 & \text{if } x(f_1(a)) = x(f_2(a)) = 0, \\ 1 & \text{if } x(f_1(a)) = 1, \\ 2 & \text{if } x(f_2(a)) = 2. \end{cases}$$

$x(f_1(a))$	$x(f_2(a))$	$\overline{x}(a)$	$x(f_1(\widetilde{a}))$	$x(f_2(\widetilde{a}))$	$\overline{x}(\widetilde{a})$
0	0	0	0	0	0
1	0	1	0	2	2
0	2	2	1	0	1

Table 3.4

We wish to show that $\overline{x} : \mathbf{A} \to \underline{\mathbf{M}}$ is a homomorphism.

Let $a \in A\backslash B$. We will use the description of the unary term functions of $\underline{\mathbf{M}}$ given in Lemma 3.3.2. First let $p, q \in M$ and assume that ppq is a term function of $\underline{\mathbf{M}}$. Using Table 3.4, we see that

$$\begin{aligned} ppq(\overline{x}(a)) &= ppq(x(f_2(a))) = x(ppq \circ 002(a)) = x(ppq(a)) \\ &= \overline{x}(ppq(a)). \end{aligned}$$

Assume that pqp is a term function of $\underline{\mathbf{M}}$. Then

$$\begin{aligned} pqp(\overline{x}(a)) &= pqp(x(f_1(a))) = x(pqp \circ 010(a)) = x(pqp(a)) \\ &= \overline{x}(pqp(a)). \end{aligned}$$

Lastly, assume that 021 is a term function of $\underline{\mathbf{M}}$, and define $\widetilde{a} := 021(a)$ in A. Then

$$\begin{aligned} x(f_1(\widetilde{a})) &= x(010 \circ 021(a)) = x(001(a)) = x(021 \circ 002(a)) \\ &= 021(x(f_2(a))) \end{aligned}$$

and

$$\begin{aligned} x(f_2(\widetilde{a})) &= x(002 \circ 021(a)) = x(020(a)) = x(021 \circ 010(a)) \\ &= 021(x(f_1(a))). \end{aligned}$$

It now follows from Table 3.4 that $021(\overline{x}(a)) = \overline{x}(\widetilde{a}) = \overline{x}(021(a))$. Thus $\overline{x}$ is a homomorphism. ∎

The next result follows at once from the previous lemma.

3.3.6 Lemma *Assume $\underline{\mathbf{M}}$ is of type $(2)_{\mathrm{R}}$. Let $\mathbf{A}$ be a subalgebra of $\underline{\mathbf{M}}^S$, for some non-empty set S, and let $x : \mathbf{A}_{\downarrow 2} \to \underline{\mathbf{M}}$ be a homomorphism. Then x extends to $\mathbf{A}$ if and only if $x(f_1(a)) = 0$ or $x(f_2(a)) = 0$, for all $a \in A\backslash A_{\downarrow 2}$. Furthermore, if x extends to $\mathbf{A}$, then x extends to $\mathbf{A}$ uniquely.*

The subalgebra $\mathbf{A}$ of $\underline{\mathbf{M}}^S$ is **hom-minimal (relative to $\underline{\mathbf{M}}$)** if every homomorphism from $\mathbf{A}$ to $\underline{\mathbf{M}}$ is the restriction of a projection. It is easy to check that every algebra in $\mathcal{A}$ is isomorphic to a hom-minimal algebra. (For instance, every algebra $\mathbf{A}$ in $\mathcal{A}$ is isomorphic to the hom-minimal subalgebra $e_{\mathbf{A}}(\mathbf{A})$ of $\underline{\mathbf{M}}^X$, where $X := \mathcal{A}(\mathbf{A}, \underline{\mathbf{M}})$.) Therefore we can further restrict our attention to those algebras in $\mathcal{A}$ that are hom-minimal.

So far in this chapter, we have twice used a particular trick for helping to establish dualisability. Given a pair of homomorphisms in the dual of an algebra, we constructed a sequence of homomorphisms, from one to the other, so that homomorphisms that were adjacent in the sequence were nearly equal. (We did this in the proofs of both Lemma 3.1.5 and Theorem 3.2.10.) We will be using the same trick to help prove that type-$(2)_{\mathrm{R}}$ algebras are dualisable.

Let $x, y \in \mathcal{A}(\mathbf{A}, \underline{\mathbf{M}})$ with $x \neq y$. We say that the homomorphisms x and y are **almost equal** if there is a petal $\mathbf{P}$ of $\mathbf{A}_{\downarrow 2}$ such that x and y agree on $A_{\downarrow 2} \backslash P$.

3.3.7 Lemma *Assume $\underline{\mathbf{M}}$ is of type $(2)_{\mathrm{R}}$ and define $\mathcal{A} := \mathbb{ISP}(\underline{\mathbf{M}})$. Let $\mathbf{A}$ be a hom-minimal subalgebra of $\underline{\mathbf{M}}^S$, for some finite non-empty set S, and let $x, y \in \mathcal{A}(\mathbf{A}, \underline{\mathbf{M}})$ with $x \neq y$. Then there is a sequence $x = x_0, x_1, \dots, x_n = y$ in $\mathcal{A}(\mathbf{A}, \underline{\mathbf{M}})$, for some $n \in \omega \backslash \{0\}$, such that x_i is almost equal to x_{i+1} and $A_{\downarrow 2} \cap \mathrm{eq}(x, y) \subseteq A_{\downarrow 2} \cap \mathrm{eq}(x_i, y)$, for every $i \in \{0, \dots, n-1\}$.*

Proof We want to show that, starting from x, we can define a sequence of homomorphisms in $\mathcal{A}(\mathbf{A}, \underline{\mathbf{M}})$ that are each one step closer to agreeing with y on $\mathbf{A}_{\downarrow 2}$. We begin by proving the following claim.

Claim Assume $w \in \mathcal{A}(\mathbf{A}, \underline{\mathbf{M}})$ such that $A_{\downarrow 2} \cap \mathrm{eq}(w, y) \neq A_{\downarrow 2}$. Then there exists $w' \in \mathcal{A}(\mathbf{A}, \underline{\mathbf{M}})$ such that w' is almost equal to w and $A_{\downarrow 2} \cap \mathrm{eq}(w, y)$ is a proper subset of $A_{\downarrow 2} \cap \mathrm{eq}(w', y)$.

To each petal of $\mathbf{A}_{\downarrow 2}$, we shall associate a subset of S. Let $\mathbf{P}$ be a petal of $\mathbf{A}_{\downarrow 2}$ and, for some $a \in P \backslash C_{\mathbf{A}} \subseteq M^S$, define the subset $S_{\mathbf{P}} := a^{-1}(w(a))$ of S. As $\mathbf{A}$ is hom-minimal, the homomorphism w is the restriction of a projection. By Lemma 3.3.4, all non-centre elements of $\mathbf{P}$ determine the same partition of S. This implies that the subset $S_{\mathbf{P}}$ of S is independent of our choice of a.

We are assuming that $A_{\downarrow 2} \cap \mathrm{eq}(w, y) \neq A_{\downarrow 2}$. So w and y disagree on at least one petal of $\mathbf{A}_{\downarrow 2}$. From amongst all the petals of $\mathbf{A}_{\downarrow 2}$ on which w and y disagree, choose a petal $\mathbf{P}$ such that $|S_{\mathbf{P}}|$ is minimal. We can do this because S is finite.

The algebra $\mathbf{A}_{\downarrow 2}$ is the coproduct of its petals, by Lemma 3.1.4. (We are allowed to apply this lemma as $\mathbf{A}_{\downarrow 2} \leqslant \underline{\mathbf{M}}^S$, with S non-empty.) Therefore

$$w{\restriction}_{A_{\downarrow 2} \backslash P} \cup y{\restriction}_P : \mathbf{A}_{\downarrow 2} \to \underline{\mathbf{M}}$$

is a homomorphism. We now want to show that $w{\restriction}_{A_{\downarrow 2} \backslash P} \cup y{\restriction}_P$ extends to a homomorphism w' in $\mathcal{A}(\mathbf{A}, \underline{\mathbf{M}})$.

Suppose, by way of contradiction, that $w\restriction_{A_{\downarrow 2}\backslash P} \cup y\restriction_P$ does not extend to $\mathbf{A}$. It follows by Lemma 3.3.6 that there exists $a \in A\backslash A_{\downarrow 2}$ and $\{k, \ell\} = \{1, 2\}$, with $f_k(a) \in A_{\downarrow 2}\backslash P$ and $f_\ell(a) \in P$, such that

$$w(f_k(a)) \neq 0 \quad \text{and} \quad y(f_\ell(a)) \neq 0.$$

Since $w, y \in \mathcal{A}(\mathbf{A}, \underline{\mathbf{M}})$, applying Lemma 3.3.6 again tells us that

$$w(f_\ell(a)) = 0 \quad \text{and} \quad y(f_k(a)) = 0.$$

As $|a(S)| = 3$, we know that $f_k(a), f_\ell(a) \notin C_{\mathbf{A}}$.

Now define $\mathbf{Q}$ to be the petal of $\mathbf{A}_{\downarrow 2}$ containing $f_k(a)$. We have

$$w(f_k(a)) \neq 0 = y(f_k(a)),$$

and therefore w and y disagree on $\mathbf{Q}$. Since $w(f_k(a)) \neq 0$ and w is the restriction of a projection, we must have $w(f_k(a)) = k$. It follows that

$$S_{\mathbf{Q}} = f_k(a)^{-1}(k) = a^{-1}(k) \subset a^{-1}(0) \cup a^{-1}(k) = f_\ell(a)^{-1}(0) = S_{\mathbf{P}}.$$

Note that $S_{\mathbf{Q}}$ is a proper subset of $S_{\mathbf{P}}$, as $|a(S)| = 3$ and therefore $a^{-1}(0) \neq \varnothing$. This contradicts the minimality of $|S_{\mathbf{P}}|$.

We have established that, for our chosen petal $\mathbf{P}$ of $\mathbf{A}_{\downarrow 2}$ with $w\restriction_P \neq y\restriction_P$, the homomorphism $w\restriction_{A_{\downarrow 2}\backslash P} \cup y\restriction_P : \mathbf{A}_{\downarrow 2} \to \underline{\mathbf{M}}$ extends to a homomorphism w' in $\mathcal{A}(\mathbf{A}, \underline{\mathbf{M}})$. The maps w and w' are almost equal, and $A_{\downarrow 2} \cap \operatorname{eq}(w, y)$ is a proper subset of $A_{\downarrow 2} \cap \operatorname{eq}(w', y)$.

We can now prove the lemma. First, define $x_0 := x$. As $x_0 \neq y$, it follows from Lemma 3.3.6 that $A_{\downarrow 2} \cap \operatorname{eq}(x_0, y) \neq A_{\downarrow 2}$. Using the claim, there exists x_1 in $\mathcal{A}(\mathbf{A}, \underline{\mathbf{M}})$, with x_0 and x_1 almost equal, such that $A_{\downarrow 2} \cap \operatorname{eq}(x_0, y)$ is a proper subset of $A_{\downarrow 2} \cap \operatorname{eq}(x_1, y)$. If $A_{\downarrow 2} \cap \operatorname{eq}(x_1, y) \neq A_{\downarrow 2}$, then we can use the claim again to find $x_2 \in \mathcal{A}(\mathbf{A}, \underline{\mathbf{M}})$, with x_1 and x_2 almost equal, such that $A_{\downarrow 2} \cap \operatorname{eq}(x_1, y)$ is a proper subset of $A_{\downarrow 2} \cap \operatorname{eq}(x_2, y)$. Iterate this construction. As $A_{\downarrow 2}$ is finite, there will be some $n \in \omega\backslash\{0\}$ with $A_{\downarrow 2} \cap \operatorname{eq}(x_n, y) = A_{\downarrow 2}$. By Lemma 3.3.6, it follows that $x_n = y$. ∎

The next lemma completes the preparation for our proof that type-$(2)_{\mathrm{R}}$ algebras are dualisable. This lemma is the only place we use the fact that algebras of type $(2)_{\mathrm{R}}$ are not of type $(2)_{\mathrm{M}}$.

3.3.8 Lemma *Assume that $\underline{\mathbf{M}}$ is of type $(2)_{\mathrm{R}}$. Let $\mathbf{A}$ be a subalgebra of $\underline{\mathbf{M}}^S$, for some non-empty set S, and let $m \in M$. If A does not contain the constant map in M^S with value m, then there is a homomorphism $x : \mathbf{A} \to \underline{\mathbf{M}}$ such that $m \notin x(A)$.*

Proof Let F be the set of unary term functions of $\underline{\mathbf{M}}$. For each $m \in M$, let $\widehat{m}$ denote the constant map in M^S with value m. We must have $\widehat{0} \in A$, since $000 = f_2 \circ f_1 \in F$. So, by symmetry, we can assume that $\widehat{2} \notin A$. We want to find a homomorphism $x : \mathbf{A} \to \underline{\mathbf{M}}$ such that $2 \notin x(A)$.

Since $\widehat{2} \notin A$, we must have $222 \notin F$. First assume that $111 \notin F$. For all $u \in F$, we have $u \circ 000 \in F$, and therefore $u(0) = 0$. So the constant map $\underline{0} : \mathbf{A} \to \underline{\mathbf{M}}$ is a homomorphism, and $2 \notin \underline{0}(A)$.

Now assume that $111 \in F$. For every $u \in F$, we have $u \circ 000 \in F$ and $u \circ 111 \in F$. Since $222 \notin F$, we must have $u(0) \in \{0, 1\}$ and $u(1) \in \{0, 1\}$, for all $u \in F$. Using Lemma 3.3.2, it follows that

$$F \subseteq \{012, 010, 101\} \cup \{\, 00q, 11q \mid q \in M \,\}.$$

As $010 = f_1 \in F$ and $\underline{\mathbf{M}}$ is not of type $(2)_{\mathrm{M}}$, we know that $\{001, 110\} \nsubseteq F$. There are three cases to consider.

Case 1: $001 \in F$ and $110 \notin F$. We must have $101 \notin F$ and $112 \notin F$, since $101 \circ 001 = 010 \circ 112 = 110 \notin F$. So

$$F \subseteq \{012, 010, 111\} \cup \{\, 00q \mid q \in M \,\}.$$

Since $\widehat{2} \notin A$, the homomorphism $x : \mathbf{A} \to \underline{\mathbf{M}}$, given by $x := \underline{0}{\restriction}_{A \setminus \{\widehat{1}\}} \cup \underline{1}{\restriction}_{\{\widehat{1}\}}$, satisfies $2 \notin x(A)$.

Case 2: $110 \in F$ and $001 \notin F$. This implies that $101 \notin F$, since $101 \circ 110 = 001 \notin F$. Therefore

$$F \subseteq \{012, 010, 000, 002\} \cup \{\, 11q \mid q \in M \,\}.$$

Since f_2 fixes 0 and 2, we have $f_2(A) = A \cap \{0, 2\}^S$. As $\widehat{2} \notin A$, the homomorphism $x : \mathbf{A} \to \underline{\mathbf{M}}$, given by $x := \underline{1}{\restriction}_{A \setminus f_2(A)} \cup \underline{0}{\restriction}_{f_2(A)}$, satisfies $2 \notin x(A)$.

Case 3: $001 \notin F$ and $110 \notin F$. We have $112 \notin F$, as $010 \circ 112 = 110 \notin F$. So

$$F \subseteq \{012, 010, 101, 000, 002, 111\}.$$

Choose any $s \in S$. Then we can define the homomorphism $x : \mathbf{A} \to \underline{\mathbf{M}}$ by $x := f_1 \circ \pi_s{\restriction}_{A \setminus f_2(A)} \cup \underline{0}{\restriction}_{f_2(A)}$, and $2 \notin x(A)$. ∎

3.3.9 Theorem *All three-element unary algebras of type* $(2)_{\mathrm{R}}$ *are dualisable.*

Proof Assume that $\underline{\mathbf{M}}$ is of type $(2)_{\mathrm{R}}$ and define $\mathcal{A} := \mathbb{ISP}(\underline{\mathbf{M}})$. Define the alter ego $\underset{\sim}{\mathbf{M}} := \langle \{0, 1, 2\}; R_8, \mathcal{T} \rangle$ of $\underline{\mathbf{M}}$. (By doing a little extra work at one of the steps in this proof, we can actually get by with R_6 instead of R_8.) Let $\mathbf{A}$ be a hom-minimal subalgebra of $\underline{\mathbf{M}}^S$, for some finite non-empty set S, and let

$\alpha : \mathrm{D}(\mathbf{A}) \to \underset{\sim}{\mathrm{M}}$ be a morphism. We will show that α is an evaluation. It will then follow by the Duality Compactness Theorem, 1.4.2, that $\underset{\sim}{\mathrm{M}}$ dualises $\underline{\mathbf{M}}$.

First assume that α is constant. Let m be the value of α in M, and suppose that A does not contain the constant map $\widehat{m}$ in M^S with value m. By Lemma 3.3.8, there exists $x \in \mathcal{A}(\mathbf{A}, \underline{\mathbf{M}})$ with $m \notin x(A)$. The set $x(A)$ is a unary algebraic relation on $\underline{\mathbf{M}}$. Since α preserves $x(A)$, we have $\alpha(x) \in x(A)$. So $\alpha(x) \neq m$, which is a contradiction. We have shown that $\widehat{m} \in A$. The map α is given by evaluation at $\widehat{m}$, as each element of $\mathcal{A}(\mathbf{A}, \underline{\mathbf{M}})$ is the restriction of a projection.

Now assume that the map α is not constant. There are $v_1, v_2 \in \mathcal{A}(\mathbf{A}, \underline{\mathbf{M}})$ such that $\alpha(v_1) \neq \alpha(v_2)$. By Lemma 3.3.7, there is a sequence

$$v_1 = v_{10},\ v_{11},\ \ldots,\ v_{1n} = v_2 \ \text{ in } \mathcal{A}(\mathbf{A}, \underline{\mathbf{M}}),$$

for some $n \in \omega \backslash \{0\}$, with v_{1i} almost equal to v_{1i+1}, for all $i \in \{0, \ldots, n-1\}$. As $\alpha(v_1) \neq \alpha(v_2)$, there exists $j \in \{0, \ldots, n-1\}$ with $\alpha(v_{1j}) \neq \alpha(v_{1j+1})$. Define $y_1 := v_{1j}$ and $y_2 := v_{1j+1}$. Since y_1 and y_2 are almost equal, there is a petal $\mathbf{P}_y$ of $\mathbf{A}_{\downarrow 2}$ such that $A_{\downarrow 2} \backslash P_y \subseteq \mathrm{eq}(y_1, y_2)$. As $y_1 \neq y_2$, we must have $y_1{\restriction}_{A_{\downarrow 2}} \neq y_2{\restriction}_{A_{\downarrow 2}}$, by Lemma 3.3.6. So $y_1{\restriction}_{P_y} \neq y_2{\restriction}_{P_y}$.

Case 1: P_y is a support for α. We begin by showing that there is some $a \in P_y$ such that α is given by evaluation at a on $\{y_1, y_2\}$. Since α preserves R_8, we can use the Preservation Lemma, 1.4.4, to find some $b \in A$ such that α is given by evaluation at b on $\{y_1, y_2\}$. Assume that $b \notin P_y$. We have

$$y_1(b) = \alpha(y_1) \neq \alpha(y_2) = y_2(b).$$

As $A_{\downarrow 2} \backslash P_y \subseteq \mathrm{eq}(y_1, y_2)$, this implies that $b \notin A_{\downarrow 2}$ and therefore $|b(S)| = 3$. Since f_1 and f_2 separate the elements of M, we have

$$f_k(y_1(b)) \neq f_k(y_2(b)), \quad \text{for some } k \in \{1, 2\}.$$

Since $y_1(f_k(b)) \neq y_2(f_k(b))$ and $A_{\downarrow 2} \backslash P_y \subseteq \mathrm{eq}(y_1, y_2)$, it now follows that $f_k(b) \in P_y$. We shall show that α is given by evaluation at $f_k(b)$ on $\{y_1, y_2\}$.

As $|b(S)| = 3$, the elements $f_1(b)$ and $f_2(b)$ of $\mathbf{A}$ determine two different two-block partitions of S. So $f_1(b)$ and $f_2(b)$ belong to different petals of $\mathbf{A}_{\downarrow 2}$, by Lemma 3.3.4. Choose $\ell \in \{1, 2\}$ with $\ell \neq k$. Then $f_\ell(b) \notin P_y$ and so, as $A_{\downarrow 2} \backslash P_y \subseteq \mathrm{eq}(y_1, y_2)$, we have

$$f_\ell(y_1(b)) = y_1(f_\ell(b)) = y_2(f_\ell(b)) = f_\ell(y_2(b)).$$

Since $y_1(b) \neq y_2(b)$, this gives us $\{y_1(b), y_2(b)\} = \{0, k\}$. Consequently,

$$\alpha(y_i) = y_i(b) = f_k(y_i(b)) = y_i(f_k(b)),$$

for each $i \in \{1, 2\}$.

We have shown that α is given by evaluation at some $a \in P_y$ on $\{y_1, y_2\}$. By Lemma 3.3.4, all non-centre elements of $\mathbf{P}_y$ determine the same partition of S. Since $P_y \subseteq A_{\downarrow 2}$, this partition has at most two blocks. So there are at most two functions from P_y to M that are the restriction of a projection. Now let $x \in \mathcal{A}(\mathbf{A}, \underline{\mathbf{M}})$. As $\mathbf{A}$ is hom-minimal, the homomorphisms x, y_1 and y_2 are restrictions of projections. Since $y_1{\restriction_{P_y}} \neq y_2{\restriction_{P_y}}$, there must be some $i \in \{1, 2\}$ such that $x{\restriction_{P_y}} = y_i{\restriction_{P_y}}$. Therefore

$$\alpha(x) = \alpha(y_i) = y_i(a) = x(a),$$

as P_y is a support for α. Thus α is an evaluation.

Case 2: P_y is not a support for α. By Lemma 3.3.7, there exist almost equal homomorphisms $z_1, z_2 \in \mathcal{A}(\mathbf{A}, \underline{\mathbf{M}})$ such that

$$z_1{\restriction_{P_y}} = z_2{\restriction_{P_y}} \quad \text{and} \quad \alpha(z_1) \neq \alpha(z_2).$$

There is a petal $\mathbf{P}_z$ of $\mathbf{A}_{\downarrow 2}$ with $A_{\downarrow 2} \backslash P_z \subseteq \mathrm{eq}(z_1, z_2)$. Using Lemma 3.3.6, we must have $z_1{\restriction_{P_z}} \neq z_2{\restriction_{P_z}}$. So $\mathbf{P}_y$ and $\mathbf{P}_z$ are different petals of $\mathbf{A}_{\downarrow 2}$.

The petals $\mathbf{P}_y$ and $\mathbf{P}_z$ of $\mathbf{A}_{\downarrow 2}$ determine two partitions of S, each with at most two blocks. So there are at most four functions from $P_y \cup P_z$ to M that are the restriction of a projection. Since $\mathbf{A}$ is hom-minimal, there is a subset W of $\mathcal{A}(\mathbf{A}, \underline{\mathbf{M}})$, with $|W| \leqslant 7$, such that $\{y_1, y_2, z_1, z_2\} \subseteq W$ and

$$\{\, w{\restriction_{P_y \cup P_z}} \mid w \in W \,\} = \{\, x{\restriction_{P_y \cup P_z}} \mid x \in \mathcal{A}(\mathbf{A}, \underline{\mathbf{M}}) \,\}.$$

(We can actually choose W so that $|W| \leqslant 5$.) Define

$$A_W := \{\, a \in A \mid \alpha \text{ is given by evaluation at } a \text{ on } W \,\}.$$

Then A_W is non-empty, by the Preservation Lemma, as α preserves R_8.

We want to show that A_W has only one element. To do this, let $a_1, a_2 \in A_W$ and $x \in \mathcal{A}(\mathbf{A}, \underline{\mathbf{M}})$. Since $\mathbf{A}$ is separated by homomorphisms into $\underline{\mathbf{M}}$, it suffices to prove that $x(a_1) = x(a_2)$. Let $i \in \{1, 2\}$. As $a_i \in A_W$, the map α is given by evaluation at a_i on the set W. Therefore

$$y_1(a_i) = \alpha(y_1) \neq \alpha(y_2) = y_2(a_i)$$

and

$$z_1(a_i) = \alpha(z_1) \neq \alpha(z_2) = z_2(a_i).$$

The maps f_1 and f_2 separate the elements of M. So there is some $k \in \{1, 2\}$ with

$$y_1(f_k(a_i)) = f_k(y_1(a_i)) \neq f_k(y_2(a_i)) = y_2(f_k(a_i)).$$

Thus, as $A_{\downarrow 2}\backslash P_y \subseteq \mathrm{eq}(y_1, y_2)$, we have $f_1(a_i) \in P_y\backslash C_{\mathbf{A}}$ or $f_2(a_i) \in P_y\backslash C_{\mathbf{A}}$. Similarly, we find that $f_1(a_i) \in P_z\backslash C_{\mathbf{A}}$ or $f_2(a_i) \in P_z\backslash C_{\mathbf{A}}$. Consequently, $f_1(a_i), f_2(a_i) \in P_y \cup P_z$, as $\mathbf{P}_y$ and $\mathbf{P}_z$ are distinct petals of $\mathbf{A}_{\downarrow 2}$.

By the construction of W, there exists $w \in W$ with $x\restriction_{P_y \cup P_z} = w\restriction_{P_y \cup P_z}$. Since $f_1(a_i), f_2(a_i) \in P_y \cup P_z$, this gives us

$$f_1(x(a_i)) = f_1(w(a_i)) \quad \text{and} \quad f_2(x(a_i)) = f_2(w(a_i)).$$

As f_1 and f_2 separate the elements of M, we have

$$x(a_1) = w(a_1) = \alpha(w) = w(a_2) = x(a_2).$$

Since $\mathbf{A}$ is separated by homomorphisms into $\underline{\mathbf{M}}$, it follows that $a_1 = a_2$.

Now let a be the unique element of A_W. To see that α is given by evaluation at a, let $x \in \mathcal{A}(\mathbf{A}, \underline{\mathbf{M}})$. Since α preserves R_8, there is some $b \in A$ such that α is given by evaluation at b on $W \cup \{x\}$. As a is the only element of A_W, we must have $b = a$, and therefore $\alpha(x) = x(a)$. Thus α is an evaluation. ∎

3.4 Non-dualisable three-element unary algebras

Proofs of non-dualisability are often easier than proofs of dualisability. The ghost-element method provides an extremely elegant way to show that a finite algebra is not dualisable. Constructing a ghost-element proof can require a fair amount of inspiration. However, verifying that the construction is correct often involves only routine calculations.

For all our ghost-element proofs in this section, we will use the following refinement of the Ghost Element Theorem, 1.4.6. (The result below is the unpublished precursor to the Inherent Non-dualisability Theorem [23, 8]; see 5.2.2. We cannot use the Inherent Non-dualisability Theorem itself, since there are no inherently non-dualisable unary algebras, by Theorem 2.1.4.)

3.4.1 Non-dualisability Lemma *Let $\underline{\mathbf{M}}$ be a finite algebra and $n \in \omega\backslash\{0\}$. Assume that there is a subalgebra $\mathbf{A}$ of $\underline{\mathbf{M}}^S$, for some set S, and an infinite subset A_0 of A such that*

(i) *for each homomorphism $x : \mathbf{A} \to \underline{\mathbf{M}}$, the equivalence relation $\ker(x\restriction_{A_0})$ has a unique block of size greater than n,*

(ii) *the algebra $\mathbf{A}$ does not contain the element g of M^S that is defined by $g(s) := \rho_s(a_s)$, where a_s is any element of the unique block of $\ker(\rho_s\restriction_{A_0})$ of size greater than n.*

Then $\underline{\mathbf{M}}$ is non-dualisable.

Proof We shall use the Ghost Element Theorem, 1.4.6, to prove that $\underline{\mathbf{M}}$ is non-dualisable. Define $\mathcal{A} := \mathbb{ISP}(\underline{\mathbf{M}})$ and define the map

$$\alpha : \mathcal{A}(\mathbf{A}, \underline{\mathbf{M}}) \to M \quad \text{by} \quad \alpha(x) := x(a_x),$$

where a_x is any element of the unique block of $\ker(x{\restriction}_{A_0})$ of size greater than n. Then, for all $s \in S$, we have

$$g_\alpha(s) = \alpha(\rho_s) = \rho_s(a_s) = g(s).$$

Therefore $g_\alpha = g \notin A$. By the Ghost Element Theorem and the Brute Force Lemma, 1.4.5, it is now enough to prove that α has a finite support and that α is locally an evaluation.

Choose any finite subset B of A_0 with $|B| \geqslant n|M| + 1$. To see that B is a support for α, let $x, y \in \mathcal{A}(\mathbf{A}, \underline{\mathbf{M}})$ with $x{\restriction}_B = y{\restriction}_B$. The equivalence relation $\ker(x{\restriction}_B)$ on B has at most $|M|$ blocks. As $B \subseteq A_0$ and $|B| > n|M|$, there is a unique block of $\ker(x{\restriction}_B)$ of size greater than n. Choose some $b \in B$ that lies in this largest block of $\ker(x{\restriction}_B) = \ker(y{\restriction}_B)$. Then b belongs to the unique block of $\ker(x{\restriction}_{A_0})$ of size greater than n, and also to the unique block of $\ker(y{\restriction}_{A_0})$ of size greater than n. So

$$\alpha(x) = x(b) = y(b) = \alpha(y).$$

Thus B is a finite support for α.

Now let X be a finite subset of $\mathcal{A}(\mathbf{A}, \underline{\mathbf{M}})$. For each $x \in X$, let A_x denote the unique block of $\ker(x{\restriction}_{A_0})$ of size greater than n. For each $x \in X$, the set A_x is cofinite in A_0, since $\ker(x{\restriction}_{A_0})$ has finitely many blocks. So there is some $a \in \bigcap\{\, A_x \mid x \in X \,\}$. We have $\alpha(x) = x(a)$, for all $x \in X$. Thus α agrees with an evaluation on X. ∎

We will be applying the Non-dualisability Lemma many times throughout this text, always using the bound $n = 1$. When applying this lemma to an algebra $\underline{\mathbf{M}}$, we shall want to specify sequences in M^ω. For $a \in M$, we use $\widehat{a}$ to denote the constant sequence in M^ω with value a. Now let $k \in \omega\backslash\{0\}$, let $n_1, \dots, n_k \in \omega$ and let $a, b_1, \dots, b_k \in M$. We define the sequence $a^{b_1 \dots b_k}_{n_1 \dots n_k}$ in M^ω by

$$a^{b_1 \dots b_k}_{n_1 \dots n_k}(i) = \begin{cases} b_j & \text{if } i = n_j, \text{ for some } j \in \{1, \dots, k\}, \\ a & \text{otherwise,} \end{cases}$$

for all $i \in \omega$.

The next theorem tells us that every three-element unary algebra of type $(2)_{\mathrm{P}}$ is non-dualisable.

3.4.2 Theorem *Let $\underline{\mathbf{M}}$ be a dualisable unary algebra on the set $\{0,1,2\}$. If ppq and pqp are term functions of $\underline{\mathbf{M}}$, for some $p, q \in \{0,1,2\}$ with $p \neq q$, then so are* 010 *and* 002.

Proof We will prove that, if both *ppq* and *pqp* are term functions of $\underline{\mathbf{M}}$, for some $p, q \in M$ with $p \neq q$, then 002 is as well. The rest of the result will then follow using conjugation by 021.

Assume that *ppq* and *pqp* are term functions of $\underline{\mathbf{M}}$, for some $p, q \in M$ with $p \neq q$. Define two subsets of M^ω by

$$A_0 := \{\, 0^{21}_{0n} \mid n \in \omega\backslash\{0\} \,\}$$

and

$$B := \{\, 0^{2\,1}_{mn} \mid m, n \in \omega\backslash\{0\} \text{ and } m \neq n \,\}.$$

Let $\mathbf{A}$ denote the subalgebra of $\underline{\mathbf{M}}^\omega$ generated by $A_0 \cup B$. Now choose a homomorphism $x : \mathbf{A} \to \underline{\mathbf{M}}$. We shall show that $\ker(x{\restriction}_{A_0})$ has a unique non-trivial block.

Case 1: $2 \in x(A_0)$. There exists some $n \in \omega\backslash\{0\}$ such that $x(0^{21}_{0n}) = 2$. Let $m \in \omega\backslash\{0\}$. Then

$$0^{21}_{0n} \xrightarrow{ppq} p^{q}_{0} \xleftarrow{ppq} 0^{2\,1}_{0m}$$

in $\mathbf{A}$. Applying the homomorphism x gives us

$$\boxed{2} \xrightarrow{ppq} q \xleftarrow{ppq} x(0^{2\,1}_{0m})$$

in $\underline{\mathbf{M}}$. (The box around 2 indicates that $x(0^{21}_{0n}) = 2$ by assumption.) So $x(0^{2\,1}_{0m}) = 2$, as $p \neq q$. It follows that $x(A_0) = \{2\}$, and therefore $\ker(x{\restriction}_{A_0})$ has only one block.

Case 2: $x(A_0) \subseteq \{0,1\}$. We can assume that $x(A_0) \neq \{0\}$. So there is some $n \in \omega\backslash\{0\}$ such that $x(0^{21}_{0n}) = 1$. Let $m \in \omega\backslash\{0, n\}$. Then

$$0^{21}_{0n} \xrightarrow{pqp} p^{q}_{n} \xleftarrow{pqp} 0^{2\,1}_{mn} \xrightarrow{ppq} p^{q}_{m} \xleftarrow{pqp} 0^{2\,1}_{0m}$$

in $\mathbf{A}$. Applying x, we have

$$\boxed{1} \xrightarrow{pqp} q \xleftarrow{pqp} 1 \xrightarrow{ppq} p \xleftarrow{pqp} x(0^{2\,1}_{0m})$$

in $\underline{\mathbf{M}}$. Since $x(0^{2\,1}_{0m}) \neq 2$, this implies that $x(0^{2\,1}_{0m}) = 0$. So $A_0\backslash\{0^{21}_{0n}\}$ is the unique non-trivial block of $\ker(x{\restriction}_{A_0})$.

We have proven that $\ker(x{\restriction}_{A_0})$ has a unique non-trivial block, for each homomorphism $x : \mathbf{A} \to \underline{\mathbf{M}}$. Now define $g \in M^\omega$ by $g(n) := \rho_n(a_n)$,

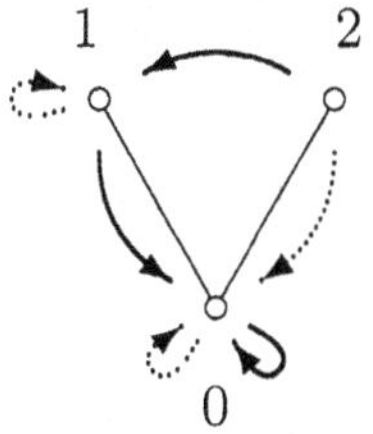

Figure 3.5 The non-dualisable algebra $\underline{\mathbf{Q}}$ has an algebraic semilattice operation

where a_n is chosen from the non-trivial block of $\ker(\rho_n{\restriction}_{A_0})$. The only block of $\ker(\rho_0{\restriction}_{A_0})$ is A_0. So $g(0) = \rho_0(0^{21}_{01}) = 2$. For each $n \in \omega\backslash\{0\}$, the unique non-trivial block of $\ker(\rho_n{\restriction}_{A_0})$ is $A_0\backslash\{0^{21}_{0n}\}$, and therefore we have $g(n) = \rho_n(0^{2\;\,1}_{0\,n+1}) = 0$. So $g = 0^2_0$. We are assuming that $\underline{\mathbf{M}}$ is dualisable and thus, by the Non-dualisability Lemma, 3.4.1, we must have $g \in A$. Therefore $0^2_0 \in \mathrm{sg}_{\underline{\mathbf{M}}^\omega}(A_0 \cup B)$, whence 002 must be a term function of $\underline{\mathbf{M}}$. ∎

Using the previous result, we can show that there are two natural generalisations of Theorem 2.1.1 that do not hold. We know that a finite algebra is dualisable if it has a pair of algebraic lattice operations. But an algebraic semilattice operation or an algebraic majority operation is not enough to guarantee the dualisability of an algebra.

3.4.3 Example *Let* $\underline{\mathbf{Q}} := \langle\{0, 1, 2\}; 001, 010\rangle$ *be the unary algebra first considered in 1.2.3. Then* $\underline{\mathbf{Q}}$ *has both an algebraic semilattice operation and an algebraic majority operation, but* $\underline{\mathbf{Q}}$ *is not dualisable.*

Proof The maps 001 and 010 are endomorphisms of the meet semilattice shown in Figure 3.5. So there is an algebraic semilattice operation on $\underline{\mathbf{Q}}$. There is an algebraic majority operation $m : \underline{\mathbf{Q}}^3 \to \underline{\mathbf{Q}}$, given by

$$m(a, b, c) = \begin{cases} \mathrm{maj}(a, b, c) & \text{if } \big|\{a, b, c\}\big| \leqslant 2, \\ 0 & \text{otherwise.} \end{cases}$$

Nevertheless, the algebra $\underline{\mathbf{Q}}$ is not dualisable, by Theorem 3.4.2. ∎

There are some semilattices and majority algebras whose endomorphisms do form dualisable unary algebras, as the following general argument shows. Consider a finite algebra $\mathbf{M}_0$ (of any type) such that $|\mathrm{Con}(\mathbf{M}_0)| \leqslant 3$, and let $F \subseteq \mathrm{End}(\mathbf{M}_0)$. Then $\langle M; F\rangle$ is a zero- or one-kernel algebra and is therefore dualisable, by Theorems 3.1.8 and 3.2.10. This argument can also be applied, for example, to the dihedral group $\mathbf{G}$ of order $2p$, for any prime p. Each set of endomorphisms of $\mathbf{G}$ forms the operations of a dualisable unary algebra.

Now we will show that every algebra of type $(2)_{\mathrm{M}}$ is non-dualisable.

3.4.4 Theorem *Let* $\underline{\mathbf{M}}$ *be a dualisable unary algebra on the set* $\{0,1,2\}$.

(i) *If* 010, 001 *and* 110 *are term functions of* $\underline{\mathbf{M}}$, *then so is* 222.

(ii) *If* 002, 020 *and* 202 *are term functions of* $\underline{\mathbf{M}}$, *then so is* 111.

Proof We will prove (i). Claim (ii) will then follow using conjugation by 021. Assume that 010, 001 and 110 are term functions of $\underline{\mathbf{M}}$. Define two subsets of M^ω by

$$A_0 := \{\, 2^{0\ \ 1}_{k\ k+1} \mid k \in \omega \text{ and } k \text{ is even} \,\}$$

and

$$B := \{\, 0^{1\,1\,2\,2}_{k\ell mn} \mid k, \ell, m, n \in \omega \text{ are distinct} \,\}.$$

Let $\mathbf{A}$ denote the subalgebra of $\underline{\mathbf{M}}^\omega$ generated by $A_0 \cup B$.

Let $x\colon \mathbf{A} \to \underline{\mathbf{M}}$ be a homomorphism. We want to show that $\ker(x{\restriction}_{A_0})$ has a unique non-trivial block. So we can assume that $x(A_0) \neq \{2\}$. There is an even number $k \in \omega$ such that $x(2^{0\ \ 1}_{k\ k+1}) \in \{0,1\}$. Let $\ell \in \omega\backslash\{k\}$ such that ℓ is even. Then k, $k+1$, ℓ and $\ell+1$ are all distinct, and

$$2^{0\ \ 1}_{k\ k+1} \xrightarrow{110} 0^{1\ \ 1}_{k\ k+1} \xleftarrow{010} 0^{1\ \ 1\ \ 2\ \ 2}_{k\ k+1\ \ell\ \ell+1} \xrightarrow{001} 0^{1\ \ 1}_{\ell\ \ell+1} \xleftarrow{110} 2^{0\ \ 1}_{\ell\ \ell+1}$$

in $\mathbf{A}$. Applying the homomorphism x gives us

$$\boxed{0,1} \xrightarrow{110} 1 \xleftarrow{010} 1 \xrightarrow{001} 0 \xleftarrow{110} 2$$

in $\underline{\mathbf{M}}$. So $x(2^{0\ \ 1}_{\ell\ \ell+1}) = 2$, and $A_0\backslash\{2^{0\ \ 1}_{k\ k+1}\}$ is the unique non-trivial block of $\ker(x{\restriction}_{A_0})$.

Define $g \in M^\omega$ by $g(n) := \rho_n(a_n)$, where a_n is any member of the unique non-trivial block of $\ker(\rho_n{\restriction}_{A_0})$. Then g is the constant sequence $\widehat{2}$. Since $\underline{\mathbf{M}}$ is dualisable, the Non-dualisability Lemma, 3.4.1, tells us that $g \in A$. Thus $\widehat{2} \in \mathrm{sg}_{\underline{\mathbf{M}}^\omega}(A_0 \cup B)$, whence 222 is a term function of $\underline{\mathbf{M}}$. ∎

It remains to prove that every three-element unary algebra with three kernels is non-dualisable. We shall obtain this as a corollary of the following stronger result.

3.4.5 Theorem *Let* $\underline{\mathbf{M}}$ *be a finite unary algebra with at least three elements. Assume that, for each* $m \in M$, *the equivalence relation coming from the two-block partition* $\{\{m\}, M\backslash\{m\}\}$ *is a kernel of* $\underline{\mathbf{M}}$. *Then* $\underline{\mathbf{M}}$ *is not dualisable.*

Proof We can assume that $M = \{0,\dots,n\}$, for some $n \in \omega$ with $n \geqslant 2$. For each $m \in M$, there is a unary term function u_m of $\underline{\mathbf{M}}$ whose kernel is the equivalence relation coming from $\{\{m\}, M\backslash\{m\}\}$. Define two subsets of M^ω by

$$A_0 := \{\, 0^{11}_{0k} \mid k \in \omega\backslash\{0\}\,\}$$

and

$$B := \{\, 0^{112}_{0k\ell} \mid k, \ell \in \omega\backslash\{0\} \text{ and } k \neq \ell\,\}.$$

Let $\mathbf{A}$ be the subalgebra of $\underline{\mathbf{M}}^\omega$ generated by $A_0 \cup B$, and let $x : \mathbf{A} \to \underline{\mathbf{M}}$ be a homomorphism. We will show that $\ker(x{\restriction}_{A_0})$ has a unique non-trivial block.

Case 1: $m \in x(B)$, for some $m \in M\backslash\{0,1,2\}$. There exist $k, \ell \in \omega\backslash\{0\}$, with $k \neq \ell$, such that $x(0^{112}_{0k\ell}) = m$. Let $j \in \omega\backslash\{0\}$. Then

$$0^{112}_{0k\ell} \xrightarrow{u_m} * \xleftarrow{u_m} 0^{11}_{0j} \qquad \overset{x}{\Longrightarrow} \qquad \boxed{m} \xrightarrow{u_m} * \xleftarrow{u_m} x(0^{11}_{0j}).$$

(Here we are using $*$ as a wild card.) This implies that $x(0^{11}_{0j}) = m$. So $x(A_0) = \{m\}$, and $\ker(x{\restriction}_{A_0})$ has only one block.

Case 2: $m \in x(A_0)$, for some $m \in M\backslash\{0,1\}$. There is some $k \in \omega\backslash\{0\}$ such that $x(0^{11}_{0k}) = m$. For all $j \in \omega\backslash\{0\}$, we have

$$0^{11}_{0k} \xrightarrow{u_m} * \xleftarrow{u_m} 0^{11}_{0j} \qquad \overset{x}{\Longrightarrow} \qquad \boxed{m} \xrightarrow{u_m} * \xleftarrow{u_m} x(0^{11}_{0j}),$$

and therefore $x(0^{11}_{0j}) = m$. So $x(A_0) = \{m\}$.

Case 3: $x(A_0) \subseteq \{0,1\}$ and $x(B) \subseteq \{0,1,2\}$. We can assume that $x{\restriction}_{A_0}$ is not constant. So there exist $k, \ell \in \omega\backslash\{0\}$ such that $x(0^{11}_{0k}) = 0$ and $x(0^{11}_{0\ell}) = 1$. Let $j \in \omega\backslash\{0,\ell\}$. We shall prove that $x(0^{11}_{0j}) = 0$. We have

$$\begin{array}{ccccccccc}
0^{11}_{0\ell} & & & & 0^{11}_{0k} & & & & 0^{11}_{0j} \\
\downarrow u_1 & & & & \downarrow u_1 & & & & \downarrow u_1 \\
* & & & & * & & & & * \\
\uparrow u_1 & & & & \uparrow u_1 & & & & \uparrow u_1 \\
0^{112}_{0\ell k} & \xrightarrow{u_0} & * & \xleftarrow{u_0} & 0^{112}_{0k\ell} & \xrightarrow{u_2} & * & \xleftarrow{u_2} & 0^{112}_{0j\ell}
\end{array}$$

in $\mathbf{A}$. As $x(0^{112}_{0k\ell}) \in \{0,1,2\}$, applying x gives us the following in $\underline{\mathbf{M}}$.

$$\begin{array}{ccccccccc}
\boxed{1} & & & & \boxed{0} & & & & x(0^{11}_{0j}) \\
\downarrow u_1 & & & & \downarrow u_1 & & & & \downarrow u_1 \\
* & & & & * & & & & * \\
\uparrow u_1 & & & & \uparrow u_1 & & & & \uparrow u_1 \\
1 & \xrightarrow{u_0} & * & \xleftarrow{u_0} & 2 & \xrightarrow{u_2} & * & \xleftarrow{u_2} & 2
\end{array}$$

Since $x(0^{11}_{0j}) \in \{0,1\}$, it follows that $x(0^{11}_{0j}) = 0$. Thus $A_0 \backslash \{0^{11}_{0\ell}\}$ is the unique non-trivial block of $\ker(x{\restriction}_{A_0})$.

Define $g \in M^\omega$ by $g(n) := \rho_n(a_n)$, where a_n is any element of the non-trivial block of $\ker(\rho_n{\restriction}_{A_0})$. Then $g = 0^1_0$. But $0^1_0 \notin \mathrm{sg}_{\underline{\mathbf{M}}^\omega}(A_0 \cup B)$, as $\underline{\mathbf{M}}$ is a unary algebra. So $g \notin A$ and therefore $\underline{\mathbf{M}}$ is not dualisable, by the Non-dualisability Lemma, 3.4.1. ∎

3.4.6 Corollary *No three-kernel three-element unary algebra is dualisable.*

Theorem 3.4.5 also has as a corollary the following result of L. Heindorf.

3.4.7 Corollary [35] *Let $\underline{\mathbf{M}}$ be a finite unary algebra with $\{0,1,2\} \subseteq M$, and assume that each map in $\{0,1\}^M$ is a term function of $\underline{\mathbf{M}}$. Then $\underline{\mathbf{M}}$ is non-dualisable.*

We have now proved the characterisation of dualisable three-element unary algebras given in the introduction to this chapter. Claim (i) of the theorem follows from Theorems 3.1.8 and 3.2.10. Claim (ii) of the theorem follows from Theorems 3.3.3, 2.1.2, 3.4.2, 3.4.4 and 3.3.9. Claim (iii) holds by Corollary 3.4.6.

3.5 Finite unars are dualisable

A **unar** is a unary algebra with only one fundamental operation. We close this chapter by proving that all finite unars are dualisable. This result will be generalised in Chapter 7. There we will consider the broader class of 'linear' unary algebras, which includes all unars. We will show that all finite linear unary algebras are strongly dualisable, but this will require much more effort. Our proof here is a straightforward combination of the results of Section 3.1 and a theorem from Chapter 2.

Assume that $\underline{\mathbf{M}} = \langle M; u \rangle$ is a finite unar. The directed graph of u will be a disjoint union of directed graphs such as that given in Figure 3.6. An element a of $\underline{\mathbf{M}}$ is **cyclic** if there is some $n \in \omega\backslash\{0\}$ such that $u^n(a) = a$. The set C of all cyclic elements of $\underline{\mathbf{M}}$ is the largest subuniverse of $\underline{\mathbf{M}}$ on which u is a permutation. Now let $\underline{\mathbf{C}}$ denote the subalgebra of $\underline{\mathbf{M}}$ with the underlying set C. Then $\underline{\mathbf{C}}$ is a zero-kernel unary algebra, and so is dualisable by Theorem 3.1.8. We will be able to use the Term Retract Theorem, 2.3.3, to lift the duality for $\mathbb{ISP}(\underline{\mathbf{C}})$ up to a duality for $\mathbb{ISP}(\underline{\mathbf{M}})$.

3.5.1 Theorem *Every finite unar is dualisable.*

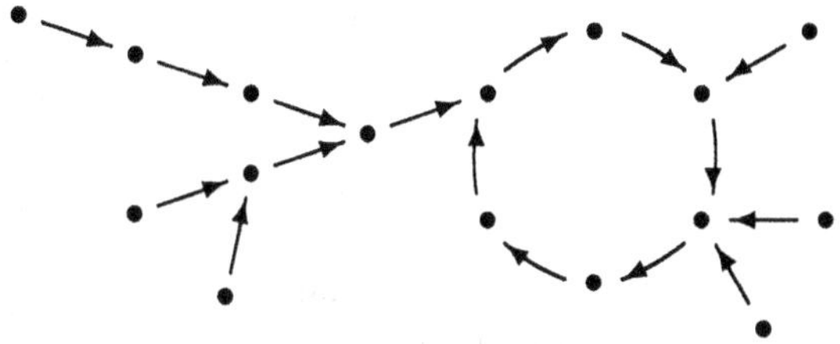

Figure 3.6 An example of a connected unar

Proof Let $\underline{\mathbf{M}} = \langle M; u \rangle$ be a finite unar. Define C to be the set of all cyclic elements of $\underline{\mathbf{M}}$, and let $\underline{\mathbf{C}}$ denote the subalgebra of $\underline{\mathbf{M}}$ with underlying set C. For each $a \in M$, define the 'distance' of a from a cycle by

$$d(a) := \min\{\, n \in \omega \mid u^n(a) \in C \,\}.$$

For each cyclic element $a \in C$, define the 'order' of a by

$$o(a) := \min\{\, n \in \omega \backslash \{0\} \mid u^n(a) = a \,\}.$$

Now set $m := \max\{\, d(a) \mid a \in M \,\}$ and $\ell := \operatorname{lcm}\{\, o(a) \mid a \in C \,\}$. Then there is a term retraction $\gamma : \underline{\mathbf{M}} \twoheadrightarrow \underline{\mathbf{C}}$, given by $\gamma := u^{\ell m}$, with $\gamma{\restriction}_C = \mathrm{id}_C$.

The operation $u{\restriction}_C : C \to C$ is a permutation, and so $\underline{\mathbf{C}}$ is a zero-kernel algebra. Thus $\underline{\mathbf{C}}$ is dualised by an alter ego of finite type, by Theorem 3.1.8. We can now assume that $M \backslash C$ is non-empty.

We will apply the Term Retract Theorem, 2.3.3, to show that $\underline{\mathbf{M}}$ is dualisable. We define the subset S of M to consist of all elements that are 'just outside' the cycles of $\underline{\mathbf{M}}$:

$$S := \{\, s \in M \mid s \notin C \text{ and } u(s) \in C \,\} = \{\, s \in M \mid d(s) = 1 \,\}.$$

The set S must be non-empty, since we are assuming that $M \backslash C$ is non-empty.

We next need to construct a set G' of binary homomorphisms of $\underline{\mathbf{M}}$. Let $s \in S$ and let $t \in M$, and define the map $g_{st} : M^2 \to M$ by

$$g_{st}(a, b) = \begin{cases} a & \text{if } (s, t) \in \mathrm{sg}_{\underline{\mathbf{M}}^2}((a, b)), \\ \gamma(a) & \text{otherwise.} \end{cases}$$

To see that g_{st} preserves u, let $a, b \in M$. First assume $(s, t) \notin \mathrm{sg}_{\underline{\mathbf{M}}^2}((a, b))$. Then $(s, t) \notin \mathrm{sg}_{\underline{\mathbf{M}}^2}((u(a), u(b)))$, and therefore

$$u(g_{st}(a, b)) = u(\gamma(a)) = \gamma(u(a)) = g_{st}(u(a), u(b)).$$

We can now assume that $(s,t) \in \mathrm{sg}_{\underline{\mathbf{M}}^2}((a,b))$ and $(s,t) \notin \mathrm{sg}_{\underline{\mathbf{M}}^2}((u(a),u(b)))$. This implies that $(s,t) = (a,b)$ and therefore, as $u(s) \in C$, we have

$$u(g_{st}(a,b)) = u(g_{st}(s,t)) = u(s) = \gamma(u(s))$$
$$= g_{st}(u(s),u(t)) = g_{st}(u(a),u(b)).$$

Thus $g_{st} : \underline{\mathbf{M}}^2 \to \underline{\mathbf{M}}$ is a binary homomorphism of $\underline{\mathbf{M}}$.

Again, let $s \in S$ and let $t \in M$. Then $s \notin C$ and $u(s) \in C$. So, for all $a,b \in M$ with $g_{st}(a,b) = s$, we must have $a = s$ and $(s,t) \in \mathrm{sg}_{\underline{\mathbf{M}}^2}((s,b))$, which implies that $b = t$. Therefore

$$g_{st}(a,b) = s \iff (a,b) = (s,t), \qquad (\pounds)_{st}$$

for all $a, b \in M$.

Now define the set G' of binary homomorphisms of $\underline{\mathbf{M}}$ by

$$G' := \{\, g_{st} \mid s \in S \text{ and } t \in M \,\}.$$

For each $s \in S$, the operation g_{ss} in G' satisfies $g_{ss}^{-1}(s) = \{(s,s)\}$, by $(\pounds)_{ss}$. So every element of S is a strong idempotent of a map in G'.

Finally, let $k \in M \backslash C$ and let $t \in M \backslash S$. Define $s_k := u^{d(k)-1}(k)$ in S. Since u is an endomorphism of $\underline{\mathbf{M}}$, we have $s_k \in S \cap \mathrm{End}(\underline{\mathbf{M}})(k)$. By $(\pounds)_{s_k t}$, we get

$$g_{s_k t}(s_k, m) = s_k \iff m = t,$$

for all $m \in M$. Thus G' and $S \cap \mathrm{End}(\underline{\mathbf{M}})(k)$ distinguish t within M. It now follows from the Term Retract Theorem, 2.3.3, that $\underline{\mathbf{M}}$ is dualisable. ∎

4

Full and strong dualisability: three-element unary algebras

We characterise the fully dualisable three-element unary algebras. Amongst the dualisable three-element unary algebras, full dualisability is equivalent to strong dualisability and to two other weak injectivity conditions.

Full dualities are more symmetric and more useful than dualities—they provide a dual equivalence, rather than just a dual representation. All the 'classical' examples of dualisable algebras are also fully dualisable: the finite cyclic groups [55, 29]; and the two-element Boolean algebra [63], lattice [56] and semilattice [37]. Indeed, the first example of a dualisable algebra that is not fully dualisable was found more than 20 years after the birth of duality theory, by Hyndman and Willard [41]. Their example was the three-element unary algebra $\langle\{0,1,2\};001,122\rangle$. In this chapter, we shall completely characterise full dualisability amongst the three-element unary algebras; thereby revealing many more examples of dualisable algebras that are not fully dualisable.

The concept of 'full duality' is very natural from a categorical viewpoint. But, from an algebraic viewpoint, the concept is not well understood and is rather difficult to work with. The stronger concept of 'strong duality' is more transparent and better behaved. For example, it is easy to prove that a duality or a strong duality is preserved when the type of the alter ego is enriched. No such result has been proven for full dualities; and it is conceivable that, by enriching the type of the alter ego, a full duality could be destroyed.

Since strong dualities are much easier to work with, every known full duality has actually been established by setting up a strong duality. The Full versus Strong Problem [8] asks: 'Is every full duality also strong?' While we don't know the answer to this question, progress has been made towards a solution in a series of papers by Davey, Haviar, Niven, Perkal and Willard [18–21].

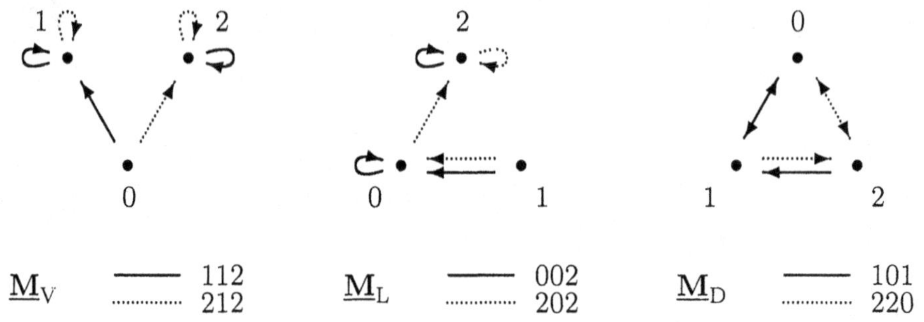

Figure 4.1 The three 'bad' three-element unary algebras

Amongst other results, it is shown that a full duality must be strong if it is based on a bounded distributive lattice, an abelian group or a semilattice. In contrast, it is also shown that every finite non-boolean bounded distributive lattice is the base for a duality that is not strong but that is full on the class of all *finite* distributive lattices.

In the preceding chapter, we uncovered some evidence that dualisability for unary algebras is complicated. The description of the dualisable three-element unary algebras found there is quite intricate. In this chapter, we shall uncover some evidence that suggests strong dualisability is not so complicated. Amongst the dualisable three-element unary algebras, the strongly dualisable algebras are easy to identify. There are three particular three-element unary algebras that, between them, capture what it is that can stop a dualisable three-element unary algebra from being strongly dualisable. These are the algebras $\underline{\mathbf{M}}_{\mathrm{V}}$, $\underline{\mathbf{M}}_{\mathrm{L}}$ and $\underline{\mathbf{M}}_{\mathrm{D}}$, drawn in Figure 4.1.

During this chapter, we prove the following theorem. We say that an algebra $\mathbf{A}$ is an **isoreduct** of an algebra $\mathbf{B}$ if $\mathbf{A}$ is isomorphic to a term reduct of $\mathbf{B}$.

4.0.1 Theorem *Let $\underline{\mathbf{M}}$ be a dualisable three-element unary algebra. Then the following are equivalent*:

(i) *$\underline{\mathbf{M}}$ has neither $\underline{\mathbf{M}}_{\mathrm{V}}$, $\underline{\mathbf{M}}_{\mathrm{L}}$ nor $\underline{\mathbf{M}}_{\mathrm{D}}$ as an isoreduct*;

(ii) *$\underline{\mathbf{M}}$ is fully dualisable*;

(iii) *$\underline{\mathbf{M}}$ is strongly dualisable*;

(iv) *$\underline{\mathbf{M}}$ has enough algebraic operations*;

(v) *$\underline{\mathbf{M}}$ is quasi-injective.*

The properties 'fully dualisable', 'strongly dualisable' and 'enough algebraic operations' in this theorem were defined in Chapter 1. The property 'quasi-injective' is a weak form of injectivity that comes from setting up distance functions on unary algebras. We shall define 'quasi-injective' in Section 4.1.

Our main theorem provides many examples of dualisable algebras that are not fully dualisable. The algebras $\underline{\mathbf{M}}_{\mathrm{V}}$ and $\underline{\mathbf{M}}_{\mathrm{L}}$ are dualisable, by the Lattice Endomorphism Theorem, 2.1.2. Nevertheless, the algebras $\underline{\mathbf{M}}_{\mathrm{V}}$ and $\underline{\mathbf{M}}_{\mathrm{L}}$ are not fully dualisable. The algebra $\underline{\mathbf{M}}_{\mathrm{D}}$ is dualisable, by Theorem 3.0.1, since 010, 002, 111 and 222 are term functions of $\underline{\mathbf{M}}_{\mathrm{D}}$. So $\underline{\mathbf{M}}_{\mathrm{D}}$ is another dualisable algebra that is not fully dualisable. More generally, any two-kernel three-element unary algebra that has $\underline{\mathbf{M}}_{\mathrm{D}}$ as a reduct is dualisable but not fully dualisable.

There are many questions related to our main theorem. Is every full duality based on a three-element unary algebra necessarily strong? Is there a fully dualisable algebra that is not strongly dualisable? Is there a strongly dualisable algebra that does not have enough algebraic operations? It is known that there is a non-quasi-injective algebra with enough algebraic operations, and a quasi-injective algebra without enough algebraic operations [40]. Even though there is no direct link between quasi-injectivity and having enough algebraic operations, the concept of quasi-injectivity will arise naturally in some of our proofs that particular algebras have enough algebraic operations.

Amongst the three-element unary algebras, the algebra $\underline{\mathbf{M}}_{\mathrm{V}}$ is an obstacle to strong dualisability. However, there is a seven-element strongly dualisable algebra that has $\underline{\mathbf{M}}_{\mathrm{V}}$ as a subalgebra [40]. In general, it is not known whether every finite unary algebra can be embedded into a strongly dualisable algebra. But every finite unary algebra can be embedded into an algebra with enough algebraic operations. Within every locally finite variety of unary algebras, each finite algebra can be embedded into a finite injective algebra (P. Berthiaume [3]). So any finite unary algebra can be embedded into a finite algebra that is injective in the quasi-variety it generates. It is easy to show that a finite injective algebra must have enough algebraic operations; use Lemma 4.1.1. (See Theorem A.7.5 in the appendix for an alternative proof of the fact that an injective dualisable algebra is strongly dualisable.)

Our main theorem has been extended by Hyndman and Pitkethly [40]. Conditions (i), (iv) and (v) are also equivalent for non-dualisable three-element unary algebras. Furthermore, I. P. Bestsennyi [4] has shown that condition (i) exactly describes the three-element unary algebras (of finite type) whose quasi-equational theory is finitely based.

Two of the implications in Theorem 4.0.1 hold in general. We know that (iv) implies (iii), by the EAO Theorem, 1.5.4, and that (iii) implies (ii), by the Full Duality Theorem, 1.5.1. The rest of the proof is spread throughout this whole chapter. In Sections 4.1 and 4.2, we shall prove that (i) implies (iv) and that (i) implies (v). In Sections 4.3 and 4.4, we will prove that ¬(i) implies ¬(ii) and that ¬(i) implies ¬(v).

The content of this chapter comes from a paper written by the first author [50]. However, the presentation of the chapter has been influenced by a paper written by the first author and J. Hyndman [40].

4.1 Zero-kernel and one-kernel unary algebras revisited

In Chapter 3, we showed that the quasi-variety generated by a finite zero- or one-kernel unary algebra is especially simple. Each such quasi-variety is encapsulated by a finite set of petals, called a finite gentle basis. We used this finiteness property to prove that every finite zero- or one-kernel unary algebra is dualisable. In this section, we use the same finiteness property to show that every finite zero- or one-kernel unary algebra has enough algebraic operations and is quasi-injective.

We require the following general lemma due to J. Hyndman.

4.1.1 Lemma [38, 2.2] *Let* $\underline{\mathbf{M}}$ *be a finite algebra and define* $\mathcal{A} := \mathbb{ISP}(\underline{\mathbf{M}})$. *Let* $\mathbf{B}$ *be a subalgebra of* $\underline{\mathbf{M}}^n$, *for some* $n \in \omega$. *There is a set of projections* $Z \subseteq \mathcal{A}(\underline{\mathbf{M}}^n, \underline{\mathbf{M}})$ *such that* $|Z| \leqslant |B| - 1$ *and* Z *separates the elements of* B.

Proof Since $\mathbf{B}$ is finite, we can write $B = \{b_1, \dots, b_k\}$, where $k := |B|$. We will construct a sequence of sets of projections $Z_1, \dots, Z_k \subseteq \mathcal{A}(\underline{\mathbf{M}}^n, \underline{\mathbf{M}})$ such that $|Z_i| \leqslant i - 1$ and Z_i separates $\{b_1, \dots, b_i\}$, for all $i \in \{1, \dots, k\}$.

Define $Z_1 := \varnothing$. Now let $i \in \{1, \dots, k-1\}$ and assume that Z_i has already been defined. We wish to construct Z_{i+1}.

Define C to be the set of all $c \in \{b_1, \dots, b_i\}$ such that Z_i does not separate c from b_{i+1}. To see that $|C| \leqslant 1$, assume that $c, d \in C$. For all $z \in Z_i$, we have $z(c) = z(b_{i+1}) = z(d)$. So Z_i does not separate $c, d \in \{b_1, \dots, b_i\}$. Thus $c = d$, whence $|C| \leqslant 1$.

If $|C| = 0$, then we can define $Z_{i+1} := Z_i$. Otherwise, let c be the unique element of C. We must have $c \neq b_{i+1}$ in $\mathbf{B}$, and so there exists a projection $z : \underline{\mathbf{M}}^n \to \underline{\mathbf{M}}$ with $z(c) \neq z(b_{i+1})$. Thus we can define $Z_{i+1} := Z_i \cup \{z\}$. The set Z_{i+1} satisfies $|Z_{i+1}| \leqslant |Z_i| + 1 \leqslant i$ and separates the elements of $\{b_1, \dots, b_{i+1}\}$. ∎

4.1.2 Lemma *Each finite zero-kernel or one-kernel unary algebra has enough algebraic operations.*

Proof Let $\underline{\mathbf{M}}$ be a finite zero- or one-kernel unary algebra. We may assume that $\underline{\mathbf{M}}$ is non-trivial. By Theorem 3.2.9, there is a finite gentle basis $\mathcal{B}$ for $\mathcal{A} := \mathbb{ISP}(\underline{\mathbf{M}})$. We can choose $k \in \omega\backslash\{0\}$ such that $k \geqslant |\mathcal{A}(\mathbf{P}, \underline{\mathbf{M}})|$, for all $\mathbf{P} \in \mathcal{B}$. Now define $f : \omega \to \omega$ by $f(m) := (1 + k^2)m$. Let $\mathbf{B} \leqslant \mathbf{A} \leqslant \underline{\mathbf{M}}^n$, for some $n \in \omega\backslash\{0\}$, and let $h : \mathbf{A} \to \underline{\mathbf{M}}$ be a homomorphism.

First assume that $n = 1$. Define the subset Y of $\mathcal{A}(\underline{\mathbf{M}}^1, \underline{\mathbf{M}})$ by $Y := \{\pi_0\}$, where $\pi_0 : \underline{\mathbf{M}}^1 \hookrightarrow\!\!\!\!\rightarrow \underline{\mathbf{M}}$ is the natural isomorphism. Then $\sqcap Y{\restriction}_A : \mathbf{A} \hookrightarrow\!\!\!\!\rightarrow \sqcap Y(\mathbf{A})$ is an isomorphism. Now define $h' : \sqcap Y(\mathbf{A}) \to \underline{\mathbf{M}}$ by $h' := h \circ (\sqcap Y{\restriction}_A)^{-1}$. This gives us $h' \circ \sqcap Y{\restriction}_B = h{\restriction}_B$, as required.

We can now assume that $n > 1$. This implies that $C_{\underline{\mathbf{M}}^n} \neq M^n$. So the set $\mathcal{P}$ of all petals of $\underline{\mathbf{M}}^n$ is non-empty. We want to construct an appropriate subset Y of $\mathcal{A}(\underline{\mathbf{M}}^n, \underline{\mathbf{M}})$.

First, choose some $\mathbf{P} \in \mathcal{P}$. There is a gentle surjection $\varphi_{\mathbf{P}} : \mathbf{P} \twoheadrightarrow \mathbf{P}^*$, for some $\mathbf{P}^* \in \mathcal{B}$. By Lemma 3.2.4, the petal $\mathbf{P}^*$ embeds into $\mathbf{P}$. Since $\mathbf{P}$ is a subalgebra of $\underline{\mathbf{M}}^n$ and $n \neq 0$, this implies that the set $\mathcal{A}(\mathbf{P}^*, \underline{\mathbf{M}})$ is non-empty. As $k \geqslant |\mathcal{A}(\mathbf{P}^*, \underline{\mathbf{M}})|$, we can write $\mathcal{A}(\mathbf{P}^*, \underline{\mathbf{M}}) = \{x_{\mathbf{P}1}, \dots, x_{\mathbf{P}k}\}$.

Now, for each petal $\mathbf{P}$ of $\underline{\mathbf{M}}^n$ and all $i, j \in \{1, \dots, k\}$, we can define the homomorphism $g_{\mathbf{P}ij} : \underline{\mathbf{M}}^n \to \underline{\mathbf{M}}$ by

$$g_{\mathbf{P}ij} := (x_{\mathbf{P}i} \circ \varphi_{\mathbf{P}}) \sqcup \bigsqcup\{\, x_{\mathbf{Q}j} \circ \varphi_{\mathbf{Q}} \mid \mathbf{Q} \in \mathcal{P}\backslash\{\mathbf{P}\} \,\},$$

as $\underline{\mathbf{M}}^n$ is the coproduct of its petals. By Lemma 4.1.1, there is a non-empty subset Z of $\mathcal{A}(\underline{\mathbf{M}}^n, \underline{\mathbf{M}})$ such that $|Z| \leqslant |B|$ and Z separates the elements of B. Let $\mathcal{P}_B$ denote the set of all petals $\mathbf{P}$ of $\underline{\mathbf{M}}^n$ such that $P \cap B \neq C_{\underline{\mathbf{M}}^n}$, and define the subset Y of $\mathcal{A}(\underline{\mathbf{M}}^n, \underline{\mathbf{M}})$ by

$$Y := Z \cup \{\, g_{\mathbf{P}ij} \mid \mathbf{P} \in \mathcal{P}_B \text{ and } i, j \in \{1, \dots, k\} \,\}.$$

Then $|Y| \leqslant |B| + k^2|B| = f(|B|)$. Define the homomorphism

$$\mu : \underline{\mathbf{M}}^n \to \underline{\mathbf{M}}^Y \quad \text{by} \quad \mu := \sqcap Y.$$

To prove that $\underline{\mathbf{M}}$ has enough algebraic operations, we shall construct a homomorphism $h' : \mu(\mathbf{A}) \to \underline{\mathbf{M}}$ such that $h' \circ \mu{\restriction}_B = h{\restriction}_B$.

First we shall express $\mu(\mathbf{A})$ as a coproduct of simpler algebras. Define the subuniverse D of $\underline{\mathbf{M}}^n$ by

$$D := \bigcup\{\, P \mid \mathbf{P} \in \mathcal{P}\backslash\mathcal{P}_B \,\}.$$

We shall prove that, if $D \cap A \neq \varnothing$, then $\mu(\mathbf{A})$ is the coproduct in $\mathcal{A}$ of the family

$$\{\, \mu(\mathbf{P} \cap \mathbf{A}) \mid \mathbf{P} \in \mathcal{P}_B \,\} \cup \{\mu(\mathbf{D} \cap \mathbf{A})\},$$

and that, if $D \cap A = \varnothing$, then $\mu(\mathbf{A})$ is the coproduct in $\mathcal{A}$ of the family

$$\{\, \mu(\mathbf{P} \cap \mathbf{A}) \mid \mathbf{P} \in \mathcal{P}_B \,\}.$$

To prove both these claims, it suffices to show that

$$\Big(\{\, \mu(P \cap A)\backslash C_{\mu(\mathbf{A})} \mid \mathbf{P} \in \mathcal{P}_B \,\} \cup \{\mu(D \cap A)\backslash C_{\mu(\mathbf{A})}\}\Big)\backslash\{\varnothing\}$$

is a partition of $\mu(A)\backslash C_{\mu(\mathbf{A})}$. The above set is certainly a cover for $\mu(A)\backslash C_{\mu(\mathbf{A})}$, since $\mu(A)\backslash C_{\mu(\mathbf{A})} \subseteq \mu(A\backslash C_{\mathbf{A}})$ and every non-centre element of $\mathbf{A}$ belongs to a petal in $\mathcal{P}$. Let $a \in (P \cap A)\backslash C_{\mathbf{A}}$ and $b \in (Q \cap A)\backslash C_{\mathbf{A}}$, for some $\mathbf{P} \in \mathcal{P}_B$ and $\mathbf{Q} \in \mathcal{P}\backslash\{\mathbf{P}\}$. It is now enough to show that $\mu(a) \neq \mu(b)$.

As $a \neq b$ in M^n, there is a homomorphism $z : \underline{\mathbf{M}}^n \to \underline{\mathbf{M}}$ with $z(a) \neq z(b)$. By Lemma 3.2.4, there is a coretraction $\psi_a : \mathbf{P}^* \hookrightarrow \mathbf{P}$ for $\varphi_{\mathbf{P}} : \mathbf{P} \twoheadrightarrow \mathbf{P}^*$ such that $\mathrm{sg}_{\mathbf{P}}(a) \subseteq \psi_a(P^*)$. It follows that $\psi_a \circ \varphi_{\mathbf{P}}(a) = a$. The map $z \circ \psi_a$ belongs to $\mathcal{A}(\mathbf{P}^*, \underline{\mathbf{M}}) = \{x_{\mathbf{P}1}, \dots, x_{\mathbf{P}k}\}$. So there exists $i \in \{1, \dots, k\}$ such that $z \circ \psi_a = x_{\mathbf{P}i}$ and hence

$$x_{\mathbf{P}i} \circ \varphi_{\mathbf{P}}(a) = z \circ \psi_a \circ \varphi_{\mathbf{P}}(a) = z(a).$$

Similarly, there is $j \in \{1, \dots, k\}$ with $x_{\mathbf{Q}j} \circ \varphi_{\mathbf{Q}}(b) = z(b)$. Thus

$$g_{\mathbf{P}ij}(a) = x_{\mathbf{P}i} \circ \varphi_{\mathbf{P}}(a) = z(a) \neq z(b) = x_{\mathbf{Q}j} \circ \varphi_{\mathbf{Q}}(b) = g_{\mathbf{P}ij}(b).$$

Since $g_{\mathbf{P}ij} \in Y$ and $\mu = \sqcap Y$, this gives us $\mu(a) \neq \mu(b)$. So we have established the desired expression of $\mu(\mathbf{A})$ as a coproduct.

Now let $\mathbf{P} \in \mathcal{P}_B$. To see that the surjection $\mu{\restriction}_P : \mathbf{P} \twoheadrightarrow \mu(\mathbf{P})$ is gentle, let $a \in P_{\text{out}}$ and let $b, c \in P_{\text{in}} \cup \mathrm{sg}_{\mathbf{P}}(a)$ with $b \neq c$. Since $\varphi_{\mathbf{P}}$ is gentle, we have $\varphi_{\mathbf{P}}(b) \neq \varphi_{\mathbf{P}}(c)$ in $\mathbf{P}^*$. As $\mathbf{P}^* \in \mathbb{ISP}(\underline{\mathbf{M}})$, there exists $i \in \{1, \dots, k\}$ with $x_{\mathbf{P}i} \circ \varphi_{\mathbf{P}}(b) \neq x_{\mathbf{P}i} \circ \varphi_{\mathbf{P}}(c)$. So $g_{\mathbf{P}i1}(b) \neq g_{\mathbf{P}i1}(c)$, and therefore $\mu(b) \neq \mu(c)$. Thus $\mu{\restriction}_P : \mathbf{P} \twoheadrightarrow \mu(\mathbf{P})$ is gentle, which implies $\mu{\restriction}_{P \cap A} : \mathbf{P} \cap \mathbf{A} \twoheadrightarrow \mu(\mathbf{P} \cap \mathbf{A})$ is gentle. Since Y separates the elements of B, the map $\mu{\restriction}_{P \cap B}$ is one-to-one. It follows, by Lemma 3.2.4, that there is a coretraction $\nu_{\mathbf{P}} : \mu(\mathbf{P} \cap \mathbf{A}) \hookrightarrow \mathbf{P} \cap \mathbf{A}$ for $\mu{\restriction}_{P \cap A}$ such that $P \cap B \subseteq \nu_{\mathbf{P}} \circ \mu(P \cap A)$.

Since $\mu(\mathbf{A}) \leqslant \underline{\mathbf{M}}^Y$ and the set Y is non-empty, there exists a homomorphism $x : \mu(\mathbf{A}) \to \underline{\mathbf{M}}$. Using our expression of $\mu(\mathbf{A})$ as a coproduct, we can now define the homomorphism $h' : \mu(\mathbf{A}) \to \underline{\mathbf{M}}$ by

$$h' := x{\restriction}_{\mu(D \cap A)} \cup \bigcup\{\, h \circ \nu_{\mathbf{P}} \mid \mathbf{P} \in \mathcal{P}_B \,\}.$$

To see that $h' \circ \mu{\restriction}_B = h{\restriction}_B$, let $\mathbf{P} \in \mathcal{P}_B$ and let $b \in P \cap B$. Since $\nu_{\mathbf{P}}$ is a coretraction for $\mu{\restriction}_{P \cap A}$ and

$$b \in P \cap B \subseteq \nu_{\mathbf{P}} \circ \mu(P \cap A),$$

we have $\nu_{\mathbf{P}} \circ \mu(b) = b$. So

$$h' \circ \mu(b) = h \circ \nu_{\mathbf{P}} \circ \mu(b) = h(b).$$

Thus $\underline{\mathbf{M}}$ has enough algebraic operations. ∎

In the previous chapter, we showed that every finite zero- or one-kernel unary algebra is dualisable; see Theorems 3.1.8 and 3.2.10. So, using the EAO Theorem, 1.5.4, we obtain the following corollary.

4.1.3 Theorem *Each finite zero-kernel or one-kernel unary algebra is strongly dualisable.*

We will now define quasi-injectivity. First let $\underline{\mathbf{M}}$ be a finite unary algebra and let $\mathbf{A}$ belong to $\mathbb{ISP}(\underline{\mathbf{M}})$. Define the directed graph

$$Q(\mathbf{A}) = \langle A; \overline{E}_{\mathbf{A}} \rangle, \quad \text{where} \quad \overline{E}_{\mathbf{A}} := \{ (a, b) \in A^2 \mid b \in \mathrm{sg}_{\mathbf{A}}(a) \}.$$

The relation $\overline{E}_{\mathbf{A}}$ on A is reflexive and transitive. So $Q(\mathbf{A})$ is a quasi-ordered set. In fact, the graph $Q(\mathbf{A})$ is the quasi-ordered set determined by the directed graph $G(\mathbf{A})$ used in the previous chapter; see page 55.

Let $Q^*(\mathbf{A})$ denote the induced subgraph of $Q(\mathbf{A})$ with vertex set $A \backslash C_{\mathbf{A}}$. For all $a, b \in A \backslash C_{\mathbf{A}}$ and $n \in \omega$, we say that there is a **fence from** a **to** b **in A of length** n if there are edges $x_1, y_1, \ldots, x_n, y_n$ of $Q^*(\mathbf{A})$ such that

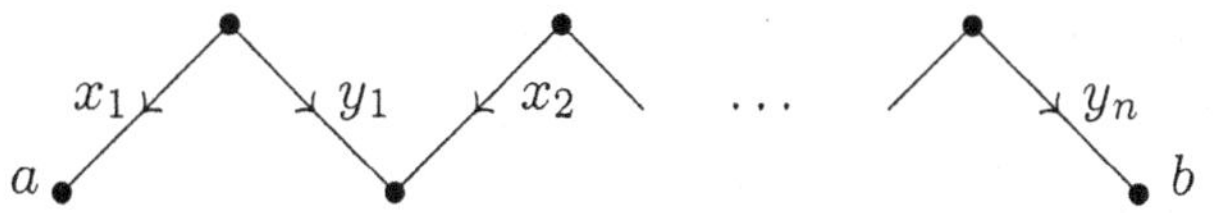

in $Q^*(\mathbf{A})$. It is easy to check that $a, b \in A \backslash C_{\mathbf{A}}$ belong to the same petal of $\mathbf{A}$ if and only if there is a fence from a to b in $\mathbf{A}$.

We can define a distance function $d_{\mathbf{A}}$ on $A \backslash C_{\mathbf{A}}$. For all $a, b \in A \backslash C_{\mathbf{A}}$, let $d_{\mathbf{A}}(a, b)$ be the length of the shortest fence from a to b in $\mathbf{A}$. (If $a, b \in A \backslash C_{\mathbf{A}}$ such that there is no fence from a to b in $\mathbf{A}$, then $d_{\mathbf{A}}(a, b) = \infty$.) The distance between two elements of $A \backslash C_{\mathbf{A}}$ is finite if and only if they belong to the same petal of $\mathbf{A}$.

Now let $n \in \omega \backslash \{0\}$. For each $a \in A \backslash C_{\mathbf{A}}$, define the ball in $\mathbf{A}$ with centre a and radius n by

$$n_{\mathbf{A}}(a) := \{ b \in A \backslash C_{\mathbf{A}} \mid d_{\mathbf{A}}(a, b) \leqslant n \} \cup C_{\mathbf{A}}.$$

For each $a \in C_{\mathbf{A}}$, we set $n_{\mathbf{A}}(a) := C_{\mathbf{A}}$. It is easy to check that, for each $a \in A$, the set $n_{\mathbf{A}}(a)$ forms a subalgebra of $\mathbf{A}$. Therefore, for each non-empty subset

B of A, the set

$$n_{\mathbf{A}}(B) := \bigcup \{ n_{\mathbf{A}}(b) \mid b \in B \}$$

forms a subalgebra $\boldsymbol{n}_{\mathbf{A}}(B)$ of $\mathbf{A}$. We say that $\underline{\mathbf{M}}$ is n-**quasi-injective** if, for all finite algebras $\mathbf{A}, \mathbf{B} \in \mathbb{ISP}(\underline{\mathbf{M}})$ such that $\mathbf{B} \leqslant \mathbf{A}$, every homomorphism $x : \mathbf{B} \to \underline{\mathbf{M}}$ that extends to $\boldsymbol{n}_{\mathbf{A}}(B)$ also extends to $\mathbf{A}$. We say that the algebra $\underline{\mathbf{M}}$ is **quasi-injective** if $\underline{\mathbf{M}}$ is n-quasi-injective, for some $n \in \omega\backslash\{0\}$.

4.1.4 Lemma *Every finite zero-kernel or one-kernel unary algebra is quasi-injective.*

Proof Let $\underline{\mathbf{M}}$ be a finite zero- or one-kernel unary algebra. Then there is a finite gentle basis $\mathcal{B}$ for $\mathcal{A} := \mathbb{ISP}(\underline{\mathbf{M}})$, by Theorem 3.2.9. For every petal $\mathbf{P}$ of $\mathcal{A}$, the distance between any two elements of $P\backslash C_{\mathbf{P}}$ is finite. So, for each finite petal $\mathbf{P}$ of $\mathcal{A}$, we can define

$$w_{\mathbf{P}} := \max\{ d_{\mathbf{P}}(a, b) \mid a, b \in P\backslash C_{\mathbf{P}} \}.$$

Now define

$$n := \max(\{ w_{\mathbf{B}} \mid \mathbf{B} \in \mathcal{B} \} \cup \{2\}).$$

We begin by proving that $w_{\mathbf{P}} \leqslant n$, for each finite petal $\mathbf{P}$ of $\mathcal{A}$.

Let $\mathbf{P}$ be a finite petal of $\mathcal{A}$. Then there is a gentle surjection $\varphi : \mathbf{P} \twoheadrightarrow \mathbf{B}$, for some $\mathbf{B} \in \mathcal{B}$. Now let $a, b \in P\backslash C_{\mathbf{P}}$ and define the subuniverse P_{ab} of $\mathbf{P}$ by $P_{ab} := P_{\text{in}} \cup \operatorname{sg}_{\mathbf{P}}(\{a, b\})$. We want to show that $d_{\mathbf{A}}(a, b) \leqslant n$.

Case 1: $\varphi{\restriction}_{P_{ab}}$ is one-to-one. Lemma 3.2.4 yields a coretraction $\psi : \mathbf{B} \hookrightarrow \mathbf{P}$ for φ with $a, b \in \psi(B)$. So $d_{\mathbf{P}}(a, b) \leqslant d_{\psi(\mathbf{B})}(a, b) \leqslant w_{\psi(\mathbf{B})} = w_{\mathbf{B}} \leqslant n$.

Case 2: $\varphi{\restriction}_{P_{ab}}$ is not one-to-one. We will show that $d_{\mathbf{P}}(a, b) \leqslant 2$. As the gentle surjection φ is not one-to-one on $P_{\text{in}} \cup \operatorname{sg}_{\mathbf{P}}(\{a, b\})$, we must have $a, b \in P_{\text{out}}$ and $\operatorname{sg}_{\mathbf{P}}(a) \neq \operatorname{sg}_{\mathbf{P}}(b)$. There exist $c_a \in \operatorname{sg}_{\mathbf{P}}(a)\backslash P_{\text{in}}$ and $c_b \in \operatorname{sg}_{\mathbf{P}}(b)\backslash P_{\text{in}}$ such that $\varphi(c_a) = \varphi(c_b)$. Since c_a and c_b are outer elements of $\mathbf{P}$, we have $\operatorname{sg}_{\mathbf{P}}(a) = \operatorname{sg}_{\mathbf{P}}(c_a)$ and $\operatorname{sg}_{\mathbf{P}}(b) = \operatorname{sg}_{\mathbf{P}}(c_b)$. The elements c_a and c_b are connected by a fence in the petal $\mathbf{P}$. As $\operatorname{sg}_{\mathbf{P}}(c_a) \neq \operatorname{sg}_{\mathbf{P}}(c_b)$, this fence must pass through $P_{\text{in}}\backslash C_{\mathbf{P}}$; see Remark 3.2.2 and Figure 3.3. So there must be a unary term function u of $\underline{\mathbf{M}}$ such that $u(c_a) \in P_{\text{in}}\backslash C_{\mathbf{P}}$. Since

$$\varphi(u(c_a)) = u(\varphi(c_a)) = u(\varphi(c_b)) = \varphi(u(c_b))$$

and φ is gentle, it follows that $u(c_a) = u(c_b)$. Therefore $d_{\mathbf{P}}(a, b) \leqslant 2 \leqslant n$, as $a \in \operatorname{sg}_{\mathbf{P}}(c_a)$ and $b \in \operatorname{sg}_{\mathbf{P}}(c_b)$.

We have shown that $w_{\mathbf{P}} \leqslant n$, for every finite petal $\mathbf{P}$ of $\mathcal{A}$. To see that the algebra $\underline{\mathbf{M}}$ is n-quasi-injective, choose finite algebras $\mathbf{A}, \mathbf{B} \in \mathcal{A}$ such that

$\mathbf{B} \leqslant \mathbf{A}$. Let $x : \mathbf{B} \to \underline{\mathbf{M}}$ be a homomorphism and assume that x extends to $n_{\mathbf{A}}(B)$. We want to show that x extends to $\mathbf{A}$, so we can assume that $\mathbf{A}$ is non-trivial. Let $\mathbf{P}$ be a petal of $\mathbf{A}$ with $P \cap B \neq C_{\mathbf{A}}$. Then $P \subseteq n_{\mathbf{A}}(B)$, since $w_{\mathbf{P}} \leqslant n$. So $x{\restriction}_{P \cap B}$ extends to the petal $\mathbf{P}$. Now let $\mathbf{Q}$ be a petal of $\mathbf{A}$ such that $Q \cap B = C_{\mathbf{A}}$. There is at least one homomorphism from $\mathbf{Q}$ to $\underline{\mathbf{M}}$, as $\mathbf{A} \in \mathbb{ISP}(\underline{\mathbf{M}})$ and $\mathbf{A}$ is non-trivial. Since $\mathbf{A}$ is the coproduct of its petals, it follows that x extends to $\mathbf{A}$. ∎

No zero- or one-kernel three-element unary algebra has $\underline{\mathbf{M}}_{\mathrm{V}}$, $\underline{\mathbf{M}}_{\mathrm{L}}$ or $\underline{\mathbf{M}}_{\mathrm{D}}$ as an isoreduct. So every zero- or one-kernel three-element unary algebra satisfies condition (i) of Theorem 4.0.1. We have shown that these algebras all satisfy conditions (iv) and (v) as well.

4.2 Nice two-kernel three-element unary algebras

The family of two-kernel three-element unary algebras is surprisingly complicated. It contains strongly dualisable algebras, dualisable algebras that are not fully dualisable, and non-dualisable algebras. In this section, we study the dualisable two-kernel three-element unary algebras that have neither $\underline{\mathbf{M}}_{\mathrm{V}}$, $\underline{\mathbf{M}}_{\mathrm{L}}$ nor $\underline{\mathbf{M}}_{\mathrm{D}}$ as an isoreduct. We shall prove that every such algebra has enough algebraic operations and is quasi-injective.

We know that every two-kernel three-element unary algebra is isomorphic to a unary algebra, on the set $\{0, 1, 2\}$, with kernels $\{01|2\}$ and $\{02|1\}$; see Lemma 3.3.1. The following lemma teases out what it means for such an algebra to have $\underline{\mathbf{M}}_{\mathrm{V}}$, $\underline{\mathbf{M}}_{\mathrm{L}}$ or $\underline{\mathbf{M}}_{\mathrm{D}}$ as an isoreduct. For each unary algebra $\underline{\mathbf{M}}$ and each permutation $v : M \to M$, we let ${}^{v}\underline{\mathbf{M}}$ denote the isomorphic copy of $\underline{\mathbf{M}}$ formed via conjugation by v; see page 70.

4.2.1 Lemma [40, 2.2] *Let $\underline{\mathbf{M}}$ be a two-kernel unary algebra, on the set $\{0, 1, 2\}$, with kernels $\{01|2\}$ and $\{02|1\}$.*

(i) *The algebra $\underline{\mathbf{M}}$ has $\underline{\mathbf{M}}_{\mathrm{V}}$ or $\underline{\mathbf{M}}_{\mathrm{L}}$ as an isoreduct if and only if ppq and qpq are term functions of $\underline{\mathbf{M}}$, for some distinct $p, q \in \{0, 1, 2\}$.*

(ii) *The algebra $\underline{\mathbf{M}}$ has $\underline{\mathbf{M}}_{\mathrm{D}}$ as an isoreduct if and only if* 101 *and* 220 *are term functions of $\underline{\mathbf{M}}$.*

Proof Define F to be the set of unary term functions of $\underline{\mathbf{M}}$. We first prove the 'only if' parts of the two claims. Assume that $\underline{\mathbf{M}}$ has $\underline{\mathbf{M}}_{\mathrm{V}}$ as an isoreduct. Then there is a permutation v of $\{0, 1, 2\}$ such that $\underline{\mathbf{M}}$ has ${}^{v}\underline{\mathbf{M}}_{\mathrm{V}}$ as a term reduct. The algebra $\underline{\mathbf{M}}$ has kernels $\{01|2\}$ and $\{02|1\}$. So ${}^{v}\underline{\mathbf{M}}_{\mathrm{V}}$ also has kernels $\{01|2\}$ and $\{02|1\}$. Since the algebra $\underline{\mathbf{M}}_{\mathrm{V}}$ has kernels $\{01|2\}$ and $\{02|1\}$, this implies that v is either 012 or 021. Therefore $\underline{\mathbf{M}}$ has

$$^{012}\underline{\mathbf{M}}_{\mathrm{V}} = \langle\{0,1,2\}; 112, 212\rangle \quad \text{or} \quad {}^{021}\underline{\mathbf{M}}_{\mathrm{V}} = \langle\{0,1,2\}; 212, 112\rangle$$

as a term reduct. Thus $\{112, 212\} \subseteq F$. Likewise, if $\underline{\mathbf{M}}_{\mathrm{L}}$ is an isoreduct of $\underline{\mathbf{M}}$, then $\{002, 202\} \subseteq F$ or $\{010, 110\} \subseteq F$; and, if $\underline{\mathbf{M}}_{\mathrm{D}}$ is an isoreduct of $\underline{\mathbf{M}}$, then $\{101, 220\} \subseteq F$.

It remains to prove the 'if' parts of the claims. Clearly, if $\{101, 220\} \subseteq F$, then $\underline{\mathbf{M}}$ has $\underline{\mathbf{M}}_{\mathrm{D}}$ as an isoreduct. So assume that there are $p, q \in M$, with $p \neq q$, such that $\{ppq, qpq\} \subseteq F$. We want to show that $\underline{\mathbf{M}}$ has $\underline{\mathbf{M}}_{\mathrm{V}}$ or $\underline{\mathbf{M}}_{\mathrm{L}}$ as an isoreduct. Since

$$\underline{\mathbf{M}}_{\mathrm{V}} = \langle\{0,1,2\}; 112, 212\rangle \quad \text{and} \quad \underline{\mathbf{M}}_{\mathrm{L}} = \langle\{0,1,2\}; 002, 202\rangle,$$

we can assume that $q \neq 2$.

First consider the case in which $p = 2$. We have

$$ppq \circ ppq = 22q \circ 22q = qq2 \quad \text{and} \quad ppq \circ qpq = 22q \circ q2q = 2q2.$$

Thus $\{qq2, 2q2\} \subseteq F$, and so $\underline{\mathbf{M}}$ has $\underline{\mathbf{M}}_{\mathrm{V}}$ or $\underline{\mathbf{M}}_{\mathrm{L}}$ as an isoreduct.

Finally, consider the case $p \neq 2$. Then $\{110, 010\} \subseteq F$ or $\{001, 101\} \subseteq F$. As $\underline{\mathbf{M}}_{\mathrm{L}}$ is isomorphic to $^{021}\underline{\mathbf{M}}_{\mathrm{L}} = \langle\{0,1,2\}; 010, 110\rangle$ and as $101 \circ 101 = 010$ and $101 \circ 001 = 110$, it follows that $\underline{\mathbf{M}}_{\mathrm{L}}$ is an isoreduct of $\underline{\mathbf{M}}$. ∎

Our next result describes four types of two-kernel three-element unary algebras. The idempotent operations $f_1 := 010$ and $f_2 := 002$ on $\{0,1,2\}$, which were important in the previous chapter, will be important in this chapter as well.

4.2.2 Theorem *Let $\underline{\mathbf{M}}$ be a two-kernel unary algebra, on the set $\{0,1,2\}$, with kernels $\{01|2\}$ and $\{02|1\}$. Let F be the set of unary term functions of $\underline{\mathbf{M}}$. Then at least one of the following is true:*

$(2)_{\mathrm{P}}$ $\{ppq, pqp\} \subseteq F$, *for some distinct* $p, q \in \{0,1,2\}$, *and* $\{f_1, f_2\} \nsubseteq F$;

$(2)_{\mathrm{VL}}$ $\{ppq, qpq\} \subseteq F$, *for some distinct* $p, q \in \{0,1,2\}$;

$(2)_{\mathrm{D}}$ $\{101, 220\} \subseteq F$;

$(2)_{\mathrm{N}}$ $\{f_1, f_2\} \subseteq F$, *and both conditions* $(2)_{\mathrm{VL}}$ *and* $(2)_{\mathrm{D}}$ *fail.*

Proof Assume that $\underline{\mathbf{M}}$ is not of type $(2)_{\mathrm{P}}$, type $(2)_{\mathrm{VL}}$ nor type $(2)_{\mathrm{D}}$. We want to show that $\{f_1, f_2\} \subseteq F$. As $\{01|2\}$ and $\{02|1\}$ are kernels of $\underline{\mathbf{M}}$, there are $p, q, r, s \in M$, with $p \neq q$ and $r \neq s$, such that $ppq \in F$ and $rsr \in F$. Since $\underline{\mathbf{M}}$ is not of type $(2)_{\mathrm{P}}$, we know that $\{f_1, f_2\} \subseteq F$ or $\{ppq, pqp\} \nsubseteq F$. So we can assume that $\{ppq, pqp\} \nsubseteq F$. As $\underline{\mathbf{M}}$ is not of type $(2)_{\mathrm{VL}}$, we have $\{ppq, qpq\} \nsubseteq F$. Thus $ppq \circ rsr \neq pqp$ and $ppq \circ rsr \neq qpq$. This implies that $s \neq 2$ and $r \neq 2$. So rsr is either 010 or 101. Since $101 \circ 101 = 010$, it follows that $f_1 = 010 \in F$. By symmetry, we have $f_2 = 002 \in F$. ∎

All the algebras of type $(2)_{\mathrm{P}}$ are non-dualisable, by Theorem 3.0.1. The algebras of type $(2)_{\mathrm{VL}}$ and the algebras of type $(2)_{\mathrm{D}}$ each have $\underline{\mathbf{M}}_{\mathrm{V}}$, $\underline{\mathbf{M}}_{\mathrm{L}}$ or $\underline{\mathbf{M}}_{\mathrm{D}}$ as an isoreduct, by Lemma 4.2.1. In this section, we will study the algebras of type $(2)_{\mathrm{N}}$. The type-$(2)_{\mathrm{N}}$ algebras are the Nice two-kernel three-element unary algebras. These algebras are all dualisable, by Theorem 3.0.1. We shall show that the algebras of type $(2)_{\mathrm{N}}$ all have enough algebraic operations and are all quasi-injective.

The algebras of type $(2)_{\mathrm{N}}$ come in three different flavours. Assume that $\underline{\mathbf{M}}$ is of type $(2)_{\mathrm{N}}$ and let F be the set of unary term functions of $\underline{\mathbf{M}}$. Then $\{101, 220\} \nsubseteq F$, since $\underline{\mathbf{M}}$ is not of type $(2)_{\mathrm{D}}$. We shall consider the two cases $101, 220 \notin F$ and $101 \in F$ separately. The third case, $220 \in F$, is symmetric under conjugation by 021 to the case $101 \in F$. To see this, assume that $220 \in F$. It is easy to check that the algebra ${}^{021}\underline{\mathbf{M}}$ is of type $(2)_{\mathrm{N}}$ and that $101 = {}^{021}220 \in {}^{021}F$.

4.2.3 Lemma *Let $\underline{\mathbf{M}}$ be a unary algebra of type $(2)_{\mathrm{N}}$.*

(i) *If neither 101 nor 220 is a term function of $\underline{\mathbf{M}}$, then all the unary term functions of $\underline{\mathbf{M}}$ belong to $\{012, 021, 001, 002, 010, 020, 000, 111, 222\}$.*

(ii) *If 101 is a term function of $\underline{\mathbf{M}}$, then all the unary term functions of $\underline{\mathbf{M}}$ belong to $\{012, 002, 010, 101, 000, 111, 222\}$.*

Proof Let F denote the set of all unary term functions of $\underline{\mathbf{M}}$. Since $\{01|2\}$ and $\{02|1\}$ are the two kernels of $\underline{\mathbf{M}}$, we know that

$$F \subseteq \{012, 021\} \cup \{\, ppq, pqp \mid p, q \in M \,\},$$

by Lemma 3.3.2. As $\underline{\mathbf{M}}$ is of type $(2)_{\mathrm{N}}$, the maps $f_1 = 010$ and $f_2 = 002$ are in F. Since $\underline{\mathbf{M}}$ is not of type $(2)_{\mathrm{VL}}$, this implies that $110 \notin F$ and $202 \notin F$. We have $010 \circ 112 = 110 \notin F$ and $221 \circ 221 = 112$. So $112, 221 \notin F$. As $002 \circ 212 = 202 \notin F$ and $121 \circ 121 = 212$, we also have $212, 121 \notin F$. Therefore

$$F \subseteq \{012, 021, 001, 002, 220, 010, 020, 101, 000, 111, 222\}.$$

Claim (i) now follows immediately. To prove (ii), assume that $101 \in F$ and $220 \notin F$. Then $021, 001 \notin F$, since $101 \circ 021 = 101 \circ 001 = 110 \notin F$, and $020 \notin F$, since $020 \circ 101 = 202 \notin F$. Thus claim (ii) holds. ∎

We now study the type-$(2)_{\mathrm{N}}$ algebras of the first flavour, those that have neither 101 nor 220 as a term function. One example of such an algebra is given in Figure 4.2, which depicts the algebra $\underline{\mathbf{M}} = \langle \{0,1,2\}; 001, 010, 002\rangle$ and an algebra $\mathbf{A}$ that belongs to $\mathbb{ISP}(\underline{\mathbf{M}})$.

Given a set S and some $m \in M$, we use $\widehat{m}$ to denote the constant map in M^S with value m.

4.2.4 Lemma *Let $\underline{\mathbf{M}}$ be a unary algebra of type $(2)_{\mathrm{N}}$ such that neither* 101 *nor* 220 *is a term function of $\underline{\mathbf{M}}$. Let $\mathbf{B} \leqslant \mathbf{A}$ in $\mathbb{ISP}(\underline{\mathbf{M}})$ and let $x : \mathbf{B} \to \underline{\mathbf{M}}$ be a homomorphism. Then the following are equivalent:*

(i) *x extends to $\mathbf{A}$;*

(ii) *x extends to $\mathbf{1}_{\mathbf{A}}(B)$;*

(iii) *for every $a \in A \backslash B$ and all unary term functions $u_1 \in \{010, 020\}$ and $u_2 \in \{002, 001\}$ of $\underline{\mathbf{M}}$ with $u_1(a)$, $u_2(a) \in B$, we have $x(u_1(a)) = 0$ or $x(u_2(a)) = 0$.*

In particular, the algebra $\underline{\mathbf{M}}$ is 1*-quasi-injective.*

Proof Define F to be the set of all unary term functions of $\underline{\mathbf{M}}$. Since $\underline{\mathbf{M}}$ is of type $(2)_{\mathrm{N}}$, the maps $f_1 = 010$ and $f_2 = 002$ belong to F. So $000 = f_1 \circ f_2$ is a constant term function of $\underline{\mathbf{M}}$.

We can assume that $\mathbf{A} \leqslant \underline{\mathbf{M}}^S$, for some set S. Clearly (i) implies (ii). To see that (ii) implies (iii), assume that $\overline{x} : \mathbf{1}_{\mathbf{A}}(B) \to \underline{\mathbf{M}}$ is an extension of x. Let $a \in A$ and let $u_1 \in F \cap \{010, 020\}$ and $u_2 \in F \cap \{002, 001\}$. Assume that $u_1(a)$, $u_2(a) \in B$ and that $x(u_1(a)) \neq 0$. We want to show that $x(u_2(a)) = 0$. First assume that $u_1(a) \in C_{\mathbf{A}} \subseteq \{\widehat{0}, \widehat{1}, \widehat{2}\}$. Since 000 is a constant term function of $\underline{\mathbf{M}}$ and $x(u_1(a)) \neq 0$, we have $u_1(a) \neq \widehat{0}$. So $a = \widehat{1}$, which implies that $x(u_2(a)) = x(\widehat{0}) = 0$. Now assume that $u_1(a) \in B \backslash C_{\mathbf{A}}$. It follows that $a \in 1_{\mathbf{A}}(B)$ and $u_1(\overline{x}(a)) = x(u_1(a)) \neq 0$. So $\overline{x}(a) = 1$, and therefore $x(u_2(a)) = u_2(\overline{x}(a)) = 0$.

It remains to show that (iii) implies (i). So assume that condition (iii) holds. By Lemma 4.2.3(i), each non-constant unary term function of $\underline{\mathbf{M}}$ preserves 0. So the set

$$A_* := A \cap \left(\{0,1\}^S \cup \{0,2\}^S\right)$$

is a subuniverse of $\mathbf{A}$. Define the extension $x_* : A_* \to M$ of the map $x{\restriction}_{B \cap A_*}$ so that, for all $a \in A_* \backslash B$, we have

$$x_*(a) = \begin{cases} 2 & \text{if } 001 \in F \text{ and } 001(a) \in x^{-1}(1), \\ 1 & \text{if } 020 \in F \text{ and } 020(a) \in x^{-1}(2), \\ 0 & \text{otherwise.} \end{cases}$$

The map x_* is well defined, by (iii). We want to show that $x_* : \mathbf{A}_* \to \underline{\mathbf{M}}$ is a homomorphism.

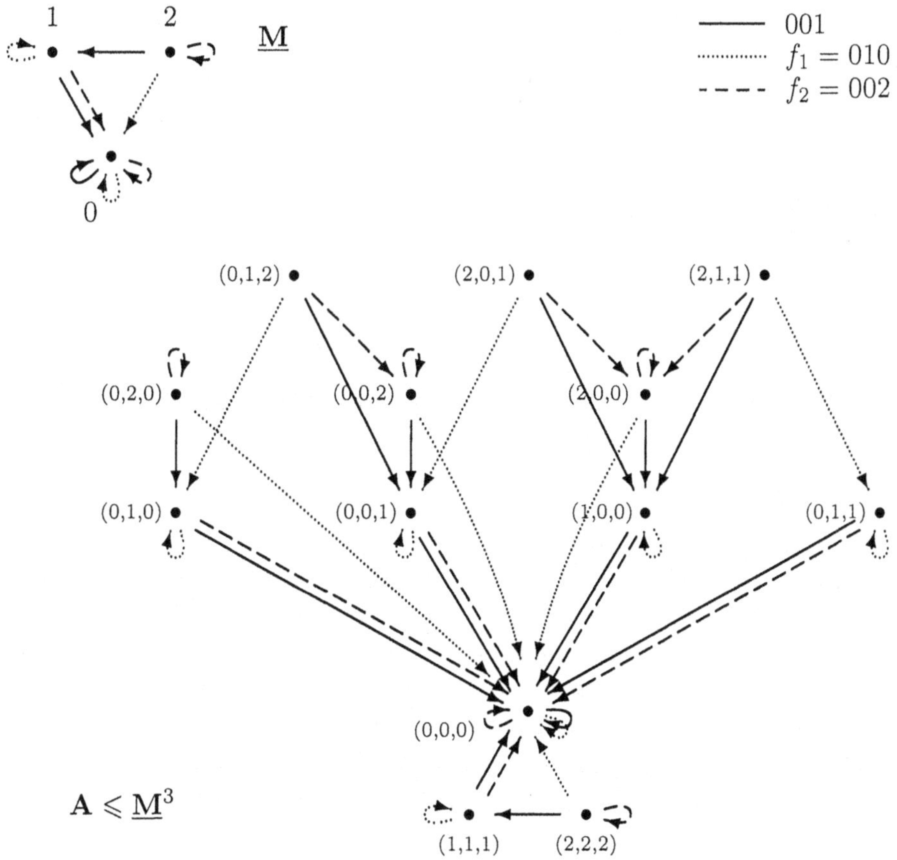

Figure 4.2 An algebra of type $(2)_{\mathrm{N}}$

Let $a \in A_* \backslash B$. We shall show that $u(x_*(a)) = x_*(u(a))$, for each non-constant term function u in F. Since x is a homomorphism, it will then follow that x_* is a homomorphism. By Lemma 4.2.3(i), the only term functions that we need to check are 010, 002, 001, 020 and 021.

Using the symmetry between 1 and 2, we can assume that $a \in \{0, 1\}^S$. Since $001(a) = \widehat{0} \notin x^{-1}(1)$, we must have $x_*(a) \in \{0, 1\}$. This gives us

$$010(x_*(a)) = x_*(a) = x_*(010(a))$$

and

$$002(x_*(a)) = 0 = x(\widehat{0}) = x_*(002(a)).$$

If $001 \in F$, then

$$001(x_*(a)) = 0 = x(\widehat{0}) = x_*(001(a)).$$

Now assume $020 \in F$. We have $020(x_*(a)) \in \{0, 2\}$. As $020 \circ 020 = 000$ and $002 \circ 020 = 020$, it follows that $x_*(020(a)) \in \{0, 2\}$. Using the fact that

$$001 \circ 020(a) = 010(a) = a \notin B,$$

we find that

$$\begin{aligned} 020(x_*(a)) = 2 &\iff x_*(a) = 1 \\ &\iff 020(a) \in B \ \& \ x(020(a)) = 2 \\ &\iff x_*(020(a)) = 2. \end{aligned}$$

Therefore $020(x_*(a)) = x_*(020(a))$. Finally, assume that $021 \in F$. Then we have $020 = 002 \circ 021 \in F$. Since $x_*(a) \in \{0, 1\}$ and $a \in \{0, 1\}^S$, we get

$$021(x_*(a)) = 020(x_*(a)) = x_*(020(a)) = x_*(021(a)).$$

Thus x_* is a homomorphism.

We shall prove that x_* extends to a homomorphism $\overline{x} : \mathbf{A} \to \underline{\mathbf{M}}$ using Lemma 3.3.5. Choose some $a \in A \backslash A_*$, and suppose that $x_*(f_1(a)) \neq 0$ and $x_*(f_2(a)) \neq 0$. Since f_1 is idempotent, we must have $x_*(f_1(a)) = 1$. So there is $u_1 \in F \cap \{010, 020\}$ such that $u_1(a) = u_1 \circ f_1(a) \in B$ and $x(u_1(a)) \neq 0$. Similarly, there is $u_2 \in F \cap \{002, 001\}$ such that $u_2(a) = u_2 \circ f_2(a) \in B$ and $x(u_2(a)) \neq 0$. But this contradicts (iii). Thus there is an extension $\overline{x} : \mathbf{A} \to \underline{\mathbf{M}}$ of x_*. By Lemma 3.3.5, the extension $x : \mathbf{B} \to \underline{\mathbf{M}}$ of $x{\restriction}_{B \cap A_*}$ is unique. Therefore $\overline{x}$ is an extension of x, since $\overline{x}{\restriction}_{B \cap A_*} = x_*{\restriction}_{B \cap A_*} = x{\restriction}_{B \cap A_*}$. ∎

4.2.5 Theorem *Let $\underline{\mathbf{M}}$ be a unary algebra of type $(2)_{\mathrm{N}}$ such that neither* 101 *nor* 220 *is a term function of $\underline{\mathbf{M}}$. Then $\underline{\mathbf{M}}$ has enough algebraic operations.*

Proof Define the map $f : \omega \to \omega$ by $f(k) := k$. Let $\mathbf{B} \leqslant \mathbf{A} \leqslant \underline{\mathbf{M}}^n$, for some $n \in \omega \backslash \{0\}$, and let $h : \mathbf{A} \to \underline{\mathbf{M}}$ be a homomorphism. We shall define some algebraic operations on $\underline{\mathbf{M}}$.

By Lemma 4.2.3(i), the set

$$f_1(M^n) \cup f_2(M^n) = \{0, 1\}^n \cup \{0, 2\}^n$$

is a subuniverse of $\underline{\mathbf{M}}^n$. Let $m \in \{1, 2\}$ and $b \in f_m(B) \backslash \{\widehat{0}\}$. We wish to define a homomorphism $g_b : \underline{\mathbf{M}}^n \to \underline{\mathbf{M}}$ such that, for all $a \in f_1(M^n) \cup f_2(M^n)$, we have

$$g_b(a) = \begin{cases} m & \text{if } a = b \text{ or } a = \widehat{m}, \\ 021(m) & \text{if } a = 021(b) \text{ or } a = 021(\widehat{m}), \\ 0 & \text{otherwise.} \end{cases}$$

To see that this is possible, first check that the specified map acts like a homomorphism on $f_1(M^n) \cup f_2(M^n)$, using Lemma 4.2.3(i), and then check that this homomorphism extends to $\underline{\mathbf{M}}^n$, using Lemma 3.3.5.

Now define the set Y of n-ary algebraic operations on $\underline{\mathbf{M}}$ by

$$Y := \{\, g_b \mid b \in (f_1(B) \cup f_2(B)) \backslash \{\widehat{0}\} \,\}.$$

Then $|Y| \leqslant |B| = f(|B|)$. Define the homomorphism $\mu : \underline{\mathbf{M}}^n \to \underline{\mathbf{M}}^Y$ by $\mu := \sqcap Y$. We will first show that $\mu\restriction_B$ is an embedding, and then show that $h \circ (\mu\restriction_B)^{-1} : \mu(\mathbf{B}) \to \underline{\mathbf{M}}$ extends to $\mu(\mathbf{A})$. It will then follow that $\underline{\mathbf{M}}$ has enough algebraic operations.

To see that $\mu\restriction_B$ is an embedding, let $b, c \in B$ with $b \neq c$. Since the maps f_1 and f_2 separate the elements of M, we must have $f_m(b) \neq f_m(c)$, for some $m \in \{1, 2\}$. We can assume that $f_m(b) \neq \widehat{0}$ and $f_m(c) \neq \widehat{m}$. So

$$g_{f_m(b)}(f_m(b)) = m \neq g_{f_m(b)}(f_m(c)),$$

which implies that $\mu(f_m(b)) \neq \mu(f_m(c))$. So $\mu(b) \neq \mu(c)$, whence $\mu\restriction_B$ is an embedding.

We shall use Lemma 4.2.4 to prove that $h \circ (\mu\restriction_B)^{-1} : \mu(\mathbf{B}) \to \underline{\mathbf{M}}$ extends to $\mu(\mathbf{A})$. Choose any $a \in A$. Let u_1 and u_2 be unary term functions of $\underline{\mathbf{M}}$, with $u_1 \in \{010, 020\}$ and $u_2 \in \{002, 001\}$, such that $u_1(\mu(a)), u_2(\mu(a)) \in \mu(B)$. Set $m_1 := u_1(1)$ and $m_2 := u_2(2)$. Then $f_{m_1} \circ u_1 = u_1$ and $f_{m_2} \circ u_2 = u_2$. So there are $b_1 \in f_{m_1}(B)$ and $b_2 \in f_{m_2}(B)$ such that $\mu(u_1(a)) = \mu(b_1)$ and $\mu(u_2(a)) = \mu(b_2)$.

We will show that $h(b_1) = 0$ or $h(b_2) = 0$. Since $000 = f_1 \circ f_2$ is a term function of $\underline{\mathbf{M}}$, we can assume that $b_1, b_2 \neq \widehat{0}$. As $\mu\restriction_B$ is one-to-one, we have $\mu(b_1), \mu(b_2) \neq \mu(\widehat{0})$. Thus $a \notin \{\widehat{2}, \widehat{1}\}$, and so $u_1(a), u_2(a) \notin \{\widehat{1}, \widehat{2}\}$. For each $i \in \{1, 2\}$, we have $m_i = g_{b_i}(b_i) = g_{b_i}(u_i(a))$ and therefore $u_i(a) = b_i$. As $h\restriction_B$ extends to $\mathbf{A}$, it follows by Lemma 4.2.4 that $h(b_1) = h(u_1(a)) = 0$ or $h(b_2) = h(u_2(a)) = 0$.

We have shown that there is some $j \in \{1, 2\}$ for which $h(b_j) = 0$. So

$$h \circ (\mu\restriction_B)^{-1}(u_j(\mu(a))) = h \circ (\mu\restriction_B)^{-1}(\mu(b_j)) = h(b_j) = 0.$$

Thus $h \circ (\mu\restriction_B)^{-1}$ extends to $\mu(\mathbf{A})$, by Lemma 4.2.4. Hence $\underline{\mathbf{M}}$ has enough algebraic operations. ∎

We now turn our attention to the type-$(2)_{\mathrm{N}}$ algebras of the remaining flavours, those with either 101 or 220 as a term function. By symmetry, we only need to consider those with 101 as a term function. Figure 4.3 gives an example of such an algebra, $\underline{\mathbf{M}} = \langle \{0, 1, 2\}; 101, 010, 002 \rangle$. Both of the constant operations

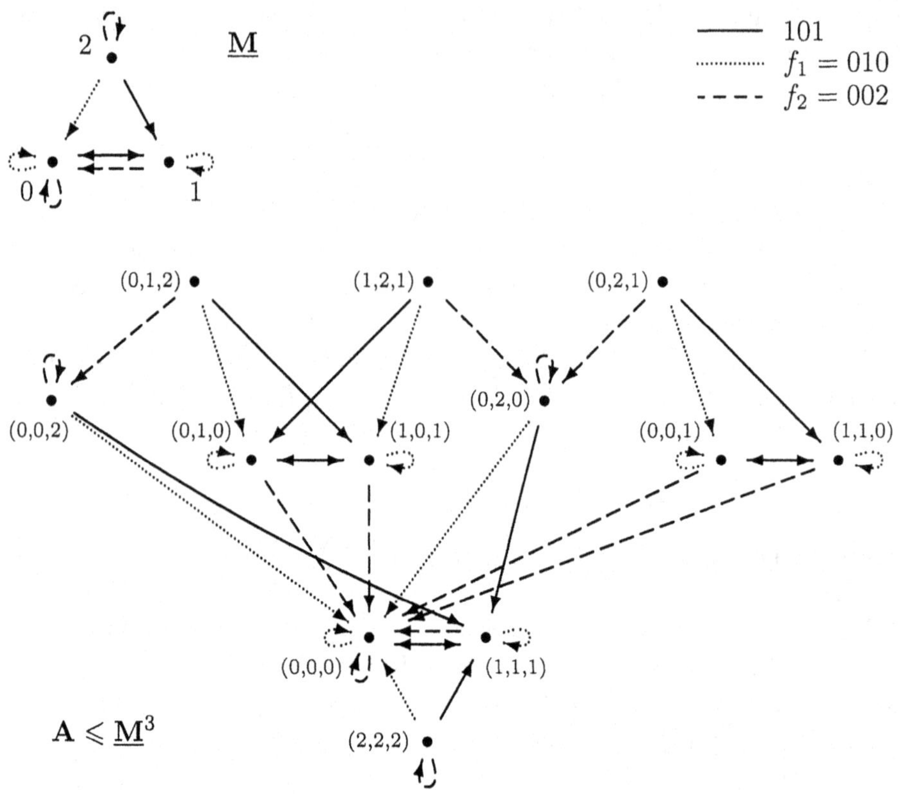

Figure 4.3 Another algebra of type $(2)_{\mathrm{N}}$

$000 = 010 \circ 002$ and $111 = 101 \circ 002$ are term functions of $\underline{\mathbf{M}}$. So it follows from Lemma 4.2.3(ii) that, up to term equivalence, there are only two different type-$(2)_{\mathrm{N}}$ algebras with 101 as a term function: the algebra $\underline{\mathbf{M}}$ from Figure 4.3, and the algebra obtained by adding the constant operation 222 to $\underline{\mathbf{M}}$.

To make the next two proofs easier to read, we introduce some notation. Assume that $\underline{\mathbf{M}}$ is of type $(2)_{\mathrm{N}}$ and let $\mathbf{A} \in \mathbb{ISP}(\underline{\mathbf{M}})$. There is a natural binary relation $\frown_{\mathbf{A}}$ on A that reflects part of the structure of $\mathbf{A}$. For all $a, b \in A$, we write $a \frown_{\mathbf{A}} b$ if and only if there is some $c \in A$ such that $a = f_1(c)$ and $b = f_2(c)$. Using the notation of Figure 4.3, we have

$a \frown_{\mathbf{A}} b$ if and only if 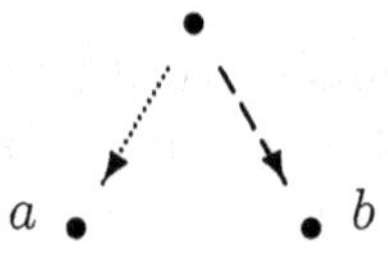 in $\mathbf{A}$,

for all $a, b \in A$.

In the algebra $\mathbf{A}$ of Figure 4.3 we have, for example, $(0,1,0) \frown_{\mathbf{A}} (0,0,2)$ and $(1,0,1) \frown_{\mathbf{A}} (0,0,0)$, but $(1,0,1) \not\frown_{\mathbf{A}} (0,0,2)$. In general, for every set S, each subalgebra $\mathbf{A}$ of $\underline{\mathbf{M}}^S$, and all $a, b \in A$, we have:

- if $a \frown_{\mathbf{A}} b$, then $a \in \{0,1\}^S$ and $b \in \{0,2\}^S$;
- if $a \frown_{\mathbf{A}} b$, then $a^{-1}(1) \subseteq b^{-1}(0)$ and $b^{-1}(2) \subseteq a^{-1}(0)$;
- if $a \in \{0,1\}^S$, then $a \frown_{\mathbf{A}} \widehat{0}$;
- if $b \in \{0,2\}^S$, then $\widehat{0} \frown_{\mathbf{A}} b$.

We will be using these four properties frequently during the next two proofs.

4.2.6 Lemma *Let $\underline{\mathbf{M}}$ be a unary algebra of type $(2)_{\mathrm{N}}$ and assume that* 101 *is a term function of $\underline{\mathbf{M}}$. Let $\mathbf{B} \leqslant \mathbf{A}$ in $\mathbb{ISP}(\underline{\mathbf{M}})$ and let $x : \mathbf{B} \to \underline{\mathbf{M}}$ be a homomorphism. Then the following are equivalent:*

(i) *x extends to $\mathbf{A}$;*

(ii) *x extends to $\mathbf{1}_{\mathbf{A}}(B)$;*

(iii) *both the following conditions hold:*

 (a) *for all $b, c \in B$ such that $b \frown_{\mathbf{A}} c$, we have $x(b) = 0$ or $x(c) = 0$;*

 (b) *for all $b, c \in B$ and all $a \in A$ such that $a \frown_{\mathbf{A}} b$ and $101(a) \frown_{\mathbf{A}} c$, we have $x(b) = 0$ or $x(c) = 0$.*

In particular, the algebra $\underline{\mathbf{M}}$ is 1*-quasi-injective.*

Proof Assume that $\mathbf{A} \leqslant \underline{\mathbf{M}}^S$, for some set S. We must have $\widehat{0}, \widehat{1} \in C_{\mathbf{A}}$, since both $000 = f_1 \circ f_2$ and $111 = 101 \circ f_2$ are constant term functions of $\underline{\mathbf{M}}$. To prove that (ii) implies (iii), assume that x extends to a homomorphism $\overline{x} : \mathbf{1}_{\mathbf{A}}(B) \to \underline{\mathbf{M}}$. We first prove two claims.

Claim 1 Let $a \in A$ and $b \in C_{\mathbf{A}}$ such that $a \frown_{\mathbf{A}} b$. Then $a = \widehat{0}$ or $b = \widehat{0}$.

There is some $c \in A$ with $f_1(c) = a$ and $f_2(c) = b$. Since $b \in C_{\mathbf{A}} \subseteq \{\widehat{0}, \widehat{1}, \widehat{2}\}$, we must have $b = \widehat{0}$ or $b = \widehat{2}$. If $b = \widehat{2}$, then $c = \widehat{2}$ and so $a = f_1(c) = \widehat{0}$.

Claim 2 Let $a \in A$ and let $b \in B \backslash C_{\mathbf{A}}$ such that $a \frown_{\mathbf{A}} b$. Then $\overline{x}(a) = 0$ or $x(b) = 0$.

There is some $c \in A$ such that $f_1(c) = a$ and $f_2(c) = b$. Since $b \notin C_{\mathbf{A}}$, we have $a, c \in 1_{\mathbf{A}}(B)$. Assume that $x(b) \neq 0$. Then $f_2(\overline{x}(c)) = \overline{x}(f_2(c)) = x(b) \neq 0$, and so $\overline{x}(c) = 2$. Therefore $\overline{x}(a) = \overline{x}(f_1(c)) = f_1(\overline{x}(c)) = 0$.

We can now show that (iii) holds. As $\widehat{0} \in C_{\mathbf{A}}$, we have $x(\widehat{0}) = 0$. So condition (a) follows straight from Claims 1 and 2. To see that condition (b) holds, let $b, c \in B$ and $a \in A$, with $a \frown_{\mathbf{A}} b$ and $101(a) \frown_{\mathbf{A}} c$, such that $x(b) \neq 0$. We wish to show that $x(c) = 0$. First assume that $b \in C_{\mathbf{A}}$. Then $b \neq \widehat{0}$, and so $a = \widehat{0}$, by Claim 1. As $101(a) = \widehat{1}$ and $101(a) \frown_{\mathbf{A}} c$, it follows that $c = \widehat{0}$. Thus $x(c) = 0$. Now assume that $b \notin C_{\mathbf{A}}$. Then $a \in 1_{\mathbf{A}}(B)$, and so $101(a) \in 1_{\mathbf{A}}(B)$. Since $x(b) \neq 0$, we must have $\overline{x}(a) = 0$, by Claim 2. Therefore $\overline{x}(101(a)) = 101(\overline{x}(a)) = 1$, which implies that $101(a) \neq \widehat{0}$. So, as $101(a) \frown_{\mathbf{A}} c$, either Claim 1 or Claim 2 tells us that $x(c) = 0$. Thus (iii) is satisfied.

To prove that (iii) implies (i), assume that (iii) holds. Since the functions f_1 and f_2 fix $\{0,1\}$ and $\{0,2\}$, respectively, we have $f_1(A) = A \cap \{0,1\}^S$ and $f_2(A) = A \cap \{0,2\}^S$. By Lemma 4.2.3(ii), the two sets

$$A_1 := f_1(A) \cup C_{\mathbf{A}} \quad \text{and} \quad A_2 := f_2(A) \cup C_{\mathbf{A}}$$

are subuniverses of $\mathbf{A}$. Let $\mathcal{T}$ be a transversal of $\{\, \{a, 101(a)\} \mid a \in f_1(A)\,\}$. Then, by using Lemma 4.2.3(ii) and referring to Figure 4.3 for help, we can define the homomorphism $x_1 : \mathbf{A}_1 \to \underline{\mathbf{M}}$ such that, for all $a \in \mathcal{T}$, we have

$$x_1(a) = \begin{cases} x(a) & \text{if } a \in B, \\ 0 & \text{if } a \notin B \text{ and } a \frown_{\mathbf{A}} b, \text{ for some } b \in B \text{ with } x(b) \neq 0, \\ 1 & \text{otherwise.} \end{cases}$$

We can also define a homomorphism $x_2 : \mathbf{A}_2 \to \underline{\mathbf{M}}$ by

$$x_2(a) = \begin{cases} x(a) & \text{if } a \in B, \\ 0 & \text{otherwise.} \end{cases}$$

The set $A_* := A_1 \cup A_2$ forms a subalgebra of $\mathbf{A}$. So we can now define the homomorphism $x_* : \mathbf{A}_* \to \underline{\mathbf{M}}$ by $x_* := x_1 \cup x_2$. The homomorphisms x_* and x agree on $B \cap A_*$.

We will use Lemma 3.3.5 to show that x_* extends to $\mathbf{A}$. Let $a \in A \backslash A_*$ with $x_*(f_2(a)) \neq 0$. Since $f_2(a) \in A_2$, we have $f_2(a) \in B$ and $x(f_2(a)) \neq 0$. We want to show that $x_*(f_1(a)) = 0$. First assume that $f_1(a) \in B$. Then $x_*(f_1(a)) = x(f_1(a)) = 0$, by (a). Now assume that $f_1(a) \notin B$ and $f_1(a) \in \mathcal{T}$. Then $x_*(f_1(a)) = x_1(f_1(a)) = 0$. Finally, assume that $f_1(a) \notin B$ and $101(a) = 101(f_1(a)) \in \mathcal{T}$. Since $101 \circ 101(a) = f_1(a)$, we have $101(a) \notin B$. We must have $x_*(101(a)) = x_1(101(a)) = 1$, by (b), and therefore

$$x_*(f_1(a)) = x_*(101 \circ 101(a)) = 101(x_*(101(a))) = 0.$$

So there is an extension $\overline{x} : \mathbf{A} \to \underline{\mathbf{M}}$ of x_*. By Lemma 3.3.5, the extension $x : \mathbf{B} \to \underline{\mathbf{M}}$ of $x{\restriction}_{B \cap A_*} = \overline{x}{\restriction}_{B \cap A_*}$ is unique. Hence $\overline{x}$ is an extension of x. ∎

4.2.7 Theorem *Let* $\underline{\mathbf{M}}$ *be a unary algebra of type* $(2)_{\mathrm{N}}$ *such that either* 101 *or* 220 *is a term function of* $\underline{\mathbf{M}}$. *Then* $\underline{\mathbf{M}}$ *has enough algebraic operations.*

Proof By symmetry, we can assume that 101 is a term function of $\underline{\mathbf{M}}$. Define $\mathcal{A} := \mathbb{ISP}(\underline{\mathbf{M}})$, and define the map $f : \omega \to \omega$ by $f(k) := 3k$. Assume that $\mathbf{B} \leqslant \mathbf{A} \leqslant \underline{\mathbf{M}}^n$, for some $n \in \omega\backslash\{0\}$, and let $h : \mathbf{A} \to \underline{\mathbf{M}}$ be a homomorphism. We will define two families of n-ary algebraic operations on $\underline{\mathbf{M}}$, each indexed by the elements of $f_2(B)\backslash\{\widehat{0}\}$.

Let $b \in f_2(B)\backslash\{\widehat{0}\}$. By Lemma 4.2.3(ii), the set $f_2(M^n) \cup C_{\underline{\mathbf{M}}^n}$ forms a subalgebra of $\underline{\mathbf{M}}^n$. We want to define a homomorphism $g_b : \underline{\mathbf{M}}^n \to \underline{\mathbf{M}}$ such that, for all $a \in f_2(M^n)$, we have

$$g_b(a) = \begin{cases} 2 & \text{if } a = b \text{ or } a = \widehat{2}, \\ 0 & \text{otherwise.} \end{cases}$$

To see that this is possible, first check that there is such a homomorphism on the set $f_2(M^n) \cup C_{\underline{\mathbf{M}}^n}$, using Lemma 4.2.3(ii) and maybe Figure 4.3. Then check that this homomorphism extends to $\underline{\mathbf{M}}^n$, using Lemma 4.2.6. Similarly, the set $f_1(M^n) \cup C_{\underline{\mathbf{M}}^n}$ forms a subalgebra of $\underline{\mathbf{M}}^n$ and, using Lemmas 4.2.3(ii) and 4.2.6 again, we can define $g'_b : \underline{\mathbf{M}}^n \to \underline{\mathbf{M}}$ such that, for $a \in f_1(M^n)$, we have

$$g'_b(a) = \begin{cases} g_b(a) & \text{if } a \frown_{\mathbf{A}} b \text{ or } 101(a) \frown_{\mathbf{A}} b, \\ 101(g_b(a)) & \text{otherwise.} \end{cases}$$

By Lemma 4.1.1, there exists a non-empty subset Z of $\mathcal{A}(\underline{\mathbf{M}}^n, \underline{\mathbf{M}})$ such that $|Z| \leqslant |B|$ and Z separates the elements of B. Now define

$$Y := Z \cup \{\, g_b, g'_b \mid b \in f_2(B)\backslash\{\widehat{0}\} \,\}.$$

Then $|Y| \leqslant |B| + 2|B| = f(|B|)$. Define the homomorphism $\mu : \underline{\mathbf{M}}^n \to \underline{\mathbf{M}}^Y$ by $\mu := \sqcap Y$. Then $\mu{\restriction}_B$ is an embedding, as Y separates the elements of B.

Claim Let $a \in A$ and $b \in B$ with $\mu(a) \frown_{\mu(\mathbf{A})} \mu(b)$. Then $f_1(a) \frown_{\mathbf{A}} b$.

Since $f_1(a) \frown_{\mathbf{A}} \widehat{0}$, we can assume that $b \neq \widehat{0}$. As $\mu(a) \frown_{\mu(\mathbf{A})} \mu(b)$, there exists $c \in A$ with $f_1(\mu(c)) = \mu(a)$ and $f_2(\mu(c)) = \mu(b)$. As f_1 is idempotent, this implies that $\mu(f_1(c)) = \mu(f_1(a))$. Since f_2 is idempotent, we also have $\mu(b) = \mu(f_2(c)) = \mu(f_2(b))$. Therefore $b \in f_2(B)$, as $\mu{\restriction}_B$ is one-to-one. Since $\mu = \sqcap Y$ and $g_b \in Y$, we now have $g_b(f_2(c)) = g_b(b) = 2$, which implies that $f_2(c) = b$ or $c = \widehat{2}$. If $f_2(c) = b$, then $f_1(c) \frown_{\mathbf{A}} b$, by definition. If $c = \widehat{2}$, then $f_1(c) = \widehat{0}$, and therefore $f_1(c) \frown_{\mathbf{A}} b$. So, in either case, we have $f_1(c) \frown_{\mathbf{A}} b$. Since $\mu(f_1(c)) = \mu(f_1(a))$, it follows that

$$g'_b(f_1(a)) = g'_b(f_1(c)) = g_b(f_1(c)) = g_b(f_1(a)).$$

As $101 \circ f_1 = 101$, this tells us that $f_1(a) \frown_{\mathbf{A}} b$ or $101(a) \frown_{\mathbf{A}} b$.

To finish the proof of the claim, it suffices to check that $101(a) \not\frown_{\mathbf{A}} b$. Since $\mu(b) = \mu(f_2(c))$, we have

$$f_2(g_b(c)) = g_b(f_2(c)) = g_b(b) = 2.$$

Therefore $g_b(c) = 2$. As $\mu(f_1(c)) = \mu(f_1(a))$, this gives us

$$\begin{aligned} g_b(101(a)) = g_b(101 \circ f_1(a)) &= 101(g_b(f_1(a))) = 101(g_b(f_1(c))) \\ &= 101 \circ f_1(g_b(c)) = 101(g_b(c)) = 101(2) = 1. \end{aligned}$$

So $g_b(101(a)) = 1$ and $g_b(b) = 2$. Since $g_b{\restriction}_A$ preserves f_1 and f_2, it follows that $101(a) \not\frown_{\mathbf{A}} b$; see Lemma 3.3.5. Thus $f_1(a) \frown_{\mathbf{A}} b$, and the claim holds.

We will use Lemma 4.2.6 to prove that $h \circ (\mu{\restriction}_B)^{-1} : \mu(\mathbf{B}) \to \underline{\mathbf{M}}$ extends to $\mu(\mathbf{A})$. To see that condition (iii)(a) of the lemma holds, let $b, c \in B$ such that $\mu(b) \frown_{\mu(\mathbf{A})} \mu(c)$. Then $f_1(b) \frown_{\mathbf{A}} c$, by the above claim. We have $f_1(b) = b$, as $f_1(\mu(b)) = \mu(b)$ and $\mu{\restriction}_B$ is one-to-one, and so $b \frown_{\mathbf{A}} c$. Therefore $h(b) = 0$ or $h(c) = 0$, by Lemma 4.2.6, since $h{\restriction}_B$ extends to $\mathbf{A}$.

To check that condition (iii)(b) of the lemma holds, let $b, c \in B$ and $a \in A$ such that $\mu(a) \frown_{\mu(\mathbf{A})} \mu(b)$ and $101(\mu(a)) \frown_{\mu(\mathbf{A})} \mu(c)$. Since $f_1 \circ 101 = 101$, we must have $f_1(a) \frown_{\mathbf{A}} b$ and $101(a) \frown_{\mathbf{A}} c$, by the above claim. This implies that $h(b) = 0$ or $h(c) = 0$, as $101 \circ f_1 = 101$ and $h{\restriction}_B$ extends to $\mathbf{A}$. It now follows that $h \circ (\mu{\restriction}_B)^{-1}$ extends to $\mu(\mathbf{A})$, whence $\underline{\mathbf{M}}$ has enough algebraic operations. ∎

By Lemmas 3.3.1 and 4.2.1 and Theorems 4.2.2 and 3.0.1, every dualisable two-kernel three-element unary algebra that has neither $\underline{\mathbf{M}}_{\mathrm{V}}$, $\underline{\mathbf{M}}_{\mathrm{L}}$ nor $\underline{\mathbf{M}}_{\mathrm{D}}$ as an isoreduct must be isomorphic to an algebra of type $(2)_{\mathrm{N}}$. We have shown that each algebra of type $(2)_{\mathrm{N}}$ has enough algebraic operations and is 1-quasi-injective. Theorem 3.0.1 tells us that there are no dualisable three-element unary algebras with three kernels. So we have now finished proving that (i) implies (iv) and that (i) implies (v) in Theorem 4.0.1.

4.3 Three-element unary algebras that are not strongly dualisable

In this section, we will show that every dualisable three-element unary algebra that has $\underline{\mathbf{M}}_{\mathrm{V}}$, $\underline{\mathbf{M}}_{\mathrm{L}}$ or $\underline{\mathbf{M}}_{\mathrm{D}}$ as a reduct is not strongly dualisable. Our proof is based on the proof used by Hyndman and Willard [41] to show that the three-element unary algebra $\langle\{0, 1, 2\}; 001, 122\rangle$ is not strongly dualisable. Actually, to finish the proof of our main theorem we need to show that every dualisable three-element unary algebra with $\underline{\mathbf{M}}_{\mathrm{V}}$, $\underline{\mathbf{M}}_{\mathrm{L}}$ or $\underline{\mathbf{M}}_{\mathrm{D}}$ as a reduct is not *fully*

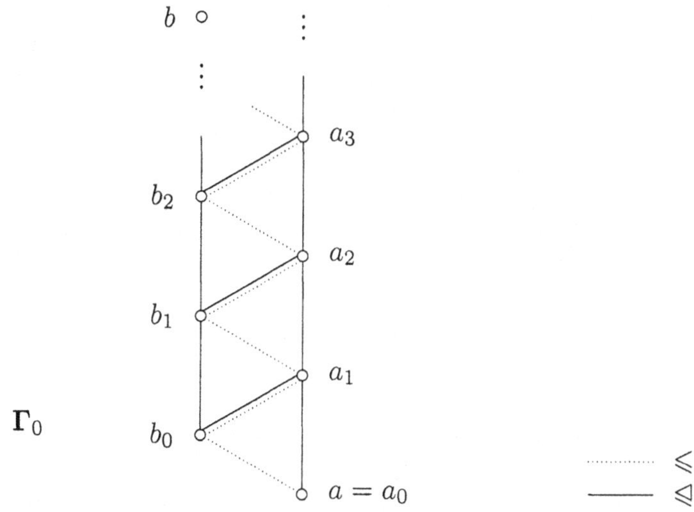

Figure 4.4 The bi-ordered set $\mathbf{\Gamma}_0 = \langle \Gamma_0; \leqslant, \trianglelefteqslant \rangle$

dualisable. We will prove this in the next section, with the help of some of the results from this section.

We use a special pair of ordered sets.

4.3.1 Lemma [41, 4.1] *There are ordered sets* $\mathbf{\Gamma} = \langle \Gamma; \leqslant \rangle$ *and* $\mathbf{\Gamma}' = \langle \Gamma; \trianglelefteqslant \rangle$ *such that*

(i) $\mathbf{\Gamma}$ *is a chain, and* $\trianglelefteqslant$ *is strictly contained in* $\leqslant$,

(ii) *for all* $c, d \in \Gamma$ *with* $c \leqslant d$ *and* $c \ntrianglelefteq d$, *there are subsets* $\{ c_n \mid n \in \omega \}$ *and* $\{ d_n \mid n \in \omega \}$ *of* Γ *such that* $c \trianglelefteqslant c_n$ *and* $d_n \trianglelefteqslant d$ *and* $c_n \leqslant d_n \leqslant c_{n+1}$, *for every* $n \in \omega$.

Proof We shall sketch the proof. Let $\mathbf{\Gamma}_0 = \langle \Gamma_0; \leqslant, \trianglelefteqslant \rangle$ be the bi-ordered set illustrated in Figure 4.4: the underlying set is

$$\Gamma_0 := \{ a_n \mid n \in \omega \} \cup \{ b_n \mid n \in \omega \} \cup \{b\},$$

the dotted lines indicate the order $\leqslant$, and the solid lines indicate the order $\trianglelefteqslant$. The ordered set $\langle \Gamma_0; \leqslant \rangle$ is an $(\omega{+}1)$-chain, with a as bottom and b as top. By construction, condition (ii) is satisfied in $\mathbf{\Gamma}_0$ for the pair $(c, d) = (a, b)$.

The failures of condition (ii) in $\mathbf{\Gamma}_0$ occur for pairs $(c, d) = (a_i, b_j)$, where $i, j \in \omega$ and $i \leqslant j$. Now fix $k \in \omega$. We can insert a copy of $\mathbf{\Gamma}_0$ between a_k and b_k, where we identify the elements a and b of the inserted copy with a_k

and b_k, respectively. This will correct all failures of condition (ii) in $\mathbf{\Gamma}_0$ for pairs $(c, d) = (a_k, b_j)$, where $k \leqslant j$.

We now define $\mathbf{\Gamma}_1$ to be the bi-ordered set obtained from $\mathbf{\Gamma}_0$ by inserting a copy of $\mathbf{\Gamma}_0$ between a_i and b_i, for all $i \in \omega$. For each $n \in \omega\backslash\{0\}$, we obtain $\mathbf{\Gamma}_{n+1}$ from $\mathbf{\Gamma}_n$ by inserting a copy of $\mathbf{\Gamma}_0$ between a_i and b_i, for all $i \in \omega$, in each of the copies of $\mathbf{\Gamma}_0$ inserted when obtaining $\mathbf{\Gamma}_n$ from $\mathbf{\Gamma}_{n-1}$. Thus we have an ω-chain of bi-ordered sets $\mathbf{\Gamma}_0 \subseteq \mathbf{\Gamma}_1 \subseteq \cdots$. Define $\mathbf{\Gamma}_\omega = \langle \Gamma_\omega; \leqslant, \trianglelefteq\rangle$ to be the union of this chain. Then $\mathbf{\Gamma} := \langle \Gamma_\omega; \leqslant\rangle$ and $\mathbf{\Gamma}' := \langle \Gamma_\omega; \trianglelefteq\rangle$ are the required ordered sets. ∎

The following result gives a general method for proving that a finite algebra is not strongly dualisable. We use $\mathcal{G}$ to denote the category of directed graphs.

4.3.2 Lemma *Let $\mathbf{\Gamma}$ and $\mathbf{\Gamma}'$ be ordered sets as in Lemma 4.3.1. Let $\underline{\mathbf{M}}$ be a finite algebra, and let $\mathbf{B} \leqslant \mathbf{A}$ in $\mathcal{A} := \mathbb{ISP}(\underline{\mathbf{M}})$ such that $\Gamma \subseteq B$. Assume that there is a chain $\mathbf{C} = \langle C; \leqslant\rangle$, with $C \subseteq M$, for which the maps $-{\restriction_\Gamma} : \mathcal{A}(\mathbf{A}, \underline{\mathbf{M}}) \to \mathcal{G}(\mathbf{\Gamma}, \mathbf{C})$ and $-{\restriction_\Gamma} : \mathcal{A}(\mathbf{B}, \underline{\mathbf{M}}) \to \mathcal{G}(\mathbf{\Gamma}', \mathbf{C})$ are well-defined bijections.*

(i) *For each alter ego $\underset{\sim}{\mathbf{M}}$ of $\underline{\mathbf{M}}$, the set $X := \{\, x{\restriction_B} \mid x \in \mathcal{A}(\mathbf{A}, \underline{\mathbf{M}})\,\}$ forms a closed substructure of $\underset{\sim}{\mathbf{M}}^B$.*

(ii) *The algebra $\underline{\mathbf{M}}$ is not strongly dualisable.*

Proof Let $\underset{\sim}{\mathbf{M}} = \langle M; G, H, R, \mathcal{T}\rangle$ be an alter ego of $\underline{\mathbf{M}}$. There are $c, d \in \Gamma$ with $c \leqslant d$ and $c \ntrianglelefteq d$. The chain $\mathbf{C}$ must be non-trivial, so there exist $0, 1 \in C$ such that $0 \neq 1$ and $0 \leqslant 1$. Define the map $w : \Gamma \to C$ by

$$w(a) = \begin{cases} 1 & \text{if } c \trianglelefteq a, \\ 0 & \text{otherwise.} \end{cases}$$

Then $w \in \mathcal{G}(\mathbf{\Gamma}', \mathbf{C})$, and so there is a homomorphism $\overline{w} \in \mathcal{A}(\mathbf{B}, \underline{\mathbf{M}})$ with $\overline{w}{\restriction_\Gamma} = w$. Since

$$c \leqslant d \quad \text{and} \quad w(c) = 1 \nleqslant 0 = w(d),$$

we know that $w \notin \mathcal{G}(\mathbf{\Gamma}, \mathbf{C})$. So $\overline{w} \notin X$, and therefore $X \neq \mathcal{A}(\mathbf{B}, \underline{\mathbf{M}})$. Since $\mathbf{A} \in \mathbb{ISP}(\underline{\mathbf{M}})$, the elements of B are separated by the homomorphisms in $\mathcal{A}(\mathbf{A}, \underline{\mathbf{M}})$. So X separates B. We shall prove that X forms a closed substructure of $\mathrm{D}(\mathbf{B}) \leqslant \underset{\sim}{\mathbf{M}}^B$. It will then follow, by Theorem 1.5.3, that $\underset{\sim}{\mathbf{M}}$ does not yield a strong duality on $\mathcal{A}$.

Let $\iota : \mathbf{B} \hookrightarrow \mathbf{A}$ denote the inclusion homomorphism. Then X is the image of the morphism $\mathrm{D}(\iota) : \mathrm{D}(\mathbf{A}) \to \mathrm{D}(\mathbf{B})$. This implies that X is topologically closed in $\mathrm{D}(\mathbf{B})$ and that X is closed under the operations in G. It remains to check that X is closed under the partial operations in H. So let h be a k-ary partial operation in H, for some $k \in \omega \backslash \{0\}$, and let $x_0, \dots, x_{k-1} \in X$ with $(x_0, \dots, x_{k-1}) \in \operatorname{dom}(h)^{\mathrm{D}(\mathbf{B})}$. We want to show that $z := h(x_0, \dots, x_{k-1})$ belongs to X.

Homomorphisms in $\mathcal{A}(\mathbf{B}, \underline{\mathbf{M}})$ are uniquely determined by their restrictions to Γ. So, to show that $z \in X$, it is enough to prove that $z{\restriction}_\Gamma \in \mathcal{G}(\mathbf{\Gamma}, \mathbf{C})$. To do this, let $c, d \in \Gamma$ with $c \leqslant d$. We now wish to show that $z(c) \leqslant z(d)$ in $\mathbf{C}$. Since $z \in \mathcal{A}(\mathbf{B}, \underline{\mathbf{M}})$, we know that $z{\restriction}_\Gamma \in \mathcal{G}(\mathbf{\Gamma}', \mathbf{C})$. So we can assume that $c \ntrianglelefteq d$. There exist subsets $\{\, c_n \mid n \in \omega \,\}$ and $\{\, d_n \mid n \in \omega \,\}$ of Γ such that $c \trianglelefteq c_n$ and $d_n \trianglelefteq d$ and $c_n \leqslant d_n \leqslant c_{n+1}$, for all $n \in \omega$. The elements of the set $\{\, c_n \mid n \in \omega \,\}$ must be pairwise distinct. Since $\underline{\mathbf{M}}$ is finite, the elements of $\{\, c_n \mid n \in \omega \,\}$ are not separated by the maps $x_0, \dots, x_{k-1}$ in $\mathcal{A}(\mathbf{B}, \underline{\mathbf{M}})$. So there exist $m, n \in \omega$, with $m < n$, such that $x_i(c_m) = x_i(c_n)$, for each $i \in \{0, \dots, k-1\}$. For all $i \in \{0, \dots, k-1\}$, we have $x_i \in X$ and therefore $x_i{\restriction}_\Gamma \in \mathcal{G}(\mathbf{\Gamma}, \mathbf{C})$. As $c_m \leqslant d_m \leqslant c_n$, it follows that $x_i(c_m) = x_i(d_m)$, for all $i \in \{0, \dots, k-1\}$, and therefore

$$z(c_m) = h(x_0, \dots, x_{k-1})(c_m) = h(x_0, \dots, x_{k-1})(d_m) = z(d_m).$$

Since $z{\restriction}_\Gamma \in \mathcal{G}(\mathbf{\Gamma}', \mathbf{C})$, and since $c \trianglelefteq c_m$ and $d_m \trianglelefteq d$, we can conclude that $z(c) \leqslant z(c_m) = z(d_m) \leqslant z(d)$, as required. Thus $h(x_0, \dots, x_{k-1}) = z \in X$, whence X forms a closed substructure of $\mathrm{D}(\mathbf{B})$. ∎

We want to show that the algebras of type $(2)_{\mathrm{VL}}$ and type $(2)_{\mathrm{D}}$ are not strongly dualisable. Let $\underline{\mathbf{M}}$ be a unary algebra on the set $\{0, 1, 2\}$. In order to apply Lemma 4.3.2, we shall introduce a method for constructing algebras in the quasi-variety $\mathcal{A} := \mathbb{ISP}(\underline{\mathbf{M}})$ from ordered sets. For the purposes of the construction, we give $\{0, 1, 2\}$ the non-standard order $2 \preccurlyeq 0 \preccurlyeq 1$.

4.3.3 Definition Let $\underline{\mathbf{M}}$ be a unary algebra on $\{0, 1, 2\}$, let $\mathbf{P} = \langle P; \leqslant \rangle$ be an ordered set, and let $\trianglelefteq$ be a reflexive subset of $\leqslant$. (More precisely, the subset $\trianglelefteq$ of $\leqslant$ must contain the diagonal relation $\{\, (a, a) \mid a \in P \,\}$.) Define the set

$$P^+ := P \cup \{\bot, \top\},$$

where we are assuming that $\bot, \top \notin P$. Now, for all $a, b \in P$ such that $a \trianglelefteq b$, define $\widehat{ab} \in M^{P^+}$ by $\widehat{ab}(\bot) = 2$, $\widehat{ab}(\top) = 1$ and, for all $c \in P$,

$$\widehat{ab}(c) = \begin{cases} 2 & \text{if } c \leqslant a, \\ 0 & \text{if } c \leqslant b \text{ and } c \nleqslant a, \\ 1 & \text{otherwise.} \end{cases}$$

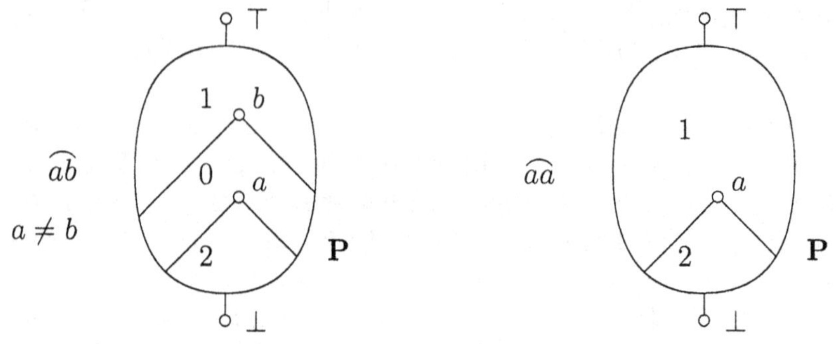

Figure 4.5 Turning a pair (a, b) in $\trianglelefteqslant$ into a map $\widehat{ab}$ in M^{P^+}

This definition is illustrated in Figure 4.5. Define the algebra

$$\widehat{\mathbf{P}}_{\trianglelefteqslant} := \mathbf{sg}_{\underline{\mathbf{M}}^{P^+}}(\{\, \widehat{ab} \mid a, b \in P \text{ and } a \trianglelefteqslant b \,\}).$$

Since the relation $\trianglelefteqslant$ on P is reflexive, we are able to define the one-to-one map $\iota_{\mathbf{P}} : P \to \widehat{P}_{\trianglelefteqslant}$ by $\iota_{\mathbf{P}}(a) := \widehat{aa}$.

The next lemma describes the structure of the algebra $\widehat{\mathbf{P}}_{\trianglelefteqslant}$ under the assumption that $\underline{\mathbf{M}}$ is a two-kernel algebra. As in the previous chapter, for each set S and $a \in M^S$, the partition of S determined by a is given by

$$\mathcal{P}(a) := \{a^{-1}(0), a^{-1}(1), a^{-1}(2)\} \backslash \{\varnothing\},$$

and, for every subalgebra $\mathbf{A}$ of $\underline{\mathbf{M}}^S$, we define the subuniverse $A_{\downarrow 2}$ of $\mathbf{A}$ by $A_{\downarrow 2} := \{\, a \in A \mid |a(S)| \leqslant 2 \,\}$.

4.3.4 Lemma *Let $\underline{\mathbf{M}}$ be a two-kernel unary algebra, on the set $\{0, 1, 2\}$, with kernels $\{01|2\}$ and $\{02|1\}$. Let $\mathbf{P} = \langle P; \leqslant \rangle$ be an ordered set, let $\trianglelefteqslant$ be a reflexive subset of $\leqslant$, and define the algebra $\mathbf{A} := \widehat{\mathbf{P}}_{\trianglelefteqslant}$.*

(i) *The set of petals of $\mathbf{A}_{\downarrow 2}$ is $\{\, \mathbf{sg}_{\mathbf{A}}(\widehat{aa}) \mid a \in P \,\}$, and distinct elements of P determine distinct petals of $\mathbf{A}_{\downarrow 2}$.*

(ii) *For all $a, b \in P$ such that $a \trianglelefteqslant b$ and $a \neq b$, we have $\widehat{ab}$, $021(\widehat{ab}) \notin A_{\downarrow 2}$ and $\mathrm{sg}_{\mathbf{A}}(\widehat{ab}) \subseteq \{\widehat{ab}, 021(\widehat{ab})\} \cup \mathrm{sg}_{\mathbf{A}}(\{\widehat{aa}, \widehat{bb}\})$.*

(iii) *For all $a, b, c, d \in P$, with $a \trianglelefteqslant b$ and $c \trianglelefteqslant d$, such that $(a, b) \neq (c, d)$, we have $\mathrm{sg}_{\mathbf{A}}(\widehat{ab}) \cap \mathrm{sg}_{\mathbf{A}}(\widehat{cd}) \subseteq \mathrm{sg}_{\mathbf{A}}(\{\widehat{aa}, \widehat{bb}\})$.*

Proof We begin by proving (ii). Let $a, b \in P$ such that $a \trianglelefteqslant b$ and $a \neq b$. For each unary term function u_1 of $\underline{\mathbf{M}}$ such that $\ker(u_1) = \{02|1\}$, we have

$u_1(\widehat{ab}) = u_1(\widehat{bb}) \in \mathrm{sg}_{\mathbf{A}}(\widehat{bb})$. For each unary term function u_2 of $\underline{\mathbf{M}}$ with $\ker(u_2) = \{01|2\}$, we have $u_2(\widehat{ab}) = u_2(\widehat{aa}) \in \mathrm{sg}_{\mathbf{A}}(\widehat{aa})$. Using Lemma 3.3.2, it follows that

$$\mathrm{sg}_{\mathbf{A}}(\widehat{ab}) \subseteq \{\widehat{ab}, 021(\widehat{ab})\} \cup \mathrm{sg}_{\mathbf{A}}(\{\widehat{aa}, \widehat{bb}\}).$$

Since $a \neq b$, the partitions $\mathcal{P}(\widehat{ab})$ and $\mathcal{P}(021(\widehat{ab}))$ of P^+ each have three blocks, and therefore $\widehat{ab}$, $021(\widehat{ab}) \notin A_{\downarrow 2}$. So claim (ii) holds.

We shall now prove (i). For each $a \in P$, the partition $\mathcal{P}(\widehat{aa})$ has two blocks. It follows from (ii) that $A_{\downarrow 2} = \bigcup\{\mathrm{sg}_{\mathbf{A}}(\widehat{aa}) \mid a \in P\}$. For every $a \in P$, the non-constant map $\widehat{aa}$ in M^{P^+} cannot belong to the centre $C_{\mathbf{A}}$. So, for all $a \in P$, the set $\mathrm{sg}_{\mathbf{A}}(\widehat{aa}) \backslash C_{\mathbf{A}}$ is non-empty and forms a connected subgraph of $G^*(\mathbf{A})$. To prove (i), it is now enough to show $\mathrm{sg}_{\mathbf{A}}(\widehat{aa}) \cap \mathrm{sg}_{\mathbf{A}}(\widehat{bb}) = C_{\mathbf{A}}$, for all $a, b \in P$ with $a \neq b$. Assume that $u(\widehat{aa}) = v(\widehat{bb})$, for some $a, b \in P$ with $a \neq b$ and some unary term functions u and v of $\underline{\mathbf{M}}$. Since the two-block partitions $\mathcal{P}(\widehat{aa})$ and $\mathcal{P}(\widehat{bb})$ of P^+ are different, we have $u(\widehat{aa}) \in \{\widehat{0}, \widehat{1}, \widehat{2}\}$. But $\widehat{aa} \in \{1, 2\}^{P^+} \backslash \{\widehat{1}, \widehat{2}\}$ and $\underline{\mathbf{M}}$ does not have any unary term functions with kernel $\{0|12\}$. So u is a constant term function of $\underline{\mathbf{M}}$, which implies that $u(\widehat{aa}) \in C_{\mathbf{A}}$.

It remains to prove (iii). Let (a, b) and (c, d) be distinct pairs in $\trianglelefteq$. Assume that there are unary term functions u and v of $\underline{\mathbf{M}}$ such that $u(\widehat{ab}) = v(\widehat{cd})$. The partitions $\mathcal{P}(\widehat{ab})$ and $\mathcal{P}(\widehat{cd})$ of P^+ each have at most three blocks. Since $(a, b) \neq (c, d)$ and $\mathbf{P}$ is an ordered set, the partitions $\mathcal{P}(\widehat{ab})$ and $\mathcal{P}(\widehat{cd})$ are different. So the partition of P^+ determined by $u(\widehat{ab}) = v(\widehat{cd})$ has at most two blocks. But this implies that $u(\widehat{ab}) \in \mathrm{sg}_{\mathbf{A}}(\{\widehat{aa}, \widehat{bb}\})$, by (ii). ∎

To illustrate Lemma 4.3.4, we consider a particular example.

4.3.5 Example Let $\underline{\mathbf{M}}$ be a two-kernel unary algebra on $\{0, 1, 2\}$, with kernels $\{01|2\}$ and $\{02|1\}$. Define the chain $\mathbf{P} = \langle\{a, b, c\}; \leqslant\rangle$, where $a \leqslant b \leqslant c$, and the ordered set $\mathbf{P}' = \langle\{a, b, c\}; \trianglelefteq\rangle$, where $\trianglelefteq = \leqslant\backslash\{(a, b)\}$. Now represent each $x \in M^{P^+}$ by the 5-tuple $(x(\bot), x(a), x(b), x(c), x(\top))$. We have

$$\widehat{ac} = (2, 2, 0, 0, 1), \qquad \widehat{bc} = (2, 2, 2, 0, 1),$$
$$\widehat{aa} = (2, 2, 1, 1, 1), \qquad \widehat{cc} = (2, 2, 2, 2, 1), \qquad \widehat{bb} = (2, 2, 2, 1, 1).$$

The structure of the algebra $\mathbf{A} := \widehat{\mathbf{P}_{\trianglelefteq}}$ is shown in Figure 4.6. The three petals of $\mathbf{A}_{\downarrow 2}$ are $\mathbf{sg}_{\mathbf{A}}(\widehat{aa})$, $\mathbf{sg}_{\mathbf{A}}(\widehat{bb})$ and $\mathbf{sg}_{\mathbf{A}}(\widehat{cc})$.

Part (ii) of the following lemma shows that, if $\underline{\mathbf{M}}$ is of type $(2)_{\mathrm{VL}}$, then the algebra $\widehat{\mathbf{P}_{\trianglelefteq}}$ and the directed graph $\mathbf{P}' = \langle P; \trianglelefteq\rangle$ are intimately connected. Let $\underline{\mathbf{2}} = \langle\{1, 2\}; \leqslant\rangle$ denote the two-element chain, where $1 \leqslant 2$.

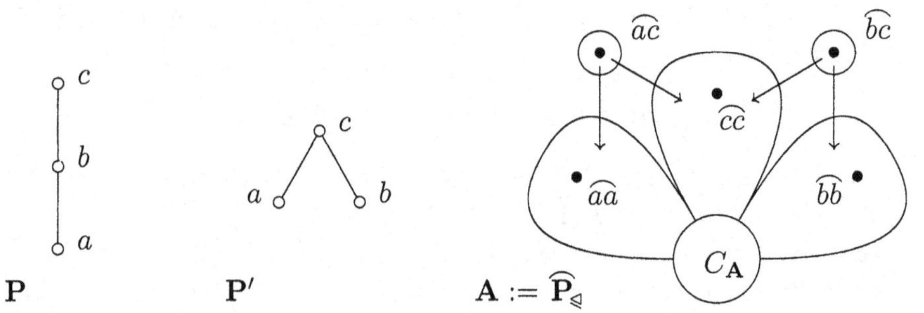

Figure 4.6 Making a unary algebra using the three-element chain

4.3.6 Lemma *Let $\underline{\mathbf{M}}$ be a unary algebra of type $(2)_{\mathrm{VL}}$ and define the quasi-variety $\mathcal{A} := \mathbb{ISP}(\underline{\mathbf{M}})$. Let $\mathbf{P} = \langle P; \leqslant\rangle$ be an ordered set, let $\trianglelefteq$ be a reflexive subset of $\leqslant$, and define $\mathbf{P}' = \langle P; \trianglelefteq\rangle$.*

(i) *For all $x \in \mathcal{A}(\widehat{\mathbf{P}}_{\trianglelefteq}, \underline{\mathbf{M}})$ and $a, b \in P$ such that $a \trianglelefteq b$, we have $x(\widehat{ab}) = 2$ if and only if $x(\widehat{aa}) = 2$, and $x(\widehat{ab}) = 1$ if and only if $x(\widehat{bb}) = 1$.*

(ii) *The map $- \circ \iota_{\mathbf{P}} : \mathcal{A}(\widehat{\mathbf{P}}_{\trianglelefteq}, \underline{\mathbf{M}}) \to \mathcal{G}(\mathbf{P}', \underline{\mathbf{2}})$ is a well-defined bijection.*

Proof As $\underline{\mathbf{M}}$ is of type $(2)_{\mathrm{VL}}$, there exist $p, q \in M$, with $p \neq q$, such that both ppq and qpq are term functions of $\underline{\mathbf{M}}$. Define $\mathbf{A} := \widehat{\mathbf{P}}_{\trianglelefteq}$. To prove (i), let $x \in \mathcal{A}(\mathbf{A}, \underline{\mathbf{M}})$ and let $a, b \in P$ with $a \trianglelefteq b$. Then

$$ppq(x(\widehat{ab})) = x(ppq(\widehat{ab})) = x(ppq(\widehat{aa})) = ppq(x(\widehat{aa})).$$

So $x(\widehat{ab}) = 2$ if and only if $x(\widehat{aa}) = 2$. Similarly, we have

$$qpq(x(\widehat{ab})) = x(qpq(\widehat{ab})) = x(qpq(\widehat{bb})) = qpq(x(\widehat{bb})),$$

and therefore $x(\widehat{ab}) = 1$ if and only if $x(\widehat{bb}) = 1$. Thus (i) holds.

We want to define the map

$$\eta : \mathcal{A}(\mathbf{A}, \underline{\mathbf{M}}) \to \mathcal{G}(\mathbf{P}', \underline{\mathbf{2}}) \;\text{ by }\; \eta(x) := x \circ \iota_{\mathbf{P}}.$$

To see that this will work, choose any $x \in \mathcal{A}(\mathbf{A}, \underline{\mathbf{M}})$. For each $a \in P$, we have $\widehat{aa} \in \{1, 2\}^{P^+}$ and therefore

$$ppq(x(\widehat{aa})) = x(ppq(\widehat{aa})) = x(qpq(\widehat{aa})) = qpq(x(\widehat{aa})),$$

which implies that $x(\widehat{aa}) \in \{1, 2\}$. So $x \circ \iota_{\mathbf{P}}(P) \subseteq \{1, 2\}$. Using (i), for all $a, b \in P$ such that $a \trianglelefteq b$ and $x(\widehat{aa}) = 2$, we have $x(\widehat{ab}) = 2$ and therefore $x(\widehat{bb}) = 2$. Thus $x \circ \iota_{\mathbf{P}} \in \mathcal{G}(\mathbf{P}', \underline{\mathbf{2}})$, and η is well defined.

To see that η is one-to-one, let $x, y \in \mathcal{A}(\mathbf{A}, \underline{\mathbf{M}})$ such that $\eta(x) = \eta(y)$. Let $a, b \in P$ with $a \trianglelefteq b$. Then

$$x(\widehat{aa}) = \eta(x)(a) = \eta(y)(a) = y(\widehat{aa}).$$

So claim (i) gives us

$$x(\widehat{ab}) = 2 \iff x(\widehat{aa}) = 2 \iff y(\widehat{aa}) = 2 \iff y(\widehat{ab}) = 2.$$

Similarly, we have $x(\widehat{bb}) = y(\widehat{bb})$ and

$$x(\widehat{ab}) = 1 \iff x(\widehat{bb}) = 1 \iff y(\widehat{bb}) = 1 \iff y(\widehat{ab}) = 1.$$

Thus $x(\widehat{ab}) = y(\widehat{ab})$. It follows that $x = y$, and so η is one-to-one.

It remains to show that η is onto. Let $z \in \mathcal{G}(\mathbf{P}', \underline{2})$. By Lemmas 4.3.4(i) and 3.1.4, we can define the homomorphism $z_* : \mathbf{A}_{\downarrow 2} \to \underline{\mathbf{M}}$ by

$$z_* := \bigsqcup \{ \pi_\top \restriction_{\mathrm{sg}_\mathbf{A}(\widehat{aa})} \mid a \in z^{-1}(1) \} \sqcup \bigsqcup \{ \pi_\bot \restriction_{\mathrm{sg}_\mathbf{A}(\widehat{aa})} \mid a \in z^{-1}(2) \}.$$

We have defined z_* so that $z_* \circ \iota_\mathbf{P} = z$. We now want to show that z_* extends to a homomorphism $\overline{z} : \mathbf{A} \to \underline{\mathbf{M}}$. To do this, it is enough to define $\overline{z}$ on the generators of $\mathbf{A}$ that do not belong to $A_{\downarrow 2}$. Let $a, b \in P$ with $a \neq b$ and $a \trianglelefteq b$. We shall find some $c \in P^+$ such that z_* agrees with π_c on $\mathrm{sg}_\mathbf{A}(\{\widehat{aa}, \widehat{bb}\})$. We will then assign $\overline{z}(\widehat{ab}) := \widehat{ab}(c)$. We are allowed to do this by Lemma 4.3.4(iii).

If $z(a) = z(b) = 1$, then z_* agrees with $\pi_\top$ on $\mathrm{sg}_\mathbf{A}(\{\widehat{aa}, \widehat{bb}\})$. Similarly, if $z(a) = z(b) = 2$, then z_* agrees with $\pi_\bot$ on $\mathrm{sg}_\mathbf{A}(\{\widehat{aa}, \widehat{bb}\})$. So we can assume that $z(a) \neq z(b)$. Therefore $z(a) = 1$ and $z(b) = 2$, as $z(a) \leqslant z(b)$. We have

$$z_*(\widehat{aa}) = \pi_\top(\widehat{aa}) = 1 = \widehat{aa}(b) \quad \text{and} \quad z_*(\widehat{bb}) = \pi_\bot(\widehat{bb}) = 2 = \widehat{bb}(b),$$

which implies that z_* agrees with π_b on $\mathrm{sg}_\mathbf{A}(\{\widehat{aa}, \widehat{bb}\})$.

It now follows that z_* extends to a homomorphism $\overline{z} : \mathbf{A} \to \underline{\mathbf{M}}$. We have $\eta(\overline{z}) = \overline{z} \circ \iota_\mathbf{P} = z_* \circ \iota_\mathbf{P} = z$. Thus η is onto, and so η is a bijection. ∎

4.3.7 Theorem *No unary algebra of type* $(2)_{\mathrm{VL}}$ *is strongly dualisable.*

Proof Let $\underline{\mathbf{M}}$ be a unary algebra of type $(2)_{\mathrm{VL}}$, and define $\mathcal{A} := \mathbb{ISP}(\underline{\mathbf{M}})$. Using Lemma 4.3.1 and Definition 4.3.3, we have algebras $\widehat{\mathbf{\Gamma}}_\trianglelefteq \leqslant \widehat{\mathbf{\Gamma}}_\leqslant$ in $\mathcal{A}$ and a one-to-one map $\iota_\mathbf{\Gamma} : \Gamma \to \widehat{\Gamma}_\trianglelefteq$. By Lemma 4.3.6(ii), the maps

$$- \circ \iota_\mathbf{\Gamma} : \mathcal{A}(\widehat{\mathbf{\Gamma}}_\leqslant, \underline{\mathbf{M}}) \to \mathcal{G}(\mathbf{\Gamma}, \underline{2}) \quad \text{and} \quad - \circ \iota_\mathbf{\Gamma} : \mathcal{A}(\widehat{\mathbf{\Gamma}}_\trianglelefteq, \underline{\mathbf{M}}) \to \mathcal{G}(\mathbf{\Gamma}', \underline{2})$$

are well-defined bijections. It follows by Lemma 4.3.2(ii) that $\underline{\mathbf{M}}$ is not strongly dualisable. ∎

We now want to show that the algebras of type $(2)_{\mathrm{D}}$ are not strongly dualisable. We will be using the most complicated algebra with kernels $\{01|2\}$ and $\{02|1\}$: the algebra $\underline{\mathbf{M}}^\sharp = \langle\{0,1,2\}; F^\sharp\rangle$, where

$$F^\sharp := \{012, 021\} \cup \{\, ppq, pqp \mid p, q \in \{0,1,2\}\,\}.$$

(See Lemma 3.3.2.) The algebra $\underline{\mathbf{M}}^\sharp$ is of type $(2)_{\mathrm{VL}}$, and so we know that $\underline{\mathbf{M}}^\sharp$ is not strongly dualisable. Certainly, every algebra of type $(2)_{\mathrm{D}}$ is a reduct of $\underline{\mathbf{M}}^\sharp$. We shall show that there is a stronger connection between $\underline{\mathbf{M}}^\sharp$ and the algebras of type $(2)_{\mathrm{D}}$.

4.3.8 Lemma *Let $\underline{\mathbf{M}}$ be a unary algebra of type $(2)_{\mathrm{D}}$. Let $\mathbf{A}$ be an algebra in $\mathcal{A}^\sharp := \mathbb{ISP}(\underline{\mathbf{M}}^\sharp)$, and let $\mathbf{A}^\flat$ be the reduct of $\mathbf{A}$ in $\mathcal{A} := \mathbb{ISP}(\underline{\mathbf{M}})$. Then the sets $\mathcal{A}(\mathbf{A}^\flat, \underline{\mathbf{M}})$ and $\mathcal{A}^\sharp(\mathbf{A}, \underline{\mathbf{M}}^\sharp)$ are equal.*

Proof The operations 101 and 220 are term functions of $\underline{\mathbf{M}}$. We begin by proving that $\underline{\mathbf{M}}^2$ is hom-minimal, that is, that the only homomorphisms from $\underline{\mathbf{M}}^2$ to $\underline{\mathbf{M}}$ are the two projections, π_0 and π_1.

Let $x : \underline{\mathbf{M}}^2 \to \underline{\mathbf{M}}$ be a homomorphism. Then $x(1,0) \in \{0,1\}$, since x preserves $010 = 101 \circ 101$. First assume that $x(1,0) = 1$. In $\underline{\mathbf{M}}^2$, we have

$$(1,0) \overset{101}{\longleftrightarrow} (0,1) \overset{101}{\longleftarrow} (1,2) \overset{220}{\longrightarrow} (2,0) \overset{220}{\longleftrightarrow} (0,2) \overset{220}{\longleftarrow} (2,1).$$

Applying the homomorphism x gives us

$$\boxed{1} \overset{101}{\longleftrightarrow} 0 \overset{101}{\longleftarrow} 1 \overset{220}{\longrightarrow} 2 \overset{220}{\longleftrightarrow} 0 \overset{220}{\longleftarrow} 2$$

in $\underline{\mathbf{M}}$. The constant operations 000, 111 and 222 are all term functions of $\underline{\mathbf{M}}$. Hence it follows that $x = \pi_0$. Now assume that $x(1,0) = 0$. Then we have $x(0,1) = 101(x(1,0)) = 1$ and, by symmetry, we conclude that $x = \pi_1$. So $\underline{\mathbf{M}}^2$ is hom-minimal.

All the constant operations on M are term functions of both $\underline{\mathbf{M}}$ and $\underline{\mathbf{M}}^\sharp$. So the lemma holds if $\mathbf{A}$ is trivial. This means that we can assume that $\mathbf{A} \leqslant (\underline{\mathbf{M}}^\sharp)^S$, for some non-empty set S.

Let $a \in A$ such that the partition $\mathcal{P}(a)$ of S has two blocks. Then

$$\mathrm{sg}_{\mathbf{A}}(a) = \{\, b \in M^S \mid \mathcal{P}(b) = \mathcal{P}(a)\,\} \cup \{\widehat{0}, \widehat{1}, \widehat{2}\},$$

as every map in $F^\sharp$ is an operation of $\underline{\mathbf{M}}^\sharp$. Let $\mathbf{sg}_{\mathbf{A}}(a)^\flat$ denote the reduct of $\mathbf{sg}_{\mathbf{A}}(a)$ in $\mathcal{A}$. Then $\underline{\mathbf{M}}^2$ is isomorphic to $\mathbf{sg}_{\mathbf{A}}(a)^\flat$, via repetition of coordinates. Since $\underline{\mathbf{M}}^2$ is hom-minimal, it follows that $\mathbf{sg}_{\mathbf{A}}(a)^\flat$ is hom-minimal.

Now let $y \in \mathcal{A}(\mathbf{A}^\flat, \underline{\mathbf{M}})$ and let $a \in A$. To prove that $y \in \mathcal{A}^\sharp(\mathbf{A}, \underline{\mathbf{M}}^\sharp)$, it suffices to show that y agrees with a projection on $\mathrm{sg}_{\mathbf{A}}(a)$. First assume that

$a \in \{\widehat{0}, \widehat{1}, \widehat{2}\}$. Each constant map on M is a term function of $\underline{\mathbf{M}}$. So y agrees with a projection on $\mathrm{sg}_{\mathbf{A}}(a) = \{\widehat{0}, \widehat{1}, \widehat{2}\}$. Now assume that $a \in A_{\downarrow 2} \backslash \{\widehat{0}, \widehat{1}, \widehat{2}\}$. The partition $\mathcal{P}(a)$ of S has two blocks. So y agrees with a projection on $\mathrm{sg}_{\mathbf{A}}(a)$, since $\mathbf{sg}_{\mathbf{A}}(a)^{\flat}$ is hom-minimal.

Finally, assume that $a \in A \backslash A_{\downarrow 2}$. As $\mathcal{P}(a)$ has three blocks, we can choose some $s \in S$ with $a(s) = y(a)$. We will show that y agrees with π_s on $\mathrm{sg}_{\mathbf{A}}(a)$. The operations

$$f_1 = 010 = 101 \circ 101 \quad \text{and} \quad f_2 = 002 = 220 \circ 220$$

are term functions of $\underline{\mathbf{M}}$. For each $m \in \{1, 2\}$, the partition $\mathcal{P}(f_m(a))$ has two blocks, and so

$$A_m := \mathrm{sg}_{\mathbf{A}}(f_m(a)) = \{\, b \in M^S \mid \mathcal{P}(b) = \mathcal{P}(f_m(a)) \,\} \cup \{\widehat{0}, \widehat{1}, \widehat{2}\}.$$

It follows that $\mathrm{sg}_{\mathbf{A}}(a) = \{a, 021(a)\} \cup A_1 \cup A_2$.

Let $m \in \{1, 2\}$. Since the algebra $\mathbf{A}_m^{\flat} := \mathbf{sg}_{\mathbf{A}}(f_m(a))^{\flat}$ is hom-minimal, we know that y agrees with a projection on A_m. We have

$$y(f_m(a)) = f_m(y(a)) = f_m(a(s)).$$

As each element of $A_m \backslash \{\widehat{0}, \widehat{1}, \widehat{2}\}$ determines the same partition of S as $f_m(a)$, it follows that y agrees with π_s on A_m. So y agrees with π_s on $A_1 \cup A_2$.

For each $m \in \{1, 2\}$, we have $f_m \circ 021(a) \in A_1 \cup A_2$ and therefore

$$f_m(y(021(a))) = y(f_m(021(a))) = f_m(021(a(s))).$$

So $y(021(a)) = 021(a(s))$, as f_1 and f_2 separate M. Thus y agrees with π_s on $\mathrm{sg}_{\mathbf{A}}(a) = \{a, 021(a)\} \cup A_1 \cup A_2$. Hence $y \in \mathcal{A}^{\sharp}(\mathbf{A}, \underline{\mathbf{M}}^{\sharp})$. ∎

4.3.9 Theorem *No unary algebra of type* $(2)_{\mathrm{D}}$ *is strongly dualisable.*

Proof Let $\underline{\mathbf{M}}$ be an algebra of type $(2)_{\mathrm{D}}$. Then $\underline{\mathbf{M}}$ is a reduct of the algebra $\underline{\mathbf{M}}^{\sharp}$ of type $(2)_{\mathrm{VL}}$. Using Lemma 4.3.1 and Definition 4.3.3, we have $\widehat{\boldsymbol{\Gamma}}_{\trianglelefteq} \leqslant \widehat{\boldsymbol{\Gamma}}_{\leqslant}$ in $\mathcal{A}^{\sharp} := \mathbb{ISP}(\underline{\mathbf{M}}^{\sharp})$, and there is a one-to-one map $\iota_{\boldsymbol{\Gamma}} : \Gamma \to \widehat{\Gamma}_{\trianglelefteq}$. The maps

$$- \circ \iota_{\boldsymbol{\Gamma}} : \mathcal{A}^{\sharp}(\widehat{\boldsymbol{\Gamma}}_{\leqslant}, \underline{\mathbf{M}}^{\sharp}) \to \mathcal{G}(\boldsymbol{\Gamma}, \underline{\mathbf{2}}) \quad \text{and} \quad - \circ \iota_{\boldsymbol{\Gamma}} : \mathcal{A}^{\sharp}(\widehat{\boldsymbol{\Gamma}}_{\trianglelefteq}, \underline{\mathbf{M}}^{\sharp}) \to \mathcal{G}(\boldsymbol{\Gamma}', \underline{\mathbf{2}})$$

are well-defined bijections, by Lemma 4.3.6(ii). Let $\widehat{\boldsymbol{\Gamma}}^{\flat}_{\leqslant}$ and $\widehat{\boldsymbol{\Gamma}}^{\flat}_{\trianglelefteq}$ denote the reducts of $\widehat{\boldsymbol{\Gamma}}_{\leqslant}$ and $\widehat{\boldsymbol{\Gamma}}_{\trianglelefteq}$ in $\mathcal{A} := \mathbb{ISP}(\underline{\mathbf{M}})$. Then, by Lemma 4.3.8, the maps

$$- \circ \iota_{\boldsymbol{\Gamma}} : \mathcal{A}(\widehat{\boldsymbol{\Gamma}}^{\flat}_{\leqslant}, \underline{\mathbf{M}}) \to \mathcal{G}(\boldsymbol{\Gamma}, \underline{\mathbf{2}}) \quad \text{and} \quad - \circ \iota_{\boldsymbol{\Gamma}} : \mathcal{A}(\widehat{\boldsymbol{\Gamma}}^{\flat}_{\trianglelefteq}, \underline{\mathbf{M}}) \to \mathcal{G}(\boldsymbol{\Gamma}', \underline{\mathbf{2}})$$

are bijections. So Lemma 4.3.2(ii) tells us that $\underline{\mathbf{M}}$ is not strongly dualisable. ∎

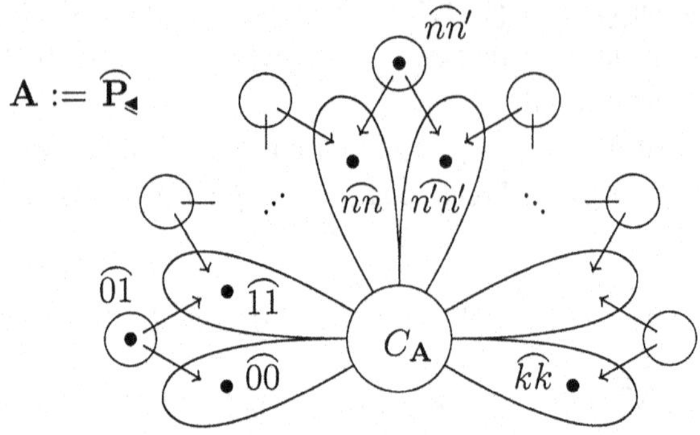

Figure 4.7 Proving non-quasi-injectivity

We finish this section by considering quasi-injectivity.

4.3.10 Lemma *No unary algebra of type* $(2)_{\mathrm{VL}}$ *or* $(2)_{\mathrm{D}}$ *is quasi-injective.*

Proof First assume that $\underline{\mathbf{M}}$ is a unary algebra of type $(2)_{\mathrm{VL}}$. Let $n \in \omega\backslash\{0\}$. We will show that $\underline{\mathbf{M}}$ is not n-quasi-injective. Define $k := 2n+1$ and let $\mathbf{P} = \langle\{0, \dots, k\}; \leqslant\rangle$ be a $(k+1)$-element chain with $0 \leqslant \dots \leqslant k$. Define the relations

$$◀ := \{\, (i,i) \mid i \in P \,\} \cup \{\, (i,i+1) \mid i \in P\backslash\{k\} \,\}$$

and $\trianglelefteq := ◀\backslash\{(n,n+1)\}$ on P. Using Definition 4.3.3, we can define the algebras $\mathbf{A} := \widehat{\mathbf{P}_{◀}}$ and $\mathbf{C} := \widehat{\mathbf{P}_{\trianglelefteq}}$ in $\mathbb{ISP}(\underline{\mathbf{M}})$, and $\mathbf{C}$ is a subalgebra of $\mathbf{A}$. Define $n' := n+1$. Then Lemma 4.3.4 tells us that the structure of the algebra $\mathbf{A}$ is as shown in Figure 4.7.

The coproduct $\mathbf{B} := \mathbf{sg}_{\mathbf{A}}(\widehat{00}) * \mathbf{sg}_{\mathbf{A}}(\widehat{kk})$ is a subalgebra of $\mathbf{A}$. Define the homomorphism

$$x : \mathbf{B} \to \underline{\mathbf{M}} \quad \text{by} \quad x := \pi_{\bot}{\restriction}_{\mathrm{sg}_{\mathbf{A}}(\widehat{00})} \sqcup \pi_{\top}{\restriction}_{\mathrm{sg}_{\mathbf{A}}(\widehat{kk})}.$$

We shall show that x extends to $\boldsymbol{n}_{\mathbf{A}}(B)$ but not to $\mathbf{A}$.

We have $x(\widehat{00}) = \widehat{00}(\bot) = 2$ and $x(\widehat{kk}) = \widehat{kk}(\top) = 1$. But, for each homomorphism $y : \mathbf{A} \to \underline{\mathbf{M}}$, we must have

$$y(\widehat{00}) = y \circ \iota_{\mathbf{P}}(0) \leqslant y \circ \iota_{\mathbf{P}}(k) = y(\widehat{kk}),$$

by Lemma 4.3.6(ii). So $x : \mathbf{B} \to \underline{\mathbf{M}}$ does not extend to $\mathbf{A}$.

Now define the map $z : P \to \{1, 2\}$ by

$$z(i) = \begin{cases} 2 & \text{if } i \in \{0, \ldots, n\}, \\ 1 & \text{if } i \in \{n+1, \ldots, k\}. \end{cases}$$

Then z is a morphism from $\langle P; \trianglelefteq \rangle$ to $\underset{\sim}{\mathbf{2}}$. Using Lemma 4.3.6(ii) again, there must be a homomorphism $\overline{z} : \mathbf{C} \to \underset{\sim}{\mathbf{M}}$ such that $\overline{z} \circ \iota_{\mathbf{P}} = z$. We have

$$\overline{z}(\widehat{00}) = z(0) = 2 = x(\widehat{00}) \quad \text{and} \quad \overline{z}(\widehat{kk}) = z(k) = 1 = x(\widehat{kk}).$$

So $\overline{z}$ is an extension of x.

We have shown that x extends to $\mathbf{C}$ but not to $\mathbf{A}$. Thus it remains to prove that $n_{\mathbf{A}}(B) \subseteq C$. By Lemma 4.3.4(ii), we have $A \backslash C \subseteq \{\widehat{nn'}, 021(\widehat{nn'})\}$. It follows from Figure 4.7 that $d_{\mathbf{A}}(a, b) \geqslant n+1$, for all $a \in A \backslash C$ and $b \in B \backslash C_{\mathbf{A}}$. Thus $n_{\mathbf{A}}(B) \subseteq C$, whence $\underset{\sim}{\mathbf{M}}$ is not n-quasi-injective.

Now assume $\underset{\sim}{\mathbf{M}}$ is of type $(2)_{\mathrm{D}}$ and let $n \in \omega \backslash \{0\}$. As $\underset{\sim}{\mathbf{M}}^{\sharp}$ is of type $(2)_{\mathrm{VL}}$, we know that $\underset{\sim}{\mathbf{M}}^{\sharp}$ is not n-quasi-injective. So there are finite algebras $\mathbf{B} \leqslant \mathbf{A}$ in $\mathbb{ISP}(\underset{\sim}{\mathbf{M}}^{\sharp})$ for which there is a homomorphism $x : \mathbf{B} \to \underset{\sim}{\mathbf{M}}^{\sharp}$ that extends to $n_{\mathbf{A}}(B)$ but not to $\mathbf{A}$. Let $\mathbf{A}^{\flat}$ and $\mathbf{B}^{\flat}$ be the reducts of $\mathbf{A}$ and $\mathbf{B}$ in $\mathbb{ISP}(\underset{\sim}{\mathbf{M}})$. The algebras $\underset{\sim}{\mathbf{M}}$ and $\underset{\sim}{\mathbf{M}}^{\sharp}$ have the same constant term functions. So $C_{\mathbf{A}^{\flat}} = C_{\mathbf{A}}$. It follows that $n_{\mathbf{A}^{\flat}}(B) \subseteq n_{\mathbf{A}}(B)$. Thus $x : \mathbf{B}^{\flat} \to \underset{\sim}{\mathbf{M}}$ extends to $n_{\mathbf{A}^{\flat}}(B)$ but not to $\mathbf{A}^{\flat}$, using Lemma 4.3.8. ∎

We know that there are no three-kernel three-element unary algebras that are dualisable, by Theorem 3.0.1. So the previous lemma tells us that every dualisable three-element unary algebra that has $\underset{\sim}{\mathbf{M}}_{\mathrm{V}}$, $\underset{\sim}{\mathbf{M}}_{\mathrm{L}}$ or $\underset{\sim}{\mathbf{M}}_{\mathrm{D}}$ as a reduct is not quasi-injective. Therefore we have shown that $\neg$(i) implies $\neg$(v) in our main theorem.

4.4 Three-element unary algebras that are not fully dualisable

In this section, we finish the proof of the main theorem by showing that each three-element unary algebra of type $(2)_{\mathrm{VL}}$ or $(2)_{\mathrm{D}}$ is not fully dualisable. This section is by far the most technical of the text, and is not called upon in any later chapters.

Our proof is an extension of that given by Hyndman and Willard [41] to show that the unary algebra $\langle \{0, 1, 2\}; 001, 122 \rangle$ is not fully dualisable. Their proof used the fact that the operations 001 and 122 preserve the total order with $0 \leqslant 1 \leqslant 2$. Our proof is more complicated since it must work, in particular, for the algebra $\underset{\sim}{\mathbf{M}}^{\sharp}$; and there is no total order on $\{0, 1, 2\}$ that is preserved by every operation in $F^{\sharp}$.

A full duality for a quasi-variety $\mathcal{A} := \mathbb{ISP}(\underline{\mathbf{M}})$ is more subtle than either a duality or a strong duality. At the moment, we have no reason to believe that, if $\underset{\sim}{\mathbf{M}}$ is a structure that yields a full duality on $\mathcal{A}$, then every enrichment of $\underset{\sim}{\mathbf{M}}$, via algebraic relations, also yields a full duality on $\mathcal{A}$. However, there are some relations that can always be added to a structure $\underset{\sim}{\mathbf{M}}$ without destroying a full duality.

The next lemma is a slightly stronger version of a result proved by Hyndman and Willard [41, 4.7]. The lemma is derived in a more general setting by Davey, Haviar and Willard [21]. Let r be an n-ary algebraic relation on $\underline{\mathbf{M}}$, for some $n \in \omega\backslash\{0\}$, and let $\mathbf{r}$ denote the subalgebra of $\underline{\mathbf{M}}^n$ on the set r. We say that the relation r is **hom-minimal (relative to $\underline{\mathbf{M}}$)** if the algebra $\mathbf{r}$ is hom-minimal, that is, if each homomorphism in $\mathcal{A}(\mathbf{r}, \underline{\mathbf{M}})$ is the restriction of a projection.

4.4.1 Lemma [41, 21] *Let $\underline{\mathbf{M}}$ be a finite algebra and assume that the alter ego $\underset{\sim}{\mathbf{M}} = \langle M; G, H, R, \mathcal{T}\rangle$ yields a full duality on $\mathbb{ISP}(\underline{\mathbf{M}})$. Let r be a hom-minimal algebraic relation on $\underline{\mathbf{M}}$. Then $\underset{\sim}{\mathbf{M}}' := \langle M; G, H, R \cup \{r\}, \mathcal{T}\rangle$ also yields a full duality on $\mathbb{ISP}(\underline{\mathbf{M}})$.*

Now assume that $\underline{\mathbf{M}}$ is a unary algebra of type $(2)_{\mathrm{VL}}$ or type $(2)_{\mathrm{D}}$. Let $\preccurlyeq$ denote the order on $\{0, 1, 2\}$ with $2 \preccurlyeq 0 \preccurlyeq 1$. We will define some algebraic relations on $\underline{\mathbf{M}}$. The definitions of these relations will depend on the type of $\underline{\mathbf{M}}$. If $\underline{\mathbf{M}}$ is of type $(2)_{\mathrm{VL}}$, then we define the algebraic relations on $\underline{\mathbf{M}}$ by

$$\preccurlyeq_n \; := \mathrm{sg}_{\underline{\mathbf{M}}^n}\Big(\big\{\, a \in M^n \mid 2 = a(0) \preccurlyeq \cdots \preccurlyeq a(n-1) = 1 \,\big\}\Big),$$

for all $n \in \omega\backslash\{0, 1\}$, and

$$\bowtie \; := \mathrm{sg}_{\underline{\mathbf{M}}^6}\Big(\big\{\, a \in M^6 \mid 2 = a(0) \preccurlyeq \cdots \preccurlyeq a(5) = 1 \,\big\}\backslash \\ \{(2,2,0,0,1,1)\}\Big).$$

If $\underline{\mathbf{M}}$ is of type $(2)_{\mathrm{D}}$ but not type $(2)_{\mathrm{VL}}$, then, since $\underline{\mathbf{M}}$ is a reduct of $\underline{\mathbf{M}}^\sharp$, we can define the algebraic relations on $\underline{\mathbf{M}}$ by

$$\preccurlyeq_n \; := \mathrm{sg}_{(\underline{\mathbf{M}}^\sharp)^n}\Big(\big\{\, a \in M^n \mid 2 = a(0) \preccurlyeq \cdots \preccurlyeq a(n-1) = 1 \,\big\}\Big),$$

for all $n \in \omega\backslash\{0, 1\}$, and

$$\bowtie \; := \mathrm{sg}_{(\underline{\mathbf{M}}^\sharp)^6}\Big(\big\{\, a \in M^6 \mid 2 = a(0) \preccurlyeq \cdots \preccurlyeq a(5) = 1 \,\big\}\backslash \\ \{(2,2,0,0,1,1)\}\Big).$$

The relations $\preccurlyeq_4$, $\preccurlyeq_6$ and $\bowtie$ will play an important role in our proof that $\underline{\mathbf{M}}$ is not fully dualisable.

4.4.2 Lemma *Let $\underline{\mathbf{M}}$ be a unary algebra of type $(2)_{\mathrm{VL}}$ or type $(2)_{\mathrm{D}}$. The relations $\preccurlyeq_4$ and $\bowtie$ on $\underline{\mathbf{M}}$ are hom-minimal.*

Proof First assume that $\underline{\mathbf{M}}$ is of type $(2)_{\mathrm{VL}}$ and define $\mathcal{A} := \mathbb{ISP}(\underline{\mathbf{M}})$. Define the three-element chain $\mathbf{C} = \langle\{0,1,2\}; \leqslant\rangle$ such that $0 \leqslant 1 \leqslant 2$. Then, using Definition 4.3.3, it is straightforward to check that $\widehat{\mathbf{C}}_{\leqslant}$ is the subalgebra of $\underline{\mathbf{M}}^{C^+}$ generated by the set

$$\{\, a \in M^{C^+} \mid 2 = a(\bot) = a(0) \preccurlyeq a(1) \preccurlyeq a(2) \preccurlyeq a(\top) = 1 \,\}.$$

Therefore $\widehat{\mathbf{C}}_{\leqslant}$ is isomorphic to the subalgebra $\preccurlyeq_4$ of $\underline{\mathbf{M}}^4$. By Lemma 4.3.6(ii), we have

$$|\mathcal{A}(\preccurlyeq_4, \underline{\mathbf{M}})| = |\mathcal{A}(\widehat{\mathbf{C}}_{\leqslant}, \underline{\mathbf{M}})| = |\mathcal{G}(\mathbf{C}, \underline{\mathbf{2}})| = 4.$$

As the relation $\preccurlyeq_4$ does not have any repeated coordinates, the four projections in $\mathcal{A}(\preccurlyeq_4, \underline{\mathbf{M}})$ are distinct. Thus $\preccurlyeq_4$ is hom-minimal.

Define the five-element chain $\mathbf{D}$ and directed graph $\mathbf{D}'$ as follows:

$$\mathbf{D} = \langle\{0,1,2,3,4\}; \leqslant\rangle, \quad \text{where} \quad 0 \leqslant \cdots \leqslant 4,$$

and

$$\mathbf{D}' = \langle\{0,1,2,3,4\}; \trianglelefteq\rangle, \quad \text{where} \quad \trianglelefteq := \leqslant \backslash \{(1,3)\}.$$

Write each element a of M^{D^+} as the 7-tuple $(a(\bot), a(0), \dots, a(4), a(\top))$. Then $\widehat{\mathbf{D}}_{\trianglelefteq}$ is the subalgebra of $\underline{\mathbf{M}}^{D^+}$ generated by the set

$$\{\, a \in M^{D^+} \mid 2 = a(\bot) = a(0) \preccurlyeq a(1) \preccurlyeq \cdots \preccurlyeq a(4) \preccurlyeq a(\top) = 1 \,\} \backslash \{(2,2,2,0,0,1,1)\},$$

and $\widehat{\mathbf{D}}_{\trianglelefteq}$ is isomorphic to the subalgebra $\bowtie$ of $\underline{\mathbf{M}}^6$. Using Lemma 4.3.6(ii), this implies that

$$|\mathcal{A}(\bowtie, \underline{\mathbf{M}})| = |\mathcal{A}(\widehat{\mathbf{D}}_{\trianglelefteq}, \underline{\mathbf{M}})| = |\mathcal{G}(\mathbf{D}', \underline{\mathbf{2}})| = 6.$$

Thus $\bowtie$ is hom-minimal.

Now assume that $\underline{\mathbf{M}}$ is of type $(2)_{\mathrm{D}}$ but not type $(2)_{\mathrm{VL}}$. The relations $\preccurlyeq_4$ and $\bowtie$, defined on $\underline{\mathbf{M}}$, are algebraic over $\underline{\mathbf{M}}^\sharp$. We have just shown that $\preccurlyeq_4$ and $\bowtie$ are hom-minimal relative to $\underline{\mathbf{M}}^\sharp$. So $\preccurlyeq_4$ and $\bowtie$ are hom-minimal relative to $\underline{\mathbf{M}}$, by Lemma 4.3.8. ∎

We shall work with the relations $\preccurlyeq_4$, $\preccurlyeq_6$ and $\bowtie$ using the properties given in the following lemma.

4.4.3 Lemma *Let* $\underline{\mathbf{M}}$ *be a unary algebra of type* $(2)_{\mathrm{VL}}$ *or type* $(2)_{\mathrm{D}}$. *Let* $m, n \in \omega\backslash\{0,1\}$ *and let* $a_0, \dots, a_{n-1} \in \{0,1,2\}$.

(i) *Let* $\sigma : \{0,\dots,m-1\} \to \{0,\dots,n-1\}$ *preserve the natural order, with* $\sigma(0) = 0$ *and* $\sigma(m-1) = n-1$. *If* $(a_0,\dots,a_{n-1}) \in \preccurlyeq_n$, *then* $(a_{\sigma(0)},\dots,a_{\sigma(m-1)}) \in \preccurlyeq_m$.

(ii) *Let* $i, j, k \in \{0,\dots,n-1\}$ *with* $i \leqslant j \leqslant k$. *If* $(a_0,\dots,a_{n-1}) \in \preccurlyeq_n$ *and* $a_i = a_k$, *then* $a_i = a_j = a_k$.

(iii) *Define* $C := \{(1,1,0,0,2,2),(2,2,0,0,1,1)\} \subseteq M^6$. *Then* $\bowtie \subseteq \preccurlyeq_6$ *and* $\preccurlyeq_6\backslash\bowtie \subseteq C \subseteq M^6\backslash\bowtie$.

Proof The first claim follows straight from the definitions. For the other claims, we also need to know that the unary term functions of the algebras $\underline{\mathbf{M}}$ and $\underline{\mathbf{M}}^\sharp$ all belong to the set $F^\sharp = \{012, 021\} \cup \{\, ppq, pqp \mid p, q \in \{0,1,2\}\,\}$; see Lemma 3.3.2. ∎

We will use a sequence of technical lemmas to prove that there are no fully dualisable algebras of type $(2)_{\mathrm{VL}}$ or type $(2)_{\mathrm{D}}$. Before starting on these lemmas, we sketch the main idea of the proof. Assume that $\underline{\mathbf{M}}$ is of type $(2)_{\mathrm{VL}}$, and suppose that $\underset{\sim}{\mathbf{M}}$ yields a full duality on $\mathcal{A} := \mathbb{ISP}(\underline{\mathbf{M}})$. We may assume that the relations $\preccurlyeq_4$ and $\bowtie$ are in the type of $\underset{\sim}{\mathbf{M}}$, by Lemmas 4.4.1 and 4.4.2. As in the previous section, we will be using the pathological algebras $\widehat{\mathbf{\Gamma}}_\leqslant$ and $\widehat{\mathbf{\Gamma}}_\trianglelefteq$ in $\mathcal{A}$, which are given by Lemma 4.3.1 and Definition 4.3.3. We know that the set $X := \{\, x{\restriction}_{\widehat{\Gamma}_\trianglelefteq} \mid x \in \mathcal{A}(\widehat{\mathbf{\Gamma}}_\leqslant, \underline{\mathbf{M}})\,\}$ forms a closed substructure $\mathbf{X}$ of $\underset{\sim}{\mathbf{M}}^{\widehat{\Gamma}_\trianglelefteq}$, by Lemmas 4.3.2(i) and 4.3.6(ii). Since we are supposing that $\underset{\sim}{\mathbf{M}}$ yields a full duality on $\mathcal{A}$, there must be an algebra $\mathbf{A}$ in $\mathcal{A}$ such that $\mathbf{X}$ is isomorphic to the dual $\mathrm{D}(\mathbf{A})$ of $\mathbf{A}$. Using our knowledge of the structure of $\mathbf{X}$, we will be able to deduce contradictory information about the structure of the putative algebra $\mathbf{A}$. The relations $\preccurlyeq_4$ and $\bowtie$ play a very important role in the proof, because they are the only part of the known structure on $\mathbf{X}$ that we can transfer across to the dual $\mathrm{D}(\mathbf{A})$.

Our proof that $\underline{\mathbf{M}}$ is not fully dualisable is rather more complicated than our proof that $\underline{\mathbf{M}}$ is not strongly dualisable. To prove that $\underline{\mathbf{M}}$ was not strongly dualisable, we only needed to study two particular algebras $\widehat{\mathbf{\Gamma}}_\leqslant$ and $\widehat{\mathbf{\Gamma}}_\trianglelefteq$ in $\mathcal{A}$. However, to prove that $\underline{\mathbf{M}}$ is not fully dualisable, we will be working with an algebra $\mathbf{A}$ from $\mathcal{A}$ that we initially know very little about.

We are now ready to begin building up the tools that will be used in our proof. Assume that $\underline{\mathbf{M}}$ is a two-kernel unary algebra on $\{0,1,2\}$, with kernels $\{01|2\}$ and $\{02|1\}$. To each subalgebra of a power of $\underline{\mathbf{M}}$, we will associate a directed graph.

Let $\mathbf{A}$ be a subalgebra of $\underline{\mathbf{M}}^S$, for some non-empty set S. For each two-block partition $\mathcal{Q}$ of S, define the **subuniverse of $\mathbf{A}_{\downarrow 2}$ determined by** $\mathcal{Q}$ to be

$$A_{\mathcal{Q}} := \{\, a \in A_{\downarrow 2} \mid \mathcal{P}(a) = \mathcal{Q} \text{ or } a \in \{\widehat{0}, \widehat{1}, \widehat{2}\}\,\}.$$

Define

$$\mathcal{P}_{\mathbf{A}} := \{\, A_{\mathcal{Q}} \mid \mathcal{Q} = \mathcal{P}(a) \text{ for some } a \in A_{\downarrow 2} \backslash \{\widehat{0}, \widehat{1}, \widehat{2}\}\,\}.$$

Then $\mathcal{P}_{\mathbf{A}}$ contains all the partition-determined subuniverses of $\mathbf{A}_{\downarrow 2}$ that do not lie in $\{\widehat{0}, \widehat{1}, \widehat{2}\}$. By assumption, there exist unary term functions u_1 and u_2 of $\underline{\mathbf{M}}$ such that $\ker(u_1) = \{02|1\}$ and $\ker(u_2) = \{01|2\}$. For each $t \in S$, we shall define the reflexive binary relation $\xrightarrow[t]{}_{\mathbf{A}}$ on $\mathcal{P}_{\mathbf{A}}$ by declaring that $P \xrightarrow[t]{}_{\mathbf{A}} Q$ if and only if $P = Q$ or there exists $a \in A \backslash A_{\downarrow 2}$ and $\{k, \ell\} = \{1, 2\}$ with

$$u_k(a) \in P,\ u_\ell(a) \in Q \text{ and } a(t) = \ell.$$

The definition of $\xrightarrow[t]{}_{\mathbf{A}}$ is independent of our choice of u_1 and u_2. Let $\xrightarrow[t]{}\!\!\!\!\rightarrow_{\mathbf{A}}$ denote the transitive closure of $\xrightarrow[t]{}_{\mathbf{A}}$. Then $\xrightarrow[t]{}\!\!\!\!\rightarrow_{\mathbf{A}}$ is a quasi-order on $\mathcal{P}_{\mathbf{A}}$. To illustrate the definition of $\xrightarrow[t]{}_{\mathbf{A}}$, we revisit the algebra constructed back in Example 4.3.5.

4.4.4 Example Let $\underline{\mathbf{M}}$ be a two-kernel unary algebra on $\{0, 1, 2\}$, with kernels $\{01|2\}$ and $\{02|1\}$. Define the ordered set $\mathbf{P}'$ and the algebra $\mathbf{A} := \widehat{\mathbf{P}}_{\trianglelefteq}$ as in Example 4.3.5. (See also Figure 4.6.) We will show that $\langle \mathcal{P}_{\mathbf{A}}; \xrightarrow[\top]{}_{\mathbf{A}} \rangle$ is isomorphic to $\mathbf{P}' = \langle P; \trianglelefteq \rangle$. Choose unary term functions u_1 and u_2 of $\underline{\mathbf{M}}$ with $\ker(u_1) = \{02|1\}$ and $\ker(u_2) = \{01|2\}$. The two-block partitions $\mathcal{P}(\widehat{aa})$, $\mathcal{P}(\widehat{bb})$ and $\mathcal{P}(\widehat{cc})$ of P^+ are distinct. So, using Lemma 4.3.4(i), we must have $\mathcal{P}_{\mathbf{A}} = \{\mathrm{sg}_{\mathbf{A}}(\widehat{aa}), \mathrm{sg}_{\mathbf{A}}(\widehat{bb}), \mathrm{sg}_{\mathbf{A}}(\widehat{cc})\}$. By Lemma 4.3.4(ii), we get

$$\widehat{ac},\ \widehat{bc} \in A \backslash A_{\downarrow 2} \quad \text{and} \quad A \backslash A_{\downarrow 2} \subseteq \{\widehat{ac}, \widehat{bc}, 021(\widehat{ac}), 021(\widehat{bc})\}.$$

Now $\widehat{ac}(\top) = 1$, with

$$u_2(\widehat{ac}) = u_2(\widehat{aa}) \in \mathrm{sg}_{\mathbf{A}}(\widehat{aa}) \quad \text{and} \quad u_1(\widehat{ac}) = u_1(\widehat{cc}) \in \mathrm{sg}_{\mathbf{A}}(\widehat{cc}),$$

which implies that $\mathrm{sg}_{\mathbf{A}}(\widehat{aa}) \xrightarrow[\top]{}_{\mathbf{A}} \mathrm{sg}_{\mathbf{A}}(\widehat{cc})$. If 021 is a term function of $\underline{\mathbf{M}}$, then we have $021(\widehat{ac})(\top) = 2$, with

$$u_1(021(\widehat{ac})) \in \mathrm{sg}_{\mathbf{A}}(\widehat{aa}) \quad \text{and} \quad u_2(021(\widehat{ac})) \in \mathrm{sg}_{\mathbf{A}}(\widehat{cc}),$$

which also implies that $\mathrm{sg}_{\mathbf{A}}(\widehat{aa}) \xrightarrow[\top]{}_{\mathbf{A}} \mathrm{sg}_{\mathbf{A}}(\widehat{cc})$. Similarly, using $\widehat{bc}$ and $021(\widehat{bc})$, we have $\mathrm{sg}_{\mathbf{A}}(\widehat{bb}) \xrightarrow[\top]{}_{\mathbf{A}} \mathrm{sg}_{\mathbf{A}}(\widehat{cc})$. Thus there is an isomorphism $\vartheta : \langle P; \trianglelefteq \rangle \to \langle \mathcal{P}_{\mathbf{A}}; \xrightarrow[\top]{}_{\mathbf{A}} \rangle$ given by $\vartheta(x) = \mathrm{sg}_{\mathbf{A}}(\widehat{xx})$.

As shown by the example above, we now have a way to recover the original relation $\trianglelefteq$ from an algebra $\widehat{\mathbf{P}}_{\trianglelefteq}$ built from that relation using Definition 4.3.3.

4.4.5 Lemma *Let $\underline{\mathbf{M}}$ be a two-kernel unary algebra on $\{0,1,2\}$, with kernels $\{01|2\}$ and $\{02|1\}$. Let $\mathbf{P} = \langle P; \leqslant \rangle$ be an ordered set, let $\trianglelefteq$ be a reflexive subset of $\leqslant$, and define $\mathbf{A} := \widehat{\mathbf{P}}_{\trianglelefteq}$. Then $\langle P; \trianglelefteq \rangle$ is isomorphic to $\langle \mathcal{P}_{\mathbf{A}}; \xrightarrow[\top]{}_{\mathbf{A}} \rangle$.*

Given any binary relation r on a set X, we shall say that a subset Z of X is r-**decreasing** if, for all $(x, y) \in r$ such that $y \in Z$, we also have $x \in Z$. Now define $\mathcal{D}_t(\mathbf{A})$ to be the set of all $\xrightarrow[t]{}_{\mathbf{A}}$-decreasing subsets of $\mathcal{P}_{\mathbf{A}}$.

4.4.6 Example We will return to the situation considered in the previous example. By Lemma 4.3.6(ii), we know that there is a bijection

$$- \circ \iota_{\mathbf{P}} : \mathcal{A}(\mathbf{A}, \underline{\mathbf{M}}) \to \mathcal{G}(\mathbf{P}', \underline{\mathbf{2}}).$$

The set $\mathcal{G}(\mathbf{P}', \underline{\mathbf{2}})$ is bijective with the set of all $\trianglelefteq$-decreasing subsets of $\mathbf{P}'$. As $\mathbf{P}' = \langle P; \trianglelefteq \rangle$ and $\langle \mathcal{P}_{\mathbf{A}}; \xrightarrow[\top]{}_{\mathbf{A}} \rangle$ are isomorphic, it now follows that $\mathcal{A}(\mathbf{A}, \underline{\mathbf{M}})$ is bijective with the set $\mathcal{D}_{\top}(\mathbf{A})$ of all $\xrightarrow[\top]{}_{\mathbf{A}}$-decreasing subsets of $\mathcal{P}_{\mathbf{A}}$. The next lemma generalises this observation.

4.4.7 Lemma *Let $\underline{\mathbf{M}}$ be a two-kernel unary algebra on $\{0,1,2\}$, with kernels $\{01|2\}$ and $\{02|1\}$, and define $\mathcal{A} := \mathbb{ISP}(\underline{\mathbf{M}})$. Let $\mathbf{A} \leqslant \underline{\mathbf{M}}^S$, for some non-empty set S, with $\mathcal{P}_{\mathbf{A}} \neq \varnothing$. Assume there are $s, t \in S$ such that $x{\restriction}_P = \pi_s{\restriction}_P$ or $x{\restriction}_P = \pi_t{\restriction}_P$, for all $x \in \mathcal{A}(\mathbf{A}, \underline{\mathbf{M}})$ and all $P \in \mathcal{P}_{\mathbf{A}}$. Then there is a bijection $\eta : \mathcal{A}(\mathbf{A}, \underline{\mathbf{M}}) \to \mathcal{D}_t(\mathbf{A})$, given by $\eta(x) := \{\, P \in \mathcal{P}_{\mathbf{A}} \mid x{\restriction}_P = \pi_t{\restriction}_P \,\}$.*

Proof By assumption, there are unary term functions u_1 and u_2 of $\underline{\mathbf{M}}$ with $\ker(u_1) = \{02|1\}$ and $\ker(u_2) = \{01|2\}$.

Claim 1 The map η is well defined.

Let $x \in \mathcal{A}(\mathbf{A}, \underline{\mathbf{M}})$. We want to show that $\eta(x)$ is a $\xrightarrow[t]{}_{\mathbf{A}}$-decreasing subset of $\mathcal{P}_{\mathbf{A}}$. So choose $P, Q \in \mathcal{P}_{\mathbf{A}}$ such that $Q \in \eta(x)$ and $P \xrightarrow[t]{}_{\mathbf{A}} Q$. We now want to check that $P \in \eta(x)$. We have $x{\restriction}_Q = \pi_t{\restriction}_Q$, and we wish to show that $x{\restriction}_P = \pi_t{\restriction}_P$. We can assume that $P \neq Q$, and so there exist $a \in A \backslash A_{\downarrow 2}$ and $\{k, \ell\} = \{1, 2\}$ such that $u_k(a) \in P$, $u_\ell(a) \in Q$ and $a(t) = \ell$. We must have

$$u_\ell(x(a)) = x(u_\ell(a)) = u_\ell(a(t)) = u_\ell(\ell),$$

since $x{\restriction}_Q = \pi_t{\restriction}_Q$. Therefore $x(a) = \ell = a(t)$, and

$$x(u_k(a)) = u_k(x(a)) = u_k(a(t)).$$

Since $|a(S)| = 3$, the partition $\mathcal{P}(u_k(a))$ of S has two blocks. We know that $u_k(a) \in P$, and we are assuming that every homomorphism in $\mathcal{A}(\mathbf{A}, \underline{\mathbf{M}})$ agrees with π_s or π_t on P. So it follows that $u_k(a)(s) \neq u_k(a)(t)$. Since

$x(u_k(a)) = u_k(a(t))$, this implies that $x{\restriction}_P = \pi_t{\restriction}_P$. Thus $\eta(x) \in \mathcal{D}_t(\mathbf{A})$, and η is well defined.

Claim 2 The map η is one-to-one.

Let $x, y \in \mathcal{A}(\mathbf{A}, \underline{\mathbf{M}})$ with $\eta(x) = \eta(y)$. Then x and y agree on $\bigcup \mathcal{P}_{\mathbf{A}} = A_{\downarrow 2}$. Now let $a \in A \backslash A_{\downarrow 2}$. For each $m \in \{1, 2\}$, we have $u_m(a) \in A_{\downarrow 2}$ and

$$u_m(x(a)) = x(u_m(a)) = y(u_m(a)) = u_m(y(a)).$$

Since u_1 and u_2 separate the elements of M, it follows that $x(a) = y(a)$. So $x = y$, whence η is one-to-one.

Claim 3 The map η is onto.

Let $Z \in \mathcal{D}_t(\mathbf{A})$. As $\{\, P \backslash \{\widehat{0}, \widehat{1}, \widehat{2}\} \mid P \in \mathcal{P}_{\mathbf{A}} \,\}$ is a partition of $A_{\downarrow 2} \backslash \{\widehat{0}, \widehat{1}, \widehat{2}\}$, we can define the homomorphism $x : \mathbf{A}_{\downarrow 2} \to \underline{\mathbf{M}}$ by

$$x := \bigcup \{\, \pi_t{\restriction}_P \mid P \in Z \,\} \cup \bigcup \{\, \pi_s{\restriction}_P \mid P \in \mathcal{P}_{\mathbf{A}} \backslash Z \,\}.$$

We wish to prove that $x : \mathbf{A}_{\downarrow 2} \to \underline{\mathbf{M}}$ has an extension $\overline{x} : \mathbf{A} \to \underline{\mathbf{M}}$. We shall show that, for each $a \in A \backslash A_{\downarrow 2}$, there is $r_a \in S$ such that x agrees with π_{r_a} on $A_{\downarrow 2} \cap \mathrm{sg}_{\mathbf{A}}(a)$. We shall then define $\overline{x}(a) := a(r_a)$, for each $a \in A \backslash A_{\downarrow 2}$.

For the sake of argument, assume that we have already set up $\overline{x}$ in this way. We will check that $\overline{x}$ is a homomorphism. For each $a \in A \backslash A_{\downarrow 2}$ and each unary term function u of $\underline{\mathbf{M}}$ that is not a permutation, we have

$$u(\overline{x}(a)) = u(a(r_a)) = \pi_{r_a}(u(a)) = x(u(a)) = \overline{x}(u(a)),$$

since $u(a) \in A_{\downarrow 2} \cap \mathrm{sg}_{\mathbf{A}}(a)$. So $\overline{x}$ preserves every unary term function of $\underline{\mathbf{M}}$ that is not a permutation. By Lemma 3.3.2, the only non-identity permutation that can be a term function of $\underline{\mathbf{M}}$ is 021. So assume that 021 is a term function of $\underline{\mathbf{M}}$. For each $a \in A \backslash A_{\downarrow 2}$ and each $m \in \{1, 2\}$, we have

$$u_m(021(\overline{x}(a))) = \overline{x}(u_m \circ 021(a)) = u_m(\overline{x}(021(a))),$$

since both $u_m \circ 021$ and u_m are unary term functions of $\underline{\mathbf{M}}$ neither of which is a permutation. As u_1 and u_2 separate the elements of M, this tells us that $021(\overline{x}(a)) = \overline{x}(021(a))$, for every $a \in A \backslash A_{\downarrow 2}$. Thus the putative map $\overline{x}$ is a homomorphism.

Now we want to show that we can set up $\overline{x}$ as desired. Let $a \in A \backslash A_{\downarrow 2}$. For each $m \in \{1, 2\}$, define Q_m to be the subuniverse of $\mathbf{A}_{\downarrow 2}$ determined by the two-block partition $\mathcal{P}(u_m(a))$ of S. The only kernels of $\underline{\mathbf{M}}$ are $\ker(u_1) = \{02|1\}$ and $\ker(u_2) = \{01|2\}$. So $A_{\downarrow 2} \cap \mathrm{sg}_{\mathbf{A}}(a) \subseteq Q_1 \cup Q_2$. We will next find some $r_a \in S$ such that x agrees with π_{r_a} on $Q_1 \cup Q_2$.

For each $m \in \{1, 2\}$, we have

$$u_m(a(s)) = u_m(a)(s) \neq u_m(a)(t) = u_m(a(t)).$$

This implies that $\{a(s), a(t)\} = \{1, 2\}$. Now define $k := a(s)$ and $\ell := a(t)$. We have $a \in A \backslash A_{\downarrow 2}$ with $u_k(a) \in Q_k$, $u_\ell(a) \in Q_\ell$ and $a(t) = \ell$. Therefore $Q_k \xrightarrow[t]{}_{\mathbf{A}} Q_\ell$. We are trying to find some $r_a \in S$ with $x{\restriction}_{Q_1 \cup Q_2} = \pi_{r_a}{\restriction}_{Q_1 \cup Q_2}$. Since Z is $\xrightarrow[t]{}_{\mathbf{A}}$-decreasing and $Q_k \xrightarrow[t]{}_{\mathbf{A}} Q_\ell$, we can assume that $Q_k \in Z$ and $Q_\ell \notin Z$. As $a \notin A_{\downarrow 2}$, we have $|a(S)| = 3$ and so there is some $r_a \in a^{-1}(0)$. We now have

$$u_k(a(t)) = u_k(\ell) = u_k(0) = u_k(a(r_a))$$

and

$$u_\ell(a(s)) = u_\ell(k) = u_\ell(0) = u_\ell(a(r_a)).$$

Each element of $Q_k \backslash \{\widehat{0}, \widehat{1}, \widehat{2}\}$ determines the same partition of S as the element $u_k(a)$. Since $Q_k \in Z$, we have $x{\restriction}_{Q_k} = \pi_t{\restriction}_{Q_k} = \pi_{r_a}{\restriction}_{Q_k}$, and similarly we have $x{\restriction}_{Q_\ell} = \pi_s{\restriction}_{Q_\ell} = \pi_{r_a}{\restriction}_{Q_\ell}$. So x agrees with π_{r_a} on $Q_1 \cup Q_2 \supseteq A_{\downarrow 2} \cap \mathrm{sg}_{\mathbf{A}}(a)$, as required.

It now follows that x extends to a homomorphism $\overline{x} \in \mathcal{A}(\mathbf{A}, \underline{\mathbf{M}})$. We must have $\eta(\overline{x}) = Z$, as $\pi_s{\restriction}_P \neq \pi_t{\restriction}_P$, for all $P \in \mathcal{P}_{\mathbf{A}}$. Thus η is onto. ∎

The subalgebra $\mathbf{A}$ of $\underline{\mathbf{M}}^S$ is **locally hom-minimal (relative to $\underline{\mathbf{M}}$)** if, for each homomorphism $x : \mathbf{A} \to \underline{\mathbf{M}}$ and each finite subset P of A, the map $x{\restriction}_P$ is the restriction of a projection. Recall that, for each subalgebra $\mathbf{B}$ of $\underline{\mathbf{M}}^S$ and each $s \in S$, we define $\rho_s : \mathbf{B} \to \underline{\mathbf{M}}$ by $\rho_s := \pi_s{\restriction}_B$.

4.4.8 Lemma *Let $\underline{\mathbf{M}}$ be a unary algebra of type $(2)_{\mathrm{VL}}$ or $(2)_{\mathrm{D}}$, and define $\mathcal{A} := \mathbb{ISP}(\underline{\mathbf{M}})$. Let $\mathbf{B} \leqslant \mathbf{A} \leqslant \underline{\mathbf{M}}^S$, for some non-empty set S, such that $A_{\downarrow 2} \subseteq B$ and $\mathbf{A}$ is locally hom-minimal. Define $X := \{\, x{\restriction}_B \mid x \in \mathcal{A}(\mathbf{A}, \underline{\mathbf{M}}) \,\}$. Assume that there are $s, t \in S$ such that the relation*

$$\leqslant := \{\, (x, y) \in X^2 \mid (\rho_s, x, y, \rho_t) \in \preccurlyeq_4 \,\}$$

on X is reflexive.

(i) *For all $x \in X$ and $P \in \mathcal{P}_{\mathbf{A}}$, we have $x{\restriction}_P = \pi_s{\restriction}_P$ or $x{\restriction}_P = \pi_t{\restriction}_P$.*

(ii) *For every $x \in X$, define the set $\eta(x) := \{\, P \in \mathcal{P}_{\mathbf{A}} \mid x{\restriction}_P = \pi_t{\restriction}_P \,\}$. Then $\eta : \langle X; \leqslant \rangle \to \langle \mathcal{D}_t(\mathbf{A}); \subseteq \rangle$ is a well-defined isomorphism.*

(iii) *For all $n \in \omega \backslash \{0\}$ and $x_1, \dots, x_n \in X$ with $x_1 \leqslant \cdots \leqslant x_n$, we have $(\rho_s, x_1, \dots, x_n, \rho_t) \in \preccurlyeq_{n+2}$.*

Proof There are unary term functions u_1 and u_2 of $\underline{\mathbf{M}}$ with $\ker(u_1) = \{02|1\}$ and $\ker(u_2) = \{01|2\}$. Since $A_{\downarrow 2} \subseteq B$, each subuniverse of $\mathbf{A}$ that belongs to $\mathcal{P}_{\mathbf{A}}$ is also a subuniverse of $\mathbf{B}$.

Proof of (i) Let $x \in X$ and let $P \in \mathcal{P}_{\mathbf{A}}$. There is some $a \in P$ such that P is the subuniverse of $\mathbf{A}_{\downarrow 2}$ determined by the two-block partition $\mathcal{P}(a)$ of S. The relation $\leqslant$ on X is reflexive, and therefore $(\rho_s, x, x, \rho_t)(a) \in \preccurlyeq_4$. Since $\mathbf{A}$ is locally hom-minimal, the map $x{\restriction}_P$ is the restriction of a projection. So $x(a) \in a(S)$. As $\mathcal{P}(a)$ has two blocks, we have $|\{a(s), x(a), a(t)\}| \leqslant 2$. By Lemma 4.4.3(ii), we must have $x(a) = a(s)$ or $x(a) = a(t)$. As $x{\restriction}_P$ is the restriction of a projection and all the elements of $P\backslash\{\widehat{0}, \widehat{1}, \widehat{2}\}$ determine the same partition of S, it now follows that $x{\restriction}_P = \pi_s{\restriction}_P$ or $x{\restriction}_P = \pi_t{\restriction}_P$. Thus (i) holds.

Proof of (ii) We can now show that the map η, given in (ii), is a well-defined bijection. First consider the trivial case that $\mathcal{P}_{\mathbf{A}} = \varnothing$. We have $A \subseteq \{\widehat{0}, \widehat{1}, \widehat{2}\}$ and so, as $\mathbf{A}$ is locally hom-minimal, we have $|X| = 1$. Therefore η is a well-defined bijection. We can now assume that $\mathcal{P}_{\mathbf{A}} \neq \varnothing$. Since $A_{\downarrow 2} \subseteq B$, it follows by Lemma 4.4.7 that $\eta : X \to \mathcal{D}_t(\mathbf{A})$ is a well-defined bijection.

We now want to show that η is an isomorphism. Let $x, y \in X$ and assume that $x \leqslant y$. To see that $\eta(x) \subseteq \eta(y)$, let $P \in \eta(x)$. Then $x{\restriction}_P = \pi_t{\restriction}_P$. There is some $a \in P$ such that P is the subuniverse of $\mathbf{A}_{\downarrow 2}$ determined by the two-block partition $\mathcal{P}(a)$. We have $(\rho_s, \rho_t, y, \rho_t)(a) = (\rho_s, x, y, \rho_t)(a) \in \preccurlyeq_4$, and so $y(a) = a(t)$, by Lemma 4.4.3(ii). Using (i), it follows that $P \in \eta(y)$, which implies that $\eta(x) \subseteq \eta(y)$.

Next assume that $\eta(x) \subseteq \eta(y)$ and choose some $b \in B$. We are aiming to prove that $(\rho_s, x, y, \rho_t)(b) \in \preccurlyeq_4$. Since $\leqslant$ is reflexive on X, we know that $(\rho_s, y, y, \rho_t)(b) \in \preccurlyeq_4$ and $(\rho_s, x, x, \rho_t)(b) \in \preccurlyeq_4$. So $(\rho_s, \rho_s, y, \rho_t)(b) \in \preccurlyeq_4$ and $(\rho_s, x, \rho_t, \rho_t)(b) \in \preccurlyeq_4$, by Lemma 4.4.3(i). We shall show that $b(s) = x(b)$ or $x(b) = y(b)$ or $y(b) = b(t)$. It will then follow that $(\rho_s, x, y, \rho_t)(b) \in \preccurlyeq_4$.

Assume that $b(s) \neq x(b)$ and $y(b) \neq b(t)$. As u_1 and u_2 separate M, there exists $k \in \{1, 2\}$ such that $x(u_k(b)) = u_k(x(b)) \neq u_k(b(s))$. As $\mathbf{A}$ is locally hom-minimal, this implies that the partition $\mathcal{P}(u_k(b))$ has two blocks. So we can define Q_k to be the subuniverse of $\mathbf{A}_{\downarrow 2}$ determined by $\mathcal{P}(u_k(b))$. We must have $x{\restriction}_{Q_k} = \pi_t{\restriction}_{Q_k}$, by (i). As $\eta(x) \subseteq \eta(y)$, this implies that $y{\restriction}_{Q_k} = \pi_t{\restriction}_{Q_k}$, and therefore

$$u_k(y(b)) = y(u_k(b)) = u_k(b(t)) = x(u_k(b)) = u_k(x(b)).$$

Let $\ell \in \{1, 2\}$ with $\ell \neq k$. Then $y(u_\ell(b)) = u_\ell(y(b)) \neq u_\ell(b(t))$, since $y(b) \neq b(t)$ and $u_k(y(b)) = u_k(b(t))$. So the partition $\mathcal{P}(u_\ell(b))$ has two blocks, and we can define Q_ℓ to be the subuniverse of $\mathbf{A}_{\downarrow 2}$ determined by $\mathcal{P}(u_\ell(b))$. Using (i), we have $y{\restriction}_{Q_\ell} = \pi_s{\restriction}_{Q_\ell}$. Since $\eta(x) \subseteq \eta(y)$, it now follows that $x{\restriction}_{Q_\ell} = \pi_s{\restriction}_{Q_\ell} = y{\restriction}_{Q_\ell}$, and therefore

$$u_\ell(x(b)) = x(u_\ell(b)) = y(u_\ell(b)) = u_\ell(y(b)).$$

We have shown that $u_k(x(b)) = u_k(y(b))$ and $u_\ell(x(b)) = u_\ell(y(b))$, whence $x(b) = y(b)$. Thus $(\rho_s, x, y, \rho_t)(b) \in \preccurlyeq_4$, and so $x \leqslant y$. Consequently, η is an isomorphism, and (ii) holds.

Proof of (iii) Let $n \in \omega\backslash\{0\}$, assume that $x_1 \leqslant \cdots \leqslant x_n$ in X and let $b \in B$. We will show that there are $j, k \in \{1, \ldots, n\}$, with $j \leqslant k$, such that

$$x_i(b) = \begin{cases} \rho_s(b) & \text{if } i < j, \\ x_j(b) & \text{if } j \leqslant i \leqslant k, \\ \rho_t(b) & \text{if } k < i, \end{cases}$$

for every $i \in \{1, \ldots, n\}$. As the relation $\leqslant$ on X is reflexive, we know that $(\rho_s, x_j, x_j, \rho_t)(b) \in \preccurlyeq_4$. So it will then follow, by Lemma 4.4.3(i), that we have $(\rho_s, x_1, \ldots, x_n, \rho_t)(b) \in \preccurlyeq_{n+2}$, as required.

First assume that $\rho_s(b) = \rho_t(b)$ and let $i \in \{1, \ldots, n\}$. Since the relation $\leqslant$ on X is reflexive, we must have $(\rho_s, x_i, x_i, \rho_t)(b) \in \preccurlyeq_4$, and consequently $\rho_s(b) = x_i(b) = \rho_t(b)$, by Lemma 4.4.3(ii). Thus we can take $j = k = n$, and we are done.

Now assume that $\rho_s(b) \neq \rho_t(b)$ and choose some $i \in \{1, \ldots, n\}$. If $i \neq n$ and $x_i(b) = \rho_t(b)$, then as $x_i \leqslant x_{i+1}$ we have

$$(\rho_s, \rho_t, x_{i+1}, \rho_t)(b) = (\rho_s, x_i, x_{i+1}, \rho_t)(b) \in \preccurlyeq_4,$$

and therefore $x_{i+1}(b) = \rho_t(b)$, by Lemma 4.4.3(ii). Similarly, if we have $i \neq 1$ and $x_i(b) = \rho_s(b)$, then $x_{i-1}(b) = \rho_s(b)$. Because the algebra $\mathbf{A}$ is locally hom-minimal, we know that $|\{\, x(b) \mid x \in X \,\}| \leqslant |b(S)| \leqslant 3$. Thus the desired $j, k \in \{1, \ldots, n\}$ must exist. Hence (iii) holds. ∎

For each ordered set $\langle X; \leqslant\rangle$, define $\mathcal{L}_\leqslant(X)$ to be the set of all elements x of X that have a unique lower cover $x^\downarrow$ in $\langle X; \leqslant\rangle$.

4.4.9 Lemma *Let $\underline{\mathbf{M}}$ be a unary algebra of type $(2)_{\mathrm{VL}}$ or $(2)_{\mathrm{D}}$, and define $\mathcal{A} := \mathbb{ISP}(\underline{\mathbf{M}})$. Let $\mathbf{B} \leqslant \mathbf{A} \leqslant \underline{\mathbf{M}}^S$, for some non-empty set S, such that $A_{\downarrow 2} \subseteq B$ and $\mathbf{A}$ is locally hom-minimal. Define $X := \{\, x{\restriction}_B \mid x \in \mathcal{A}(\mathbf{A}, \underline{\mathbf{M}}) \,\}$. Assume that there are $s, t \in S$ such that*

$$\leqslant\, := \{\, (x, y) \in X^2 \mid (\rho_s, x, y, \rho_t) \in \preccurlyeq_4 \,\}$$

is an order on X, and define

$$\leqslant^* := \{\, (x, y) \in \mathcal{L}_\leqslant(X)^2 \mid x \leqslant y \text{ and } (\rho_s, x^\downarrow, x, y^\downarrow, y, \rho_t) \notin \bowtie \,\}.$$

Then the structures $\langle \mathcal{L}_\leqslant(X); \leqslant^*, \leqslant\rangle$ and $\langle \mathcal{P}_{\mathbf{A}}; \xrightarrow[t]{}_{\mathbf{B}}, \xrightarrow[t]{}\!\!\!\!\rightarrow_{\mathbf{A}}\rangle$ are isomorphic.

Proof There are unary term functions u_1 and u_2 of $\underline{\mathbf{M}}$ with $\ker(u_1) = \{02|1\}$ and $\ker(u_2) = \{01|2\}$. Since $B \subseteq A$ and $A_{\downarrow 2} \subseteq B$, we have $A_{\downarrow 2} = B_{\downarrow 2}$ and therefore $\mathcal{P}_{\mathbf{A}} = \mathcal{P}_{\mathbf{B}}$. Lemma 4.4.8 tells us that $x{\restriction}_P = \pi_s{\restriction}_P$ or $x{\restriction}_P = \pi_t{\restriction}_P$, for all $x \in X$ and all $P \in \mathcal{P}_{\mathbf{A}}$, and moreover that there is an isomorphism $\eta : \langle X; \leqslant\rangle \to \langle \mathcal{D}_t(\mathbf{A}); \subseteq\rangle$ given by $\eta(x) := \{\, P \in \mathcal{P}_{\mathbf{A}} \mid x{\restriction}_P = \pi_t{\restriction}_P \,\}$.

For each $P \in \mathcal{P}_{\mathbf{A}}$, the smallest $\underset{t}{\longrightarrow}_{\mathbf{A}}$-decreasing subset of $\mathcal{P}_{\mathbf{A}}$ that contains P is

$$Z_P := \{\, Q \in \mathcal{P}_{\mathbf{A}} \mid Q \underset{t}{\twoheadrightarrow}_{\mathbf{A}} P \,\}.$$

We must have $\mathcal{L}_{\subseteq}\big(\mathcal{D}_t(\mathbf{A})\big) = \{\, Z_P \mid P \in \mathcal{P}_{\mathbf{A}} \,\}$. Since η is an isomorphism, we can define

$$\zeta : \mathcal{P}_{\mathbf{A}} \to \mathcal{L}_{\leqslant}(X) \quad \text{by} \quad \zeta(P) := \eta^{-1}(Z_P),$$

and ζ is onto. We shall establish, via a series of four claims, that ζ is an isomorphism from $\langle \mathcal{P}_{\mathbf{A}}; \underset{t}{\longrightarrow}_{\mathbf{B}}, \underset{t}{\twoheadrightarrow}_{\mathbf{A}}\rangle$ onto $\langle \mathcal{L}_{\leqslant}(X); \leqslant^*, \leqslant\rangle$.

For every $P \in \mathcal{P}_{\mathbf{A}}$, we will use x_P to denote the homomorphism $\zeta(P)$ in the set $\mathcal{L}_{\leqslant}(X) \subseteq \mathcal{A}(\mathbf{B}, \underline{\mathbf{M}})$.

Claim 1 The map ζ is a bijection.

We have already observed that the map ζ is onto. To see that ζ is one-to-one, let $P, Q \in \mathcal{P}_{\mathbf{A}}$ with $x_P = x_Q$. Then $Z_P = Z_Q$, since η is an isomorphism. For all $x \in X$, we have

$$x{\restriction}_{P \cup Q} = \pi_s{\restriction}_{P \cup Q} \quad \text{or} \quad x{\restriction}_{P \cup Q} = \pi_t{\restriction}_{P \cup Q},$$

as $\eta(x)$ is a $\underset{t}{\longrightarrow}_{\mathbf{A}}$-decreasing subset of $\mathcal{P}_{\mathbf{A}}$. So P and Q must determine the same two-block partition of S, and therefore $P = Q$. Thus ζ is a bijection.

Claim 2 For all $P, Q \in \mathcal{P}_{\mathbf{A}}$, we have $P \underset{t}{\twoheadrightarrow}_{\mathbf{A}} Q$ if and only if $x_P \leqslant x_Q$.

Let $P, Q \in \mathcal{P}_{\mathbf{A}}$. Then

$$P \underset{t}{\twoheadrightarrow}_{\mathbf{A}} Q \iff Z_P \subseteq Z_Q \iff \eta^{-1}(Z_P) \leqslant \eta^{-1}(Z_Q) \iff x_P \leqslant x_Q,$$

as required.

Claim 3 For all $P, Q \in \mathcal{P}_{\mathbf{A}}$ with $P \underset{t}{\longrightarrow}_{\mathbf{B}} Q$, we have $x_P \leqslant^* x_Q$.

Let $P, Q \in \mathcal{P}_{\mathbf{A}}$ with $P \underset{t}{\longrightarrow}_{\mathbf{B}} Q$. As the relation $\leqslant^*$ on $\mathcal{L}_{\leqslant}(X)$ is reflexive, by Lemma 4.4.3(ii), (iii), we can assume that $P \neq Q$. So there exists $b \in B \backslash B_{\downarrow 2}$ and $\{k, \ell\} = \{1, 2\}$ such that

$$u_k(b) \in P, \quad u_\ell(b) \in Q \quad \text{and} \quad b(t) = \ell.$$

We know that $x{\restriction}_P = \pi_s{\restriction}_P$ or $x{\restriction}_P = \pi_t{\restriction}_P$, for every $x \in X$. Therefore $u_k(b(s)) \neq u_k(b(t)) = u_k(\ell)$, which implies that $b(s) = k$. Since $b \in A \backslash A_{\downarrow 2}$, we also have $P \xrightarrow[t]{}\!\!\!\rightarrow_{\mathbf{A}} Q$, and therefore $x_P \leqslant x_Q$, by Claim 2. We will show that $x_P \leqslant^* x_Q$ by checking that $(\rho_s, x_P^{\downarrow}, x_P, x_Q^{\downarrow}, x_Q, \rho_t)(b) \notin \bowtie$.

By the definition of the map ζ, we have $\eta(x_P) = Z_P$ and $\eta(x_Q) = Z_Q$. As $\leqslant$ is an order on $\mathcal{L}_{\leqslant}(X)$, it follows that $\xrightarrow[t]{}\!\!\!\rightarrow_{\mathbf{A}}$ is an order on $\mathcal{P}_{\mathbf{A}}$, by Claim 2. So we also have $\eta(x_P^{\downarrow}) = Z_P^{\downarrow} = Z_P \backslash \{P\}$ and $\eta(x_Q^{\downarrow}) = Z_Q^{\downarrow} = Z_Q \backslash \{Q\}$. As $u_k(b) \in P \notin \eta(x_P^{\downarrow})$ and $u_\ell(b) \in Q \notin \eta(x_P^{\downarrow})$, it follows that

$$u_k(x_P^{\downarrow}(b)) = x_P^{\downarrow}(u_k(b)) = \pi_s(u_k(b)) = u_k(b(s)) = u_k(k)$$

and

$$u_\ell(x_P^{\downarrow}(b)) = x_P^{\downarrow}(u_\ell(b)) = \pi_s(u_\ell(b)) = u_\ell(b(s)) = u_\ell(k).$$

Since u_1 and u_2 separate the elements of M, this implies that $x_P^{\downarrow}(b) = k$. Similarly, we have

$$\begin{aligned} u_k(x_P(b)) &= u_k(b(t)) = u_k(0), & u_\ell(x_P(b)) &= u_\ell(b(s)) = u_\ell(0),\\ u_k(x_Q^{\downarrow}(b)) &= u_k(b(t)) = u_k(0), & u_\ell(x_Q^{\downarrow}(b)) &= u_\ell(b(s)) = u_\ell(0),\\ u_k(x_Q(b)) &= u_k(b(t)) = u_k(\ell), & u_\ell(x_Q(b)) &= u_\ell(b(t)) = u_\ell(\ell), \end{aligned}$$

giving $(\rho_s, x_P^{\downarrow}, x_P, x_Q^{\downarrow}, x_Q, \rho_t)(b) = (k, k, 0, 0, \ell, \ell)$. Hence, it follows that $(\rho_s, x_P^{\downarrow}, x_P, x_Q^{\downarrow}, x_Q, \rho_t)(b) \notin \bowtie$, by Lemma 4.4.3(iii). Thus $x_P \leqslant^* x_Q$.

Claim 4 For all $P, Q \in \mathcal{P}_{\mathbf{A}}$ with $x_P \leqslant^* x_Q$, we have $P \xrightarrow[t]{}_{\mathbf{B}} Q$.

Let $P, Q \in \mathcal{P}_{\mathbf{A}}$ such that $x_P \leqslant^* x_Q$. Then $x_P \leqslant x_Q$. Since the relation $\xrightarrow[t]{}_{\mathbf{B}}$ on $\mathcal{P}_{\mathbf{A}}$ is reflexive, we can assume that $P \neq Q$. As ζ is one-to-one, by Claim 1, we must have $x_P \neq x_Q$ and therefore $x_P^{\downarrow} \leqslant x_P \leqslant x_Q^{\downarrow} \leqslant x_Q$. This implies that $(\rho_s, x_P^{\downarrow}, x_P, x_Q^{\downarrow}, x_Q, \rho_t) \in \preccurlyeq_6$, by Lemma 4.4.8(iii). As $(\rho_s, x_P^{\downarrow}, x_P, x_Q^{\downarrow}, x_Q, \rho_t) \notin \bowtie$, there is some $b \in B$ and $\{k, \ell\} = \{1, 2\}$ such that $(\rho_s, x_P^{\downarrow}, x_P, x_Q^{\downarrow}, x_Q, \rho_t)(b) = (k, k, 0, 0, \ell, \ell)$, by Lemma 4.4.3(iii). Since $\mathbf{A}$ is locally hom-minimal, we must have $b \notin B_{\downarrow 2}$. Now

$$x_P^{\downarrow}(u_k(b)) = u_k(x_P^{\downarrow}(b)) = u_k(k) = u_k(\rho_s(b)) = u_k(b(s))$$

and

$$x_P(u_k(b)) = u_k(x_P(b)) = u_k(0) = u_k(\ell) = u_k(\rho_t(b)) = u_k(b(t)).$$

Thus the subuniverse of $\mathbf{A}_{\downarrow 2}$ determined by $\mathcal{P}(u_k(b))$ belongs to $\eta(x_P) = Z_P$ but not to $\eta(x_P^{\downarrow}) = Z_P \backslash \{P\}$, and so $u_k(b) \in P$. Similarly, we have

$$x_Q^{\downarrow}(u_\ell(b)) = u_\ell(0) = u_\ell(b(s)) \quad \text{and} \quad x_Q(u_\ell(b)) = u_\ell(\ell) = u_\ell(b(t)),$$

which implies that $u_\ell(b) \in Q$. We have shown that $b \in B \backslash B_{\downarrow 2}$ such that $u_k(b) \in P$, $u_\ell(b) \in Q$ and $b(t) = \rho_t(b) = \ell$. Thus $P \xrightarrow[t]{}_{\mathbf{B}} Q$, as required. ∎

The next lemma completes the preparation for our proof that the algebras of types $(2)_{\mathrm{VL}}$ and $(2)_{\mathrm{D}}$ are not fully dualisable.

4.4.10 Lemma *Let $\underline{\mathbf{M}}$ be a unary algebra of type $(2)_{\mathrm{VL}}$. Let $\widehat{\mathbf{\Gamma}}_{\leqslant}$ and $\widehat{\mathbf{\Gamma}}_{\trianglelefteqslant}$ be algebras in $\mathcal{A} := \mathbb{ISP}(\underline{\mathbf{M}})$ coming from Lemma 4.3.1 and Definition 4.3.3.*

(i) *We have $\widehat{\mathbf{\Gamma}}_{\trianglelefteqslant} \leqslant \widehat{\mathbf{\Gamma}}_{\leqslant} \leqslant \underline{\mathbf{M}}^{\Gamma^+}$ such that $(\widehat{\Gamma}_{\leqslant})_{\downarrow 2} \subseteq \widehat{\Gamma}_{\trianglelefteqslant}$ and the algebra $\widehat{\mathbf{\Gamma}}_{\leqslant}$ is locally hom-minimal.*

(ii) *Define the subset $X := \{\, x{\restriction}_{\widehat{\Gamma}_{\trianglelefteqslant}} \mid x \in \mathcal{A}(\widehat{\mathbf{\Gamma}}_{\leqslant}, \underline{\mathbf{M}}) \,\}$ of $\mathcal{A}(\widehat{\mathbf{\Gamma}}_{\trianglelefteqslant}, \underline{\mathbf{M}})$. Then $\leqslant \,:= \{\, (x, y) \in X^2 \mid (\rho_\perp, x, y, \rho_\top) \in \preccurlyeq_4 \,\}$ is a reflexive relation on X.*

(iii) *The structures $\langle \Gamma; \trianglelefteqslant, \leqslant \rangle$ and $\langle \mathcal{P}_{\widehat{\mathbf{\Gamma}}_{\leqslant}}; \underset{\top}{\longrightarrow}_{\widehat{\mathbf{\Gamma}}_{\trianglelefteqslant}}, \underset{\top}{\twoheadrightarrow}_{\widehat{\mathbf{\Gamma}}_{\leqslant}} \rangle$ are isomorphic.*

Proof We have $(\widehat{\Gamma}_{\leqslant})_{\downarrow 2} \subseteq \widehat{\Gamma}_{\trianglelefteqslant}$, by Lemma 4.3.4(i). To see that $\widehat{\mathbf{\Gamma}}_{\leqslant}$ is locally hom-minimal, let $x \in \mathcal{A}(\widehat{\mathbf{\Gamma}}_{\leqslant}, \underline{\mathbf{M}})$ and let B be a finite subset of $\widehat{\Gamma}_{\leqslant}$. We want to show that $x{\restriction}_B$ is the restriction of a projection. There is a finite subset Γ' of Γ such that

$$B \subseteq \mathrm{sg}_{\widehat{\mathbf{\Gamma}}_{\leqslant}}(\{\, \widehat{ab} \mid a, b \in \Gamma' \text{ and } a \leqslant b \,\}).$$

Using Lemma 4.3.6(ii), we know that $x \circ \iota_{\mathbf{\Gamma}} \in \mathcal{G}(\mathbf{\Gamma}, \underline{\mathbf{2}})$.

First assume that $x \circ \iota_{\mathbf{\Gamma}}(b) = 1$, for all $b \in \Gamma'$. Then, if $a, b \in \Gamma'$ with $a \leqslant b$, we have $x(\widehat{bb}) = x \circ \iota_{\mathbf{\Gamma}}(b) = 1$ and so $x(\widehat{ab}) = 1 = \widehat{ab}(\top)$, by Lemma 4.3.6(i). Therefore $x{\restriction}_B = \pi_\top{\restriction}_B$. As Γ' is finite and $\mathbf{\Gamma}$ is a chain, we can now assume that there is a minimum element c of Γ' in $\mathbf{\Gamma}$ such that $x \circ \iota_{\mathbf{\Gamma}}(c) = 2$. Since $x \circ \iota_{\mathbf{\Gamma}} \in \mathcal{G}(\mathbf{\Gamma}, \underline{\mathbf{2}})$, it follows by Lemma 4.3.6(i) that, for all $a, b \in \Gamma'$ with $a \leqslant b$, we have

$$\begin{aligned} x(\widehat{ab}) &= \begin{cases} 2 & \text{if } x(\widehat{aa}) = 2, \\ 0 & \text{if } x(\widehat{aa}) = 1 \text{ and } x(\widehat{bb}) = 2, \\ 1 & \text{if } x(\widehat{bb}) = 1, \end{cases} \\ &= \begin{cases} 2 & \text{if } c \leqslant a, \\ 0 & \text{if } a < c \leqslant b, \\ 1 & \text{if } b < c, \end{cases} \\ &= \widehat{ab}(c). \end{aligned}$$

So $x{\restriction}_B = \pi_c{\restriction}_B$. Thus $\widehat{\mathbf{\Gamma}}_{\leqslant}$ is locally hom-minimal, whence (i) holds.

To check (ii), let $x \in X$ and let $a, b \in \Gamma$ with $a \trianglelefteqslant b$. Then

$$\rho_\perp(\widehat{ab}) = 2 \preccurlyeq x(\widehat{ab}) \preccurlyeq 1 = \rho_\top(\widehat{ab}),$$

and therefore $(\rho_\perp, x, x, \rho_\top)(\widehat{ab}) \in \preccurlyeq_4$. So the relation $\leqslant$ on X is reflexive, and (ii) holds. Claim (iii) holds, by Lemma 4.4.5. ∎

4.4.11 Theorem *No unary algebra of type* $(2)_{\mathrm{VL}}$ *or* $(2)_{\mathrm{D}}$ *is fully dualisable.*

Proof First assume that $\underline{\mathbf{M}}$ is of type $(2)_{\mathrm{VL}}$. Suppose there is an alter ego $\underset{\sim}{\mathbf{M}}$ of $\underline{\mathbf{M}}$ that yields a full duality on $\mathcal{A} := \mathbb{ISP}(\underline{\mathbf{M}})$. By Lemmas 4.4.1 and 4.4.2, we can assume that $\preccurlyeq_4$ and $\bowtie$ are in the type of $\underset{\sim}{\mathbf{M}}$. Using Lemma 4.3.1 and Definition 4.3.3, there are algebras $\widehat{\mathbf{\Gamma}}_{\leqslant}$ and $\widehat{\mathbf{\Gamma}}_{\trianglelefteq}$ in $\mathcal{A}$. By Lemmas 4.3.2(i) and 4.3.6(ii), the set

$$X := \{\, x{\restriction}_{\widehat{\Gamma}_{\trianglelefteq}} \mid x \in \mathcal{A}(\widehat{\mathbf{\Gamma}}_{\leqslant}, \underline{\mathbf{M}}) \,\}$$

forms a closed substructure $\mathbf{X}$ of $\underset{\sim}{\mathbf{M}}^{\widehat{\Gamma}_{\trianglelefteq}}$. Since $\underset{\sim}{\mathbf{M}}$ yields a full duality on $\mathcal{A}$, there is an isomorphism $\varphi : \mathbf{X} \hookrightarrow\!\!\!\!\rightarrow \mathrm{D}(\mathbf{A})$, for some $\mathbf{A} \in \mathcal{A}$.

For each set S, an algebra $\mathbf{B} \leqslant \underline{\mathbf{M}}^S$ is said to be hom-minimal if every homomorphism from $\mathbf{B}$ to $\underline{\mathbf{M}}$ is the restriction of a projection. Every algebra in $\mathcal{A}$ is isomorphic to a hom-minimal algebra; see page 75. So we can assume that our algebra $\mathbf{A}$ is a hom-minimal subalgebra of $\underline{\mathbf{M}}^S$, for some set S.

The two projections $\rho_\perp : \widehat{\mathbf{\Gamma}}_{\trianglelefteq} \to \underline{\mathbf{M}}$ and $\rho_\top : \widehat{\mathbf{\Gamma}}_{\trianglelefteq} \to \underline{\mathbf{M}}$ belong to X. As we are assuming that the algebra $\mathbf{A}$ is hom-minimal, there must exist $s, t \in S$ such that $\varphi(\rho_\perp) = \rho_s$ and $\varphi(\rho_\top) = \rho_t$, where $\rho_s, \rho_t : \mathbf{A} \to \underline{\mathbf{M}}$ are projections. By Lemmas 4.4.10 and 4.4.8(ii), the relation

$$\leqslant \,:=\, \{\, (x, y) \in X^2 \mid (\rho_\perp, x, y, \rho_\top) \in \,\preccurlyeq_4 \,\}$$

is an order on X. Since φ is an isomorphism and $\preccurlyeq_4$ is in the type of $\underset{\sim}{\mathbf{M}}$, we have

$$\varphi(\leqslant) = \{\, (x, y) \in \mathcal{A}(\mathbf{A}, \underline{\mathbf{M}})^2 \mid (\rho_s, x, y, \rho_t) \in \,\preccurlyeq_4 \,\}.$$

Define

$$\leqslant^* := \{\, (x, y) \in \mathcal{L}_{\leqslant}(X)^2 \mid x \leqslant y \text{ and } (\rho_\perp, x^\downarrow, x, y^\downarrow, y, \rho_\top) \notin \bowtie \}.$$

Then $\varphi(\leqslant^*)$ is equal to

$$\{\, (x, y) \in \mathcal{L}_{\varphi(\leqslant)}\big(\mathcal{A}(\mathbf{A}, \underline{\mathbf{M}})\big)^2 \mid x \;\varphi(\leqslant)\; y \text{ and } (\rho_s, x^\downarrow, x, y^\downarrow, y, \rho_t) \notin \bowtie \}.$$

Using Lemma 4.4.10 and Lemma 4.4.9 twice, it follows that

$$\begin{aligned} \langle \mathcal{P}_{\widehat{\mathbf{\Gamma}}_{\leqslant}}; \xrightarrow[\top]{}_{\widehat{\mathbf{\Gamma}}_{\trianglelefteq}}, \xrightarrow[\top]{}\!\!\!\!\rightarrow_{\widehat{\mathbf{\Gamma}}_{\leqslant}} \rangle &\cong \langle \mathcal{L}_{\leqslant}(X); \leqslant^*, \leqslant \rangle \\ &\cong \big\langle \mathcal{L}_{\varphi(\leqslant)}\big(\mathcal{A}(\mathbf{A}, \underline{\mathbf{M}})\big); \varphi(\leqslant^*), \varphi(\leqslant) \big\rangle \\ &\cong \langle \mathcal{P}_{\mathbf{A}}; \xrightarrow[t]{}_{\mathbf{A}}, \xrightarrow[t]{}\!\!\!\!\rightarrow_{\mathbf{A}} \rangle. \end{aligned}$$

So $\langle \Gamma; \trianglelefteq, \leqslant \rangle$ is isomorphic to $\langle \mathcal{P}_{\mathbf{A}}; \xrightarrow[t]{}_{\mathbf{A}}, \xrightarrow[t]{}\!\!\!\!\rightarrow_{\mathbf{A}} \rangle$, by Lemma 4.4.10(iii). But this implies that $\leqslant$ is the transitive closure of $\trianglelefteq$, which is a contradiction. Thus $\underline{\mathbf{M}}$ is not fully dualisable.

Now assume that $\underline{\mathbf{M}}$ is of type $(2)_{\mathrm{D}}$ but not of type $(2)_{\mathrm{VL}}$. The algebra $\underline{\mathbf{M}}^{\sharp} = \langle\{0,1,2\}; F^{\sharp}\rangle$ was defined on page 116. We can show that $\underline{\mathbf{M}}$ is not fully dualisable, using Lemma 4.3.8, by following the proof given above with the algebras $\widehat{\mathbf{\Gamma}}^{\flat}_{\leqslant}$ and $\widehat{\mathbf{\Gamma}}^{\flat}_{\trianglelefteq}$ in $\mathcal{A}$, which are the reducts of the algebras $\widehat{\mathbf{\Gamma}}_{\leqslant}$ and $\widehat{\mathbf{\Gamma}}_{\trianglelefteq}$ in $\mathbb{ISP}(\underline{\mathbf{M}}^{\sharp})$. ∎

We have shown that every dualisable three-element unary algebra that has $\underline{\mathbf{M}}_{\mathrm{V}}$, $\underline{\mathbf{M}}_{\mathrm{L}}$ or $\underline{\mathbf{M}}_{\mathrm{D}}$ as a reduct is not fully dualisable. So $\neg$(i) implies $\neg$(ii) in the main theorem, completing the proof.

5

Dualisability and algebraic constructions

We show that there are many natural algebraic constructions under which dualisability is not always preserved. In particular, we find two dualisable algebras whose product is not dualisable.

Dualisability is such a natural algebraic property that it is tempting to suppose that it might interact well with natural algebraic constructions. This idea is supported by what is known about dualisability and one-point extensions. Davey and Knox [25] have shown that the one-point extensions of certain dualisable non-unary algebras are also dualisable. In this chapter, we shall prove that the one-point extension of a dualisable unary algebra is also dualisable.

Following this line of investigation, it is natural to ask whether each subalgebra of a dualisable algebra must be dualisable, whether each homomorphic image of a dualisable algebra must be dualisable, and whether a product of dualisable algebras must be dualisable. The subalgebra question was answered in the negative in Theorem 2.1.4: every non-dualisable unary algebra is a subalgebra of a dualisable algebra. In relation to the product question, it is known that each finite power of a dualisable algebra is dualisable. By the Independence Theorem, 1.4.1, any two finite algebras that generate the same quasi-variety must either both be dualisable or both be non-dualisable. The quasi-varieties $\mathbb{ISP}(\underline{\mathbf{M}})$ and $\mathbb{ISP}(\underline{\mathbf{M}}^n)$ coincide for each finite algebra $\underline{\mathbf{M}}$ and each $n \in \omega\backslash\{0\}$. So it follows that every finite power of a dualisable algebra is also dualisable.

The range of examples of dualisable and non-dualisable algebras that are useful for testing conjectures is growing. But there are presently not many naturally occurring varieties, containing both dualisable and non-dualisable algebras, for which there is a complete characterisation of dualisability. There are many varieties in which *every* finite algebra is known to be dualisable.

Every finite lattice is dualisable [29]. Similarly, each finite abelian group, Boolean algebra and semilattice is dualisable [17]. We also have complete descriptions of dualisability for some classes that are not varieties: for example, the class of graph algebras [23], and the class of three-element unary algebras (Theorem 3.0.1).

One variety of algebras for which there is an interesting characterisation of dualisability is that of commutative rings with identity. Consider an arbitrary finite commutative ring with identity $\underline{\mathbf{R}} = \langle R; +, \cdot, ^-, 0, 1\rangle$. An element $r \in R$ is said to be nilpotent if $r^n = 0$, for some $n \in \omega\backslash\{0\}$. The set J of all nilpotent elements of $\underline{\mathbf{R}}$ coincides with the Jacobson radical of $\underline{\mathbf{R}}$. We say that J is self annihilating if $rs = 0$, for all $r, s \in J$. Clark, Idziak, Sabourin, Szabó and Willard [14] have proven that the ring $\underline{\mathbf{R}}$ is dualisable if and only if its Jacobson radical J is self annihilating. It is now easy to check that the class of all dualisable commutative rings with identity is closed under taking subalgebras and finite products. It is also possible, though not quite so easy, to prove that every homomorphic image of a dualisable commutative ring with identity is dualisable.

Quackenbush and Szabó [59, 60] have studied dualisability within the variety of groups. A finite group is dualisable if all of its Sylow subgroups are cyclic. A finite group is non-dualisable if it has a non-abelian Sylow subgroup. These two results do not provide us with any examples of non-dualisable products of dualisable groups, or of dualisable groups with non-dualisable homomorphic images.

Within any congruence-distributive variety, a finite algebra is dualisable if and only if it has a near-unanimity term [29, 22]. However, within most naturally occurring congruence-distributive varieties, either all algebras have a near-unanimity term (for example, lattice-based varieties), or no non-trivial algebra has a near-unanimity term (for example, implication algebras [49]). There may be congruence-distributive varieties in which dualisability is not always preserved by taking products. In this chapter, we choose to focus our efforts instead on unary algebras and p-semilattices.

A **p-semilattice** is a bounded meet semilattice $\mathbf{P} = \langle P; \wedge, ^*, 0, 1\rangle$ with a unary operation * such that

$$a^* = \max\{\, b \in P \mid a \wedge b = 0 \,\},$$

for every $a \in P$. The class of p-semilattices forms a variety: a finite equational basis was found by Balbes and Horn [2]. A p-semilattice is said to be **boolean** if it satisfies the equation $x^{**} \approx x$. We will prove that a finite p-semilattice is dualisable if and only if it is boolean. So the class of dual-

construction	*preserves dualisability?*	*references*
subalgebra	✗	2.1.4, 5.3.2, 7.1.2
homomorphic image	✗	5.3.3, 5.3.4
retract	✗	5.3.3, 5.3.4
term retract	✓	5.2.1
finite power	✓	[17, 61, 30]
finite product	✗	5.3.8
finite coproduct	✗	5.3.8
one-point extension	?	[25], 5.1.11

Table 5.1

construction on unary algebras	*preserves dualisability?*	*references*
finite disjoint union	✗	5.3.8
one-point extension	✓	5.1.11
pointed one-point extension	✓	6.3.2
disjoint union with self	✓	5.1.10
disjoint union with finite algebra in q-v	✓	5.1.10
finite distant union	?	6.1.1

Table 5.2

isable p-semilattices is closed under taking subalgebras, homomorphic images and finite products. The class of non-dualisable p-semilattices is closed under taking finite products. But, as the following example shows, it is not closed under taking non-trivial subalgebras or non-trivial homomorphic images. The two-element p-semilattice $\underline{\mathbf{P}}_2 = \langle \{0, 1\}; \wedge, {}^*, 0, 1\rangle$ is dualisable, and the three-element p-semilattice $\underline{\mathbf{P}}_3 = \langle \{0, \frac{1}{2}, 1\}; \wedge, {}^*, 0, 1\rangle$ is not dualisable. However, there is a retraction $\gamma : \underline{\mathbf{P}}_3 \to \underline{\mathbf{P}}_2$, given by $\gamma(a) := a^{**}$.

In general, the dualisability of an algebra depends on the structure of the quasi-variety it generates. If $\underline{\mathbf{M}}$ and $\underline{\mathbf{N}}$ are arbitrary algebras of the same type, then the quasi-variety $\mathbb{ISP}(\underline{\mathbf{M}} \times \underline{\mathbf{N}})$ may be quite different from $\mathbb{ISP}(\underline{\mathbf{M}})$ and $\mathbb{ISP}(\underline{\mathbf{N}})$. So the algebra $\underline{\mathbf{M}} \times \underline{\mathbf{N}}$ might be non-dualisable, even if both $\underline{\mathbf{M}}$ and $\underline{\mathbf{N}}$ are dualisable. In this chapter, we find two dualisable unary algebras whose product is not dualisable. We also determine whether or not dualisability is preserved by various other algebraic constructions. The results are summarised in Tables 5.1 and 5.2, which also list some relevant results from other chapters. During our investigations, we demonstrate that algebras which generate the same variety do not have to share dualisability or non-dualisability.

Section 5.2 is based on a paper by both authors [26]. Section 5.3 is based on part of a paper by the first author [51], and Section 5.1 is an extension of another part of this paper.

5.1 Coproducts of unary algebras

In this chapter, we will show that the coproduct of two dualisable algebras can be non-dualisable. Before doing this, we need to clarify what we mean by the coproduct of two algebras of the same type. Let $\mathbf{A}$ and $\mathbf{B}$ be algebras of type F. There are three natural classes of algebras within which we can look for a coproduct of $\mathbf{A}$ and $\mathbf{B}$:

- the class of all algebras of type F;
- the variety generated by $\mathbf{A}$ and $\mathbf{B}$;
- the quasi-variety generated by $\mathbf{A}$ and $\mathbf{B}$.

We shall show how to construct coproducts of unary algebras within varieties. The following two general lemmas are part of the folklore of algebra.

5.1.1 Lemma *Let $\mathcal{A}$ be a quasi-variety of algebras, and let $\mathbf{A}$ be any algebra of the same type. There is a largest homomorphic image of $\mathbf{A}$ in $\mathcal{A}$.*

5.1.2 Lemma *Let $\mathcal{C}$ be a class of algebras of the same type, let $\mathcal{A}$ be a quasi-variety contained in $\mathcal{C}$ and let $\mathbf{A}, \mathbf{B} \in \mathcal{A}$. Assume that $\mathbf{C}$ is the coproduct of $\mathbf{A}$ and $\mathbf{B}$ in $\mathcal{C}$. Then the coproduct of $\mathbf{A}$ and $\mathbf{B}$ in $\mathcal{A}$ exists and it is the largest homomorphic image of $\mathbf{C}$ that belongs to $\mathcal{A}$.*

In fact, the previous lemma is a special case of a very general result from category theory: left adjoints preserve colimits [46].

Now assume that $\mathbf{A}$ and $\mathbf{B}$ are unary algebras of type F. The **disjoint union** $\mathbf{A} \dot\cup \mathbf{B}$ is obtained by putting $\mathbf{A}$ and $\mathbf{B}$ next to each other:

$$\mathbf{A} \dot\cup \mathbf{B} := \big\langle\, (A \times \{0\}) \cup (B \times \{1\});\ F^{\mathbf{A}\dot\cup\mathbf{B}} \,\big\rangle,$$

where

$$u^{\mathbf{A}\dot\cup\mathbf{B}}\big((a,0)\big) := (u^{\mathbf{A}}(a), 0) \quad\text{and}\quad u^{\mathbf{A}\dot\cup\mathbf{B}}\big((b,1)\big) := (u^{\mathbf{B}}(b), 1),$$

for all $u \in F$, $a \in A$ and $b \in B$.

Next, let $\mathbf{A}' \leqslant \mathbf{A}$ and $\mathbf{B}' \leqslant \mathbf{B}$, and let $\alpha : \mathbf{A}' \twoheadrightarrow \mathbf{D}$ and $\beta : \mathbf{B}' \twoheadrightarrow \mathbf{D}$ be surjective homomorphisms. The **amalgamated union** $\mathbf{A} \cup_{\alpha\beta} \mathbf{B}$ is obtained by pasting $\mathbf{A}$ and $\mathbf{B}$ together on $\mathbf{D}$:

$$\mathbf{A} \cup_{\alpha\beta} \mathbf{B} := (\mathbf{A} \mathbin{\dot\cup} \mathbf{B})/\theta_{\alpha\beta},$$

where $\theta_{\alpha\beta}$ is the congruence on $\mathbf{A} \mathbin{\dot\cup} \mathbf{B}$ whose non-trivial blocks are precisely those of the form

$$\big(\alpha^{-1}(d) \times \{0\}\big) \cup \big(\beta^{-1}(d) \times \{1\}\big),$$

for some $d \in D$.

The algebra built most freely from $\mathbf{A}$ and $\mathbf{B}$ is the disjoint union $\mathbf{A} \mathbin{\dot\cup} \mathbf{B}$. Clearly, $\mathbf{A} \mathbin{\dot\cup} \mathbf{B}$ is the coproduct of $\mathbf{A}$ and $\mathbf{B}$ in the class of all algebras of type F. Now let $\mathcal{V}$ be any variety containing $\mathbf{A}$ and $\mathbf{B}$. We have to be a little more careful when constructing the coproduct of $\mathbf{A}$ and $\mathbf{B}$ in $\mathcal{V}$. The disjoint union $\mathbf{A} \mathbin{\dot\cup} \mathbf{B}$ will not belong to $\mathcal{V}$ if there is a constant equation $\tau(x) \approx \tau(y)$ in the theory of $\mathcal{V}$, for some unary term τ of type F. We shall see that the coproduct of $\mathbf{A}$ and $\mathbf{B}$ in $\mathcal{V}$ is generally an amalgamated union of $\mathbf{A}$ and $\mathbf{B}$.

The variety $\mathcal{V}$ is determined by the one-variable equations $\sigma(x) \approx \tau(x)$ and the constant equations $\tau(x) \approx \tau(y)$ in its theory. So the equational theory of $\mathcal{V}$ is the same as the equational theory of the free algebra $\mathbf{F}_{\mathcal{V}}(2)$. Define $F_{\mathcal{V}}(0)$ to be the set of all $k \in F_{\mathcal{V}}(2)$ such that k is the value of a constant term function of $\mathbf{F}_{\mathcal{V}}(2)$. Then $F_{\mathcal{V}}(0)$ is a subuniverse of $\mathbf{F}_{\mathcal{V}}(2)$. If $F_{\mathcal{V}}(0)$ is non-empty, then the algebra $\mathbf{F}_{\mathcal{V}}(0)$ is an initial object in the category $\mathcal{V}$: for all $\mathbf{C} \in \mathcal{V}$, there is a unique homomorphism $\imath_{\mathbf{C}} : \mathbf{F}_{\mathcal{V}}(0) \to \mathbf{C}$.

5.1.3 Lemma *Let $\mathcal{V}$ be a variety of unary algebras and let $\mathbf{A}, \mathbf{B} \in \mathcal{V}$.*

(i) *Assume that $F_{\mathcal{V}}(0)$ is empty. Then the disjoint union $\mathbf{A} \mathbin{\dot\cup} \mathbf{B}$ is the coproduct of $\mathbf{A}$ and $\mathbf{B}$ in $\mathcal{V}$.*

(ii) *Assume that $F_{\mathcal{V}}(0)$ is non-empty. Define the congruence θ on the algebra $\mathbf{F}_{\mathcal{V}}(0)$ by $\theta := \ker(\imath_{\mathbf{A}}) \vee \ker(\imath_{\mathbf{B}})$. Let $\alpha : \imath_{\mathbf{A}}\big(\mathbf{F}_{\mathcal{V}}(0)\big) \twoheadrightarrow \mathbf{F}_{\mathcal{V}}(0)/\theta$ and $\beta : \imath_{\mathbf{B}}\big(\mathbf{F}_{\mathcal{V}}(0)\big) \twoheadrightarrow \mathbf{F}_{\mathcal{V}}(0)/\theta$ be the natural homomorphisms. Then the amalgamated union $\mathbf{A} \cup_{\alpha\beta} \mathbf{B}$ is the coproduct of $\mathbf{A}$ and $\mathbf{B}$ in $\mathcal{V}$.*

Proof Define $\mathrm{Th}(\mathcal{V})$ to be the equational theory of $\mathcal{V}$. We will be using two easy facts. One, the variety $\mathcal{V}$ is determined by the one-variable equations and the constant equations in $\mathrm{Th}(\mathcal{V})$. Two, an algebra satisfies the one-variable equations in $\mathrm{Th}(\mathcal{V})$ if and only if each of its one-generated subalgebras belongs to $\mathcal{V}$.

First assume that $F_{\mathcal{V}}(0)$ is empty. Since $\mathbf{A}, \mathbf{B} \in \mathcal{V}$, every one-generated subalgebra of $\mathbf{A} \mathbin{\dot\cup} \mathbf{B}$ belongs to $\mathcal{V}$. So $\mathbf{A} \mathbin{\dot\cup} \mathbf{B}$ satisfies all the one-variable equations in $\mathrm{Th}(\mathcal{V})$. As there are no constant equations in $\mathrm{Th}(\mathcal{V})$, it follows that $\mathbf{A} \,\dot{\cup}\, \mathbf{B}$ is a member of $\mathcal{V}$. So $\mathbf{A} \,\dot{\cup}\, \mathbf{B}$ is the coproduct of $\mathbf{A}$ and $\mathbf{B}$ in $\mathcal{V}$.

Now assume that $F_{\mathcal{V}}(0)$ is non-empty. First we want to check that the amalgamated union $\mathbf{C} := \mathbf{A} \cup_{\alpha\beta} \mathbf{B}$ belongs to $\mathcal{V}$. Every one-generated subalgebra of $\mathbf{C}$ is a subalgebra of a homomorphic image of $\mathbf{A}$ or $\mathbf{B}$. So $\mathbf{C}$ satisfies all the one-variable equations in $\mathrm{Th}(\mathcal{V})$.

Now choose a unary term τ such that $\tau(x) \approx \tau(y)$ belongs to $\mathrm{Th}(\mathcal{V})$. We want to show that $\tau^{\mathbf{C}}$ is a constant term function of $\mathbf{C}$. There is some $k \in F_{\mathcal{V}}(0)$ such that k is the value of the constant term function $\tau^{\mathbf{F}_{\mathcal{V}}(2)}$ of $\mathbf{F}_{\mathcal{V}}(2)$. Since $\mathbf{A}, \mathbf{B} \in \mathcal{V}$, we know that $\tau^{\mathbf{A}}$ and $\tau^{\mathbf{B}}$ are constant, with values $\imath_{\mathbf{A}}(k)$ and $\imath_{\mathbf{B}}(k)$, respectively. To see that $\tau^{\mathbf{C}}$ is constant, it is enough to show that $\imath_{\mathbf{A}}(k)$ and $\imath_{\mathbf{B}}(k)$ are identified in $\mathbf{C} = \mathbf{A} \cup_{\alpha\beta} \mathbf{B}$. More precisely, we want to show that $(\imath_{\mathbf{A}}(k), 0)\ \theta_{\alpha\beta}\ (\imath_{\mathbf{B}}(k), 1)$. But this holds, since $\mathbf{F}_{\mathcal{V}}(0)/\theta \in \mathcal{V}$, and therefore

$$\begin{aligned}
\alpha\big(\imath_{\mathbf{A}}(k)\big) &= \alpha\big(\tau^{\mathbf{A}}(\imath_{\mathbf{A}}(k))\big) = \tau^{\mathbf{F}_{\mathcal{V}}(0)/\theta}\big(\alpha(\imath_{\mathbf{A}}(k))\big) \\
&= \tau^{\mathbf{F}_{\mathcal{V}}(0)/\theta}\big(\beta(\imath_{\mathbf{B}}(k))\big) = \beta\big(\tau^{\mathbf{B}}(\imath_{\mathbf{B}}(k))\big) \\
&= \beta\big(\imath_{\mathbf{B}}(k)\big).
\end{aligned}$$

So $\tau^{\mathbf{C}}$ is a constant term function of $\mathbf{C}$, whence $\mathbf{C}$ satisfies all the constant equations in $\mathrm{Th}(\mathcal{V})$. We have shown that $\mathbf{C}$ belongs to $\mathcal{V}$.

We shall now show that $\mathbf{C}$ is the largest homomorphic image of $\mathbf{A} \,\dot{\cup}\, \mathbf{B}$ that belongs to $\mathcal{V}$. By Lemma 5.1.2, it will then follow that $\mathbf{C}$ is the coproduct of $\mathbf{A}$ and $\mathbf{B}$ in $\mathcal{V}$. Let $\varphi : \mathbf{A} \,\dot{\cup}\, \mathbf{B} \twoheadrightarrow \mathbf{D}$ be a surjective homomorphism such that $\mathbf{D} \in \mathcal{V}$. We just need to check that $\theta_{\alpha\beta} \leqslant \ker(\varphi)$ in the congruence lattice $\mathbf{Con}(\mathbf{A} \,\dot{\cup}\, \mathbf{B})$.

Let $\eta_{\mathbf{A}} : \mathbf{A} \hookrightarrow \mathbf{A} \,\dot{\cup}\, \mathbf{B}$ and $\eta_{\mathbf{B}} : \mathbf{B} \hookrightarrow \mathbf{A} \,\dot{\cup}\, \mathbf{B}$ be the natural embeddings. Since $\varphi \circ \eta_{\mathbf{A}} \circ \imath_{\mathbf{A}} : \mathbf{F}_{\mathcal{V}}(0) \to \mathbf{D}$ is a homomorphism and $\mathbf{D} \in \mathcal{V}$, we have $\imath_{\mathbf{D}} = \varphi \circ \eta_{\mathbf{A}} \circ \imath_{\mathbf{A}}$. This implies that $\ker(\imath_{\mathbf{A}}) \leqslant \ker(\imath_{\mathbf{D}})$ and, using a similar argument, $\ker(\imath_{\mathbf{B}}) \leqslant \ker(\imath_{\mathbf{D}})$. So we have

$$\theta = \ker(\imath_{\mathbf{A}}) \vee \ker(\imath_{\mathbf{B}}) \leqslant \ker(\imath_{\mathbf{D}}).$$

To prove that $\theta_{\alpha\beta} \leqslant \ker(\varphi)$, we will show that every non-trivial block of $\theta_{\alpha\beta}$ is contained in a block of $\ker(\varphi)$. Let S be a non-trivial block of $\theta_{\alpha\beta}$. Then

$$S = \big(\alpha^{-1}(k/\theta) \times \{0\}\big) \cup \big(\beta^{-1}(k/\theta) \times \{1\}\big),$$

for some $k \in F_{\mathcal{V}}(0)$. Now let $a \in \alpha^{-1}(k/\theta)$. There exists $\ell \in F_{\mathcal{V}}(0)$ such that $a = \imath_{\mathbf{A}}(\ell)$. Since $\alpha : \imath_{\mathbf{A}}(\mathbf{F}_{\mathcal{V}}(0)) \twoheadrightarrow \mathbf{F}_{\mathcal{V}}(0)/\theta$ is the natural homomorphism, we have

$$\ell/\theta = \alpha(\imath_{\mathbf{A}}(\ell)) = \alpha(a) = k/\theta.$$

We have already shown that $\theta \leqslant \ker(\imath_{\mathbf{D}})$. So this implies that $\imath_{\mathbf{D}}(\ell) = \imath_{\mathbf{D}}(k)$. Since $\varphi \circ \eta_{\mathbf{A}} \circ \imath_{\mathbf{A}} = \imath_{\mathbf{D}}$, it follows that

$$\varphi\big((a, 0)\big) = \varphi \circ \eta_{\mathbf{A}} \circ \imath_{\mathbf{A}}(\ell) = \imath_{\mathbf{D}}(\ell) = \imath_{\mathbf{D}}(k).$$

So we obtain

$$\varphi\big(\alpha^{-1}(k/\theta) \times \{0\}\big) = \{\imath_{\mathbf{D}}(k)\} \quad \text{and} \quad \varphi\big(\beta^{-1}(k/\theta) \times \{1\}\big) = \{\imath_{\mathbf{D}}(k)\},$$

using symmetry. Thus $|\varphi(S)| = 1$, and therefore $\theta_{\alpha\beta} \leqslant \ker(\varphi)$. Hence the algebra $\mathbf{C} = \mathbf{A} \cup_{\alpha\beta} \mathbf{B}$ is the coproduct of $\mathbf{A}$ and $\mathbf{B}$ in $\mathcal{V}$. ∎

In Section 5.3, we shall find a pair of dualisable unary algebras $\underline{\mathbf{K}}$ and $\underline{\mathbf{L}}$, with $\underline{\mathbf{L}} \in \mathbb{HSP}(\underline{\mathbf{K}})$, such that the disjoint union $\underline{\mathbf{K}} \dot\cup \underline{\mathbf{L}}$ is non-dualisable. To finish this section, we give an example of a similar coproduct construction on unary algebras that *does* preserve dualisability.

Consider a finite unary algebra $\underline{\mathbf{M}}$, and let $\underline{\mathbf{N}}$ be a finite algebra in $\mathbb{ISP}(\underline{\mathbf{M}})$. The quasi-varieties $\mathbb{ISP}(\underline{\mathbf{M}})$ and $\mathbb{ISP}(\underline{\mathbf{M}} \dot\cup \underline{\mathbf{N}})$ are not necessarily the same. Indeed, they must be different if $\underline{\mathbf{M}}$ has a constant term function. Nevertheless, we shall prove that, if $\underline{\mathbf{M}}$ is dualisable, then the disjoint union $\underline{\mathbf{M}} \dot\cup \underline{\mathbf{N}}$ is also dualisable. This is true even if $\underline{\mathbf{N}}$ is non-dualisable.

One important special case of this result is when $|N| = 1$. Let $\underline{\mathbf{1}}$ be a one-element algebra of the same type as $\underline{\mathbf{M}}$. Then the **one-point extension of $\underline{\mathbf{M}}$** is the disjoint union $\underline{\mathbf{M}} \dot\cup \underline{\mathbf{1}}$. So it will follow that the one-point extension of a dualisable unary algebra is also dualisable. A similar result, that the one-point extension of a finitely dualisable unary algebra is also finitely dualisable, was proved directly by Clark, Davey and Pitkethly [13].

We begin by investigating the precise difference between the quasi-varieties $\mathbb{ISP}(\underline{\mathbf{M}})$ and $\mathbb{ISP}(\underline{\mathbf{M}} \dot\cup \underline{\mathbf{N}})$.

5.1.4 Lemma *Let $\underline{\mathbf{M}}$ and $\underline{\mathbf{N}}$ be finite unary algebras, with $\underline{\mathbf{N}} \in \mathbb{ISP}(\underline{\mathbf{M}})$. Then every connected algebra in $\mathbb{ISP}(\underline{\mathbf{M}} \dot\cup \underline{\mathbf{N}})$ belongs to $\mathbb{ISP}(\underline{\mathbf{M}})$.*

Proof Assume that $M \cap N = \varnothing$. Then we can work with the union $\underline{\mathbf{M}} \cup \underline{\mathbf{N}}$, rather than the disjoint union. Let $\mathbf{C} \in \mathbb{ISP}(\underline{\mathbf{M}} \cup \underline{\mathbf{N}})$, with $\mathbf{C}$ connected. We want to show that $\mathbf{C}$ is separated by homomorphisms into $\underline{\mathbf{M}}$. So assume that $a, b \in C$ with $a \neq b$. Since $\mathbf{C}$ is separated by homomorphisms into $\underline{\mathbf{M}} \cup \underline{\mathbf{N}}$, there is a homomorphism $x : \mathbf{C} \to \underline{\mathbf{M}} \cup \underline{\mathbf{N}}$ such that $x(a) \neq x(b)$. As $\mathbf{C}$ is connected, we must have $x(C) \subseteq M$ or $x(C) \subseteq N$. We can assume that $x(C) \subseteq N$. Since $\underline{\mathbf{N}} \in \mathbb{ISP}(\underline{\mathbf{M}})$, there exists a homomorphism $y : \underline{\mathbf{N}} \to \underline{\mathbf{M}}$ with $y(x(a)) \neq y(x(b))$. Thus $y \circ x : \mathbf{C} \to \underline{\mathbf{M}}$ separates a and b. Using the $\mathbb{ISP}$ Theorem, 1.1.1, it now follows that $\mathbf{C} \in \mathbb{ISP}(\underline{\mathbf{M}})$. ∎

5.1.5 Corollary *Let $\underline{\mathbf{M}}$ be a finite unary algebra and let $\underline{\mathbf{1}}$ be a one-element algebra of the same type as $\underline{\mathbf{M}}$.*

(i) *An algebra $\mathbf{A}$ of the same type as $\underline{\mathbf{M}}$ belongs to $\mathbb{ISP}((\underline{\mathbf{M}} \mathbin{\dot\cup} \underline{\mathbf{1}}) \mathbin{\dot\cup} \underline{\mathbf{1}})$ if and only if every connected component of $\mathbf{A}$ belongs to $\mathbb{ISP}(\underline{\mathbf{M}})$.*

(ii) *The class $\mathbb{ISP}((\underline{\mathbf{M}} \mathbin{\dot\cup} \underline{\mathbf{1}}) \mathbin{\dot\cup} \underline{\mathbf{1}})$ is the largest quasi-variety that is generated by a disjoint union of algebras from $\mathbb{ISP}(\underline{\mathbf{M}})$.*

Proof Let $\mathbf{A}$ be of the same type as $\underline{\mathbf{M}}$. Assume that $\mathbf{A} \in \mathbb{ISP}((\underline{\mathbf{M}} \mathbin{\dot\cup} \underline{\mathbf{1}}) \mathbin{\dot\cup} \underline{\mathbf{1}})$. Using the previous lemma twice, every connected component of $\mathbf{A}$ belongs to $\mathbb{ISP}(\underline{\mathbf{M}})$. Conversely, if every connected component of $\mathbf{A}$ belongs to $\mathbb{ISP}(\underline{\mathbf{M}})$, then $\mathbf{A}$ is separated by homomorphisms into $(\underline{\mathbf{M}} \mathbin{\dot\cup} \underline{\mathbf{1}}) \mathbin{\dot\cup} \underline{\mathbf{1}}$. So (i) holds.

For (ii), let $\mathbf{B}$ be a disjoint union of algebras from $\mathbb{ISP}(\underline{\mathbf{M}})$. Then $\mathbf{B}$ belongs to $\mathbb{ISP}((\underline{\mathbf{M}} \mathbin{\dot\cup} \underline{\mathbf{1}}) \mathbin{\dot\cup} \underline{\mathbf{1}})$, by (i). So the quasi-variety generated by $\mathbf{B}$ is contained in the quasi-variety $\mathbb{ISP}((\underline{\mathbf{M}} \mathbin{\dot\cup} \underline{\mathbf{1}}) \mathbin{\dot\cup} \underline{\mathbf{1}})$. ∎

The following lemma compares quasi-varieties that are generated by disjoint unions of $\underline{\mathbf{M}}$'s and $\underline{\mathbf{1}}$'s.

5.1.6 Lemma *Let $\underline{\mathbf{M}}$ be a finite unary algebra and let $\underline{\mathbf{1}}$ be a one-element algebra of the same type as $\underline{\mathbf{M}}$. Then*

$$\mathbb{ISP}(\underline{\mathbf{M}}) \subseteq \mathbb{ISP}(\underline{\mathbf{M}} \mathbin{\dot\cup} \underline{\mathbf{M}}) \subseteq \mathbb{ISP}(\underline{\mathbf{M}} \mathbin{\dot\cup} \underline{\mathbf{1}}) \subseteq \mathbb{ISP}((\underline{\mathbf{M}} \mathbin{\dot\cup} \underline{\mathbf{1}}) \mathbin{\dot\cup} \underline{\mathbf{1}}), \quad \textit{where}$$

(i) *the first inclusion is an equality if and only if there is no element of M that is fixed by every endomorphism of $\underline{\mathbf{M}}$,*

(ii) *each of the second and third inclusions is an equality if and only if $\underline{\mathbf{M}}$ has a one-element subalgebra,*

(iii) *all four quasi-varieties are equal if and only if $\underline{\mathbf{M}}$ has at least two one-element subalgebras.*

Proof This lemma is easy to prove using the following consequence of the $\mathbb{ISP}$ Theorem, 1.1.1: for all finite algebras $\mathbf{A}$ and $\mathbf{B}$, we have $\mathbb{ISP}(\mathbf{A}) \subseteq \mathbb{ISP}(\mathbf{B})$ if and only if $\mathbf{A}$ is separated by homomorphisms into $\mathbf{B}$. ∎

The next easy lemma slots $\mathbb{ISP}(\underline{\mathbf{M}} \mathbin{\dot\cup} \underline{\mathbf{N}})$ into the chain of quasi-varieties given in the preceding lemma, in the case that $|N| > 1$.

5.1.7 Lemma *Let $\underline{\mathbf{M}}$ be a finite unary algebra and let $\underline{\mathbf{N}}$ be a finite non-trivial algebra in $\mathbb{ISP}(\underline{\mathbf{M}})$. Then $\mathbb{ISP}(\underline{\mathbf{M}}) \subseteq \mathbb{ISP}(\underline{\mathbf{M}} \mathbin{\dot\cup} \underline{\mathbf{N}}) \subseteq \mathbb{ISP}(\underline{\mathbf{M}} \mathbin{\dot\cup} \underline{\mathbf{M}})$.*

To illustrate the previous collection of results, we consider a particular three-element unary algebra.

5.1.8 Example Define the unary algebra

$$\underline{\mathbf{M}} := \langle \{0,1,2\}; 001, 111 \rangle,$$

and let $\underline{\mathbf{N}}$ be the subalgebra of $\underline{\mathbf{M}}$ on the set $N := \{0,1\}$. Then there is no homomorphism from $\underline{\mathbf{M}}$ into $\underline{\mathbf{N}}$; indeed, the only endomorphism of $\underline{\mathbf{M}}$ is the identity. With the help of the previous two lemmas, it is now easy to check that

$$\begin{aligned} \mathbb{ISP}(\underline{\mathbf{M}}) \subset \mathbb{ISP}(\underline{\mathbf{M}} \,\dot\cup\, \underline{\mathbf{N}}) &\subset \mathbb{ISP}(\underline{\mathbf{M}} \,\dot\cup\, \underline{\mathbf{M}}) \\ &\subset \mathbb{ISP}(\underline{\mathbf{M}} \,\dot\cup\, \underline{\mathbf{1}}) \subset \mathbb{ISP}((\underline{\mathbf{M}} \,\dot\cup\, \underline{\mathbf{1}}) \,\dot\cup\, \underline{\mathbf{1}}), \end{aligned}$$

where every inclusion is proper. To see the differences between these quasi-varieties, consider an algebra $\mathbf{A}$ of the same type as $\underline{\mathbf{M}}$. By Corollary 5.1.5, we know that $\mathbf{A} \in \mathbb{ISP}((\underline{\mathbf{M}} \,\dot\cup\, \underline{\mathbf{1}}) \,\dot\cup\, \underline{\mathbf{1}})$ if and only if every connected component of $\mathbf{A}$ belongs to $\mathbb{ISP}(\underline{\mathbf{M}})$. Now assume that $\mathbf{A} \in \mathbb{ISP}((\underline{\mathbf{M}} \,\dot\cup\, \underline{\mathbf{1}}) \,\dot\cup\, \underline{\mathbf{1}})$. Then, as explained below, we have:

(i) $\mathbf{A} \in \mathbb{ISP}(\underline{\mathbf{M}} \,\dot\cup\, \underline{\mathbf{1}})$ if and only if at most one connected component of $\mathbf{A}$ has only one element;

(ii) $\mathbf{A} \in \mathbb{ISP}(\underline{\mathbf{M}} \,\dot\cup\, \underline{\mathbf{M}})$ if and only if $\mathbf{A}$ is trivial or $\mathbf{A}$ has no one-element connected components;

(iii) $\mathbf{A} \in \mathbb{ISP}(\underline{\mathbf{M}} \,\dot\cup\, \underline{\mathbf{N}})$ if and only if $\mathbf{A} \in \mathbb{ISP}(\underline{\mathbf{M}} \,\dot\cup\, \underline{\mathbf{M}})$ and at most one connected component of $\mathbf{A}$ has a subalgebra isomorphic to $\underline{\mathbf{M}}$;

(iv) $\mathbf{A} \in \mathbb{ISP}(\underline{\mathbf{M}})$ if and only if $\mathbf{A}$ is connected.

Claims (i) and (ii) follow since $\underline{\mathbf{M}}$ has no one-element subalgebras. Claim (iv) follows since $\underline{\mathbf{M}}$ has a constant term function. Claim (iii) requires some knowledge of the structure of the algebras in $\mathbb{ISP}(\underline{\mathbf{M}})$. In particular, claim (iii) uses the fact: for every non-trivial algebra $\mathbf{C} \in \mathbb{ISP}(\underline{\mathbf{M}})$, if $\mathbf{C}$ does not have a subalgebra isomorphic to $\underline{\mathbf{M}}$, then there exists a homomorphism from $\mathbf{C}$ into $\underline{\mathbf{N}}$.

We now turn to proving that the finite unary algebra $\underline{\mathbf{M}} \,\dot\cup\, \underline{\mathbf{N}}$ is dualisable, whenever $\underline{\mathbf{M}}$ is dualisable and $\underline{\mathbf{N}} \in \mathbb{ISP}(\underline{\mathbf{M}})$. The following preliminary lemma is similar to Lemma 3.1.5, which was used to obtain the Petal Duality Lemma, 3.1.6.

5.1.9 Lemma *Let $\underline{\mathbf{M}}$ be a finite unary algebra and let $\mathbf{A}$ belong to the quasi-variety $\mathcal{A} := \mathbb{ISP}(\underline{\mathbf{M}})$. Assume that $\alpha : \mathcal{A}(\mathbf{A}, \underline{\mathbf{M}}) \to M$ has a finite support and that α agrees with an evaluation on each subset of $\mathcal{A}(\mathbf{A}, \underline{\mathbf{M}})$ with at most four elements. Then there is a connected component $\mathbf{C}$ of $\mathbf{A}$ such that C is a support for α.*

Proof There is a finite non-empty support S for α. Let $\mathbf{C}_1, \dots, \mathbf{C}_n$ be the connected components of $\mathbf{A}$ that contain a member of S, where $n \in \omega \backslash \{0\}$, and define the subuniverse B of $\mathbf{A}$ by $B := C_1 \cup \cdots \cup C_n$.

We can assume that the map α is not constant. So there are $y, z \in \mathcal{A}(\mathbf{A}, \underline{\mathbf{M}})$ such that $\alpha(y) \neq \alpha(z)$. Define the sequence $y_0, \dots, y_n$ of homomorphisms in $\mathcal{A}(\mathbf{A}, \underline{\mathbf{M}})$ by

$$y_0 := y \quad \text{and} \quad y_{i+1} := y_i{\restriction}_{A \backslash C_{i+1}} \cup z{\restriction}_{C_{i+1}},$$

for all $i \in \{0, \dots, n-1\}$. Then $y_n = y{\restriction}_{A \backslash B} \cup z{\restriction}_B$. As $S \subseteq B$ and S is a support for α, we have

$$\alpha(y_0) = \alpha(y) \neq \alpha(z) = \alpha(y_n).$$

This implies that $\alpha(y_j) \neq \alpha(y_{j+1})$, for some $j \in \{0, \dots, n-1\}$.

To prove that C_{j+1} is a support for α, let $w, x \in \mathcal{A}(\mathbf{A}, \underline{\mathbf{M}})$ such that $w{\restriction}_{C_{j+1}} = x{\restriction}_{C_{j+1}}$. There is some $a \in A$ such that α is given by evaluation at a on $\{w, x, y_j, y_{j+1}\}$. So

$$y_j(a) = \alpha(y_j) \neq \alpha(y_{j+1}) = y_{j+1}(a),$$

and therefore $a \in C_{j+1}$. Thus

$$\alpha(w) = w(a) = x(a) = \alpha(x),$$

whence C_{j+1} is a support for α. ∎

5.1.10 Theorem *Let $\underline{\mathbf{M}}$ and $\underline{\mathbf{N}}$ be finite unary algebras, with $\underline{\mathbf{N}} \in \mathbb{ISP}(\underline{\mathbf{M}})$. If $\underline{\mathbf{M}}$ is dualisable, then $\underline{\mathbf{M}} \dot\cup \underline{\mathbf{N}}$ is dualisable.*

Proof Assume that $M \cap N = \varnothing$ and that $\underline{\mathbf{M}}$ is dualisable. We shall show that the union $\underline{\mathbf{M}} \cup \underline{\mathbf{N}}$ is dualisable. Define the two quasi-varieties

$$\mathcal{A} := \mathbb{ISP}(\underline{\mathbf{M}}) \quad \text{and} \quad \mathcal{B} := \mathbb{ISP}(\underline{\mathbf{M}} \cup \underline{\mathbf{N}}).$$

Then $\mathcal{A}$ is contained in $\mathcal{B}$.

Let $\mathcal{A}(\underline{\mathbf{N}}, \underline{\mathbf{M}}) = \{e_1, \dots, e_k\}$, where $k \in \omega$. For every $i \in \{1, \dots, k\}$, define the endomorphism

$$\overline{e}_i : \underline{\mathbf{M}} \cup \underline{\mathbf{N}} \to \underline{\mathbf{M}} \cup \underline{\mathbf{N}} \quad \text{by} \quad \overline{e}_i := \mathrm{id}_M \cup e_i.$$

Then $\overline{e}_i(M \cup N) \subseteq M$, for each $i \in \{1, \dots, k\}$. Since $\underline{\mathbf{N}} \in \mathbb{ISP}(\underline{\mathbf{M}})$, the endomorphisms $\overline{e}_1, \dots, \overline{e}_k$ of $\underline{\mathbf{M}} \cup \underline{\mathbf{N}}$ separate the elements of N. So, in the case that $|N| > 1$, we have $k > 0$.

Now let $\mathbf{B} \in \mathcal{B}$ and let $\beta : \mathcal{B}(\mathbf{B}, \underline{\mathbf{M}} \cup \underline{\mathbf{N}}) \to M \cup N$ be a brute-force morphism. Then β preserves every endomorphism of $\underline{\mathbf{M}} \cup \underline{\mathbf{N}}$, since the graph of an endomorphism is an algebraic relation on $\underline{\mathbf{M}} \cup \underline{\mathbf{N}}$. Using the Brute Force Lemma, 1.4.5, we know that β has a finite support and is locally an evaluation.

We will know that $\underline{\mathbf{M}} \cup \underline{\mathbf{N}}$ is dualisable once we have proved that β is an evaluation. The proof splits up into three cases.

Case 1: β is constant and $|N| = 1$. Let 0 denote the unique element of $\underline{\mathbf{N}}$, and let $z : \mathbf{B} \to \underline{\mathbf{M}} \cup \underline{\mathbf{N}}$ denote the constant homomorphism with value 0. The brute-force morphism β must preserve the unary algebraic relation $N = \{0\}$ on $\underline{\mathbf{M}} \cup \underline{\mathbf{N}}$. So $\beta(z) = 0$, and therefore 0 is the value of the constant map β.

The map β preserves the unary algebraic relation M on $\underline{\mathbf{M}} \cup \underline{\mathbf{N}}$. Since the value of β is $0 \notin M$, this implies that $x(B) \nsubseteq M$, for all $x \in \mathcal{B}(\mathbf{B}, \underline{\mathbf{M}} \cup \underline{\mathbf{N}})$. So there are no homomorphisms from $\mathbf{B}$ into $\underline{\mathbf{M}}$. But we know that every connected component of $\mathbf{B}$ belongs to $\mathbb{ISP}(\underline{\mathbf{M}})$, by Lemma 5.1.4. It follows that there is some $b \in B$ that determines a one-element connected component of $\mathbf{B}$, and that $\underline{\mathbf{M}}$ has no one-element subalgebras. For all $x \in \mathcal{B}(\mathbf{B}, \underline{\mathbf{M}} \cup \underline{\mathbf{N}})$, we have $\beta(x) = 0 = x(b)$. Thus β is an evaluation.

Case 2: β is constant and $|N| > 1$. We must have $k > 0$. Let m denote the value of β in $M \cup N$. For any $x \in \mathcal{B}(\mathbf{B}, \underline{\mathbf{M}} \cup \underline{\mathbf{N}})$, we have

$$m = \beta(\bar{e}_1 \circ x) = \bar{e}_1(\beta(x)) \in M,$$

since β preserves the endomorphism $\bar{e}_1$ of $\underline{\mathbf{M}} \cup \underline{\mathbf{N}}$. Thus the value m of β belongs to M.

Define the set B_N to be the union of all the connected components of $\mathbf{B}$ that have at least one homomorphism into $\underline{\mathbf{N}}$. If B_N is non-empty, then it forms a subalgebra $\mathbf{B}_N$ of $\mathbf{B}$. In this case, there is a homomorphism $z_N : \mathbf{B}_N \to \underline{\mathbf{N}}$. If B_N is empty, we will just define $z_N : B_N \to N$ to be the empty map. As β preserves the unary relation N and has constant value $m \in M$, there is no homomorphism from $\mathbf{B}$ into $\underline{\mathbf{N}}$. So $B_N \neq B$.

Now define $B_M := B \backslash B_N$. Then B_M is the union of all the connected components of $\mathbf{B}$ that do not have any homomorphisms into $\underline{\mathbf{N}}$. Since $B_N \neq B$, the set B_M is the universe of a subalgebra $\mathbf{B}_M$ of $\mathbf{B}$. As $\mathbf{B}$ is separated by homomorphisms into $\underline{\mathbf{M}} \cup \underline{\mathbf{N}}$, it follows that $\mathbf{B}_M$ is separated by homomorphisms into $\underline{\mathbf{M}}$. So $\mathbf{B}_M \in \mathbb{ISP}(\underline{\mathbf{M}}) = \mathcal{A}$, by the $\mathbb{ISP}$ Theorem, 1.1.1.

Now define the constant map $\alpha : \mathcal{A}(\mathbf{B}_M, \underline{\mathbf{M}}) \to M$ with value m. We shall prove that α is a brute-force morphism. Since $\underline{\mathbf{M}}$ is dualisable, it will then follow that α is given by evaluation at some element of $B_M \subseteq B$. We shall then show that β is given by evaluation at the same element.

Clearly, the constant map α has a finite support. To prove that α is locally an evaluation, let Y be a finite non-empty subset of $\mathcal{A}(\mathbf{B}_M, \underline{\mathbf{M}})$. Define

$$Y^+ := \{y \cup z_N \mid y \in Y\} \subseteq \mathcal{B}(\mathbf{B}, \underline{\mathbf{M}} \cup \underline{\mathbf{N}}).$$

Since β is locally an evaluation, there is some $b \in B$ such that β is given by evaluation at b on Y^+. For all $y \in Y$, we have

$$(y \cup z_N)(b) = \beta(y \cup z_N) = m \in M.$$

Since $z_N(B_N) \subseteq N$, this tells us that $b \in B_M$. Now, to see that α is given by evaluation at b on Y, let $y \in Y$. Then

$$\alpha(y) = m = \beta(y \cup z_N) = (y \cup z_N)(b) = y(b).$$

So α is locally an evaluation. Thus α is a brute-force morphism, by the Brute Force Lemma.

As $\underline{\mathbf{M}}$ is dualisable, it now follows that $\alpha : \mathcal{A}(\mathbf{B}_M, \underline{\mathbf{M}}) \to M$ is given by evaluation at some $a \in B_M \subseteq B$. There are no homomorphisms from any of the connected components of $\mathbf{B}_M$ into $\underline{\mathbf{N}}$. So, for all $x \in \mathcal{B}(\mathbf{B}, \underline{\mathbf{M}} \cup \underline{\mathbf{N}})$, we have $x{\restriction}_{B_M} \in \mathcal{A}(\mathbf{B}_M, \underline{\mathbf{M}})$ and therefore

$$\beta(x) = m = \alpha(x{\restriction}_{B_M}) = x{\restriction}_{B_M}(a) = x(a).$$

So β is an evaluation.

Case 3: β is not constant. There are $v, w \in \mathcal{B}(\mathbf{B}, \underline{\mathbf{M}} \cup \underline{\mathbf{N}})$ with $\beta(v) \neq \beta(w)$. By Lemma 5.1.9, there is a connected component $\mathbf{C}$ of $\mathbf{B}$ such that C is a support for β. Define the map

$$\gamma : \mathcal{B}(\mathbf{C}, \underline{\mathbf{M}} \cup \underline{\mathbf{N}}) \to M \cup N \quad \text{by} \quad \gamma(y) := \beta(y \cup w{\restriction}_{B \backslash C}).$$

Then γ has a finite support, since β does.

To check that γ is locally an evaluation, let Y be a finite subset of the hom-set $\mathcal{B}(\mathbf{C}, \underline{\mathbf{M}} \cup \underline{\mathbf{N}})$. As β is locally an evaluation, there is some $b \in B$ such that β is given by evaluation at b on the finite set

$$\{\, y \cup w{\restriction}_{B \backslash C} \mid y \in Y \,\} \cup \{v{\restriction}_C \cup w{\restriction}_{B \backslash C}, w\} \subseteq \mathcal{B}(\mathbf{B}, \underline{\mathbf{M}} \cup \underline{\mathbf{N}}).$$

Since C is a support for β, we have

$$(v{\restriction}_C \cup w{\restriction}_{B \backslash C})(b) = \beta(v{\restriction}_C \cup w{\restriction}_{B \backslash C}) = \beta(v) \neq \beta(w) = w(b),$$

and therefore $b \in C$. For all $y \in Y$, we now have

$$\gamma(y) = \beta(y \cup w{\restriction}_{B \backslash C}) = (y \cup w{\restriction}_{B \backslash C})(b) = y(b).$$

So γ is given by evaluation at b on Y. We have shown that the map γ has a finite support and is locally an evaluation.

The connected algebra $\mathbf{C}$ belongs to $\mathcal{A} = \mathbb{ISP}(\underline{\mathbf{M}})$, by Lemma 5.1.4. For all $z \in \mathcal{A}(\mathbf{C}, \underline{\mathbf{M}})$, we have $z \in \mathcal{B}(\mathbf{C}, \underline{\mathbf{M}} \cup \underline{\mathbf{N}})$ and $\gamma(z) \in M$, since γ agrees with an evaluation on the set $\{z\}$. We can now define

$$\gamma' : \mathcal{A}(\mathbf{C}, \underline{\mathbf{M}}) \to M \quad \text{by} \quad \gamma' := \gamma{\restriction}_{\mathcal{A}(\mathbf{C},\underline{\mathbf{M}})}.$$

The map γ' has a finite support and is locally an evaluation. Therefore γ' is a brute-force morphism, by the Brute Force Lemma. As $\underline{\mathbf{M}}$ is dualisable, the map γ' is given by evaluation at some $c \in C \subseteq B$.

We shall complete the proof that $\underline{\mathbf{M}} \cup \underline{\mathbf{N}}$ is dualisable by showing that β is also given by evaluation at c. To this end, choose any $x \in \mathcal{B}(\mathbf{B}, \underline{\mathbf{M}} \cup \underline{\mathbf{N}})$. Since the algebra $\mathbf{C}$ is connected, we have either $x(C) \subseteq M$ or $x(C) \subseteq N$.

Case 3.1: $x(C) \subseteq M$. We must have

$$\beta(x) = \beta(x{\restriction}_C \cup w{\restriction}_{B\setminus C}) = \gamma(x{\restriction}_C) = \gamma'(x{\restriction}_C) = x(c),$$

as C is a support for β. So β is given by evaluation at c.

Case 3.2: $x(C) \subseteq N$. First we will check that $\beta(x) \in N$. Since β is locally an evaluation, the map β is given by evaluation at some $a \in B$ on the finite set $\{x, w, v{\restriction}_C \cup w{\restriction}_{B\setminus C}\}$. We know that $a \in C$, because

$$(v{\restriction}_C \cup w{\restriction}_{B\setminus C})(a) = \beta(v{\restriction}_C \cup w{\restriction}_{B\setminus C}) = \beta(v) \neq \beta(w) = w(a).$$

Therefore $\beta(x) = x(a) \in N$, as we are assuming that $x(C) \subseteq N$.

We now have $\beta(x) \in N$ and $x(c) \in N$. So, if $|N| = 1$, then $\beta(x) = x(c)$. We can now further assume that $|N| > 1$, giving $k > 0$. Let $i \in \{1, \dots, k\}$. The homomorphism $\overline{e}_i \circ x \in \mathcal{B}(\mathbf{B}, \underline{\mathbf{M}} \cup \underline{\mathbf{N}})$ satisfies $\overline{e}_i \circ x(C) \subseteq M$. Thus $\beta(\overline{e}_i \circ x) = \overline{e}_i \circ x(c)$, since Case 3.1 applies to $\overline{e}_i \circ x$. This gives us

$$\overline{e}_i(\beta(x)) = \beta(\overline{e}_i \circ x) = \overline{e}_i \circ x(c) = \overline{e}_i(x(c)),$$

as β preserves the endomorphism $\overline{e}_i$ of $\underline{\mathbf{M}} \cup \underline{\mathbf{N}}$. Since $\beta(x), x(c) \in N$ and $\overline{e}_1, \dots, \overline{e}_k$ separate the elements of N, it follows that $\beta(x) = x(c)$. Thus β is given by evaluation at c. ∎

5.1.11 Corollary *Let* $\underline{\mathbf{M}}$ *be a finite unary algebra. If* $\underline{\mathbf{M}}$ *is dualisable, then the one-point extension of* $\underline{\mathbf{M}}$ *is also dualisable.*

The following result gives us a partial converse for the previous theorem.

5.1.12 Theorem *Let* $\underline{\mathbf{M}}$ *and* $\underline{\mathbf{N}}$ *be finite unary algebras, with* $\underline{\mathbf{N}} \in \mathbb{ISP}(\underline{\mathbf{M}})$. *If* $\underline{\mathbf{M}} \dot\cup \underline{\mathbf{N}}$ *is finitely dualisable, then* $\underline{\mathbf{M}}$ *is finitely dualisable.*

Proof Assume that $M \cap N = \varnothing$ and that $\underline{\mathbf{M}} \cup \underline{\mathbf{N}}$ is finitely dualisable. We want to show that $\underline{\mathbf{M}}$ is finitely dualisable. Define

$$\mathcal{A} := \mathbb{ISP}(\underline{\mathbf{M}}) \quad \text{and} \quad \mathcal{B} := \mathbb{ISP}(\underline{\mathbf{M}} \cup \underline{\mathbf{N}}).$$

As in the proof of the previous theorem, there is a collection of endomorphisms $\overline{e}_1, \dots, \overline{e}_k$ of $\underline{\mathbf{M}} \cup \underline{\mathbf{N}}$, for some $k \in \omega$, such that $\overline{e}_1, \dots, \overline{e}_k$ each map into M and together separate the elements of N.

Since $\underline{\mathbf{M}} \cup \underline{\mathbf{N}}$ is finitely dualisable, there exists some $n \in \omega \backslash \{0\}$ such that $R_n(\underline{\mathbf{M}} \cup \underline{\mathbf{N}})$, the set of all n-ary algebraic relations on $\underline{\mathbf{M}} \cup \underline{\mathbf{N}}$, yields a duality on each algebra in $\mathcal{B}$. Define $m := \max(n, nk)$. We shall show that $R_m(\underline{\mathbf{M}})$ yields a duality on each finite connected algebra in $\mathcal{A}$. Each petal of $\mathcal{A}$ is connected. So it will then follow, by the Petal Duality Lemma, 3.1.6, that $\underline{\mathbf{M}}$ is finitely dualisable.

Let $\mathbf{C}$ be a finite connected algebra in $\mathcal{A}$, and let $\alpha : \mathcal{A}(\mathbf{C}, \underline{\mathbf{M}}) \to M$ preserve $R_m(\underline{\mathbf{M}})$. We wish to prove that α is an evaluation. Since $\mathcal{A} \subseteq \mathcal{B}$, we know that $\mathbf{C} \in \mathcal{B}$. We want to define $\beta : \mathcal{B}(\mathbf{C}, \underline{\mathbf{M}} \cup \underline{\mathbf{N}}) \to M \cup N$ by

$$\beta(x) = \begin{cases} \alpha(x) & \text{if } x(C) \subseteq M, \\ x(c) & \text{if } x(C) \subseteq N, \text{ where } c \text{ is any element of } C \text{ such that} \\ & \alpha \text{ is given by evaluation at } c \text{ on } \{\overline{e}_1 \circ x, \dots, \overline{e}_k \circ x\}, \end{cases}$$

for all $x \in \mathcal{B}(\mathbf{C}, \underline{\mathbf{M}} \cup \underline{\mathbf{N}})$. In the following two claims, we establish that β is a well-defined map that preserves $R_n(\underline{\mathbf{M}} \cup \underline{\mathbf{N}})$.

Claim 1 The map β is well defined.

For each $x \in \mathcal{B}(\mathbf{C}, \underline{\mathbf{M}} \cup \underline{\mathbf{N}})$, we have $x(C) \subseteq M$ or $x(C) \subseteq N$, as $\mathbf{C}$ is connected. Now let $x \in \mathcal{B}(\mathbf{C}, \underline{\mathbf{M}} \cup \underline{\mathbf{N}})$ with $x(C) \subseteq N$. Since α preserves $R_m(\underline{\mathbf{M}})$, the Preservation Lemma, 1.4.4, tells us that the map α agrees with an evaluation on every subset of $\mathcal{A}(\mathbf{C}, \underline{\mathbf{M}})$ with at most m elements. Therefore α agrees with an evaluation on $\{\overline{e}_1 \circ x, \dots, \overline{e}_k \circ x\} \subseteq \mathcal{A}(\mathbf{C}, \underline{\mathbf{M}})$, as $k \leqslant m$.

To see that β is well defined, let $c, d \in C$ such that α is given by evaluation at both c and d on $\{\overline{e}_1 \circ x, \dots, \overline{e}_k \circ x\}$. Then, for all $i \in \{1, \dots, k\}$, we have

$$\overline{e}_i(x(c)) = \overline{e}_i \circ x(c) = \alpha(\overline{e}_i \circ x) = \overline{e}_i \circ x(d) = \overline{e}_i(x(d)).$$

Since $x(c), x(d) \in x(C) \subseteq N$ and $\overline{e}_1, \dots, \overline{e}_k$ separate N, this implies that $x(c) = x(d)$. Thus β is well defined.

Claim 2 The map β preserves $R_n(\underline{\mathbf{M}} \cup \underline{\mathbf{N}})$.

We will use the Preservation Lemma. Let X be a subset of $\mathcal{B}(\mathbf{C}, \underline{\mathbf{M}} \cup \underline{\mathbf{N}})$ with at most n elements. Define the subset X' of $\mathcal{A}(\mathbf{C}, \underline{\mathbf{M}})$ by

$$X' := \{\, x \in X \mid x(C) \subseteq M \,\} \\ \cup \{\, \overline{e}_i \circ x \mid i \in \{1, \dots, k\} \text{ and } x \in X \text{ with } x(C) \subseteq N \,\}.$$

The size of X' is at most $\max(n, nk) = m$. Since α preserves $R_m(\underline{\mathbf{M}})$, we know that α is given by evaluation at some $a \in C$ on X'. We shall show that β is given by evaluation at a on X.

Let $x \in X$. If $x(C) \subseteq M$, then $x \in X'$ and so $\beta(x) = \alpha(x) = x(a)$. So we can assume that $x(C) \subseteq N$. For each $i \in \{1, \dots, k\}$, we have $\overline{e}_i \circ x \in X'$ and therefore $\alpha(\overline{e}_i \circ x) = \overline{e}_i \circ x(a)$. Thus α is given by evaluation at a on $\{\overline{e}_1 \circ x, \dots, \overline{e}_k \circ x\}$. This implies that $\beta(x) = x(a)$, by the definition of β. Thus β is given by evaluation at a on X.

We have proved that $\beta : \mathcal{B}(\mathbf{C}, \underline{\mathbf{M}} \cup \underline{\mathbf{N}}) \to M \cup N$ preserves $R_n(\underline{\mathbf{M}} \cup \underline{\mathbf{N}})$. As $R_n(\underline{\mathbf{M}} \cup \underline{\mathbf{N}})$ yields a duality on $\mathcal{B}$, there exists $b \in C$ such that β is given by evaluation at b. For every $y \in \mathcal{A}(\mathbf{C}, \underline{\mathbf{M}})$, we have $y \in \mathcal{B}(\mathbf{C}, \underline{\mathbf{M}} \cup \underline{\mathbf{N}})$ with $y(C) \subseteq M$, and therefore

$$\alpha(y) = \beta(y) = y(b).$$

Hence α is also given by evaluation at b, whence $R_m(\underline{\mathbf{M}})$ yields a duality on every finite connected algebra in $\mathcal{A}$. ∎

The corollary below was proved directly by Clark, Davey and Pitkethly [13].

5.1.13 Corollary *Let $\underline{\mathbf{M}}$ be a finite unary algebra. If the one-point extension of $\underline{\mathbf{M}}$ is finitely dualisable, then $\underline{\mathbf{M}}$ is also finitely dualisable.*

We do not know at present whether or not there is a non-dualisable unary algebra $\underline{\mathbf{M}}$ whose one-point extension $\underline{\mathbf{M}} \dot{\cup} \underline{\mathbf{1}}$ is dualisable (but not finitely dualisable). Finding such a pathological unary algebra $\underline{\mathbf{M}}$ would also solve the Finite Type Problem.

5.2 Term retractions and p-semilattices

In Chapter 2, we used term retractions to help lift some dualities for small algebras up to dualities for bigger algebras. In this section, we study the general relationship between term retractions and dualisability. We shall prove that a term retract of a dualisable algebra must also be dualisable, but that a term retract of a non-dualisable algebra is not necessarily non-dualisable.

The proof that term retractions preserve dualisability is very easy.

5.2.1 Theorem *A term retract of a dualisable algebra is also dualisable.*

Proof Let $\underset{\sim}{\mathbf{N}}$ be a dualisable algebra and let $\gamma : \mathbf{N} \twoheadrightarrow \mathbf{M}$ be a term retraction. We can assume that γ fixes each element of M, by Lemma 2.3.1. Define the two quasi-varieties $\mathcal{A} := \mathbb{ISP}(\mathbf{M})$ and $\mathcal{B} := \mathbb{ISP}(\mathbf{N})$. Then $\mathcal{A}$ is contained in $\mathcal{B}$. Let $\mathbf{A} \in \mathcal{A}$ and let $\alpha : \mathcal{A}(\mathbf{A}, \mathbf{M}) \to M$ be a brute-force morphism. In order to prove that $\underset{\sim}{\mathbf{M}}$ is dualisable, we must show that α is an evaluation.

As $\mathbf{A} \in \mathcal{B}$ and $\gamma : \mathbf{N} \to \mathbf{M}$ is a homomorphism, we can define the map $\beta : \mathcal{B}(\mathbf{A}, \mathbf{N}) \to N$ by $\beta(x) := \alpha(\gamma \circ x)$. We want to prove that β is a brute-force morphism. Since α is a brute-force morphism, the Brute Force Lemma, 1.4.5, tells us that α has a finite support and is locally an evaluation.

Let S be a finite support for α. To prove that S is also a support for β, let $x, y \in \mathcal{B}(\mathbf{A}, \mathbf{N})$ such that $x{\restriction}_S = y{\restriction}_S$. Then $(\gamma \circ x){\restriction}_S = (\gamma \circ y){\restriction}_S$, and therefore

$$\beta(x) = \alpha(\gamma \circ x) = \alpha(\gamma \circ y) = \beta(y).$$

So S is a finite support for β.

To see that β is locally an evaluation, let X be a finite subset of $\mathcal{B}(\mathbf{A}, \mathbf{N})$. As α is locally an evaluation, the map α is given by evaluation at some $a \in A$ on the finite subset $\{\, \gamma \circ x \mid x \in X \,\}$ of $\mathcal{A}(\mathbf{A}, \mathbf{M})$. As $\mathbf{A} \in \mathcal{B} = \mathbb{ISP}(\mathbf{N})$, there is a term function $\gamma^{\mathbf{A}} : A \to A$ of $\mathbf{A}$ corresponding to the term function $\gamma : N \to N$ of $\mathbf{N}$. For all $x \in X$, we have

$$\beta(x) = \alpha(\gamma \circ x) = (\gamma \circ x)(a) = x(\gamma^{\mathbf{A}}(a)).$$

Thus β is given by evaluation at $\gamma^{\mathbf{A}}(a)$ on X.

We have shown that β has a finite support and is locally an evaluation. So, by the Brute Force Lemma, the map $\beta : \mathcal{B}(\mathbf{A}, \mathbf{N}) \to N$ is a brute-force morphism. Since $\underset{\sim}{\mathbf{N}}$ is dualisable, the map β is given by evaluation at some $b \in A$. For all $z \in \mathcal{A}(\mathbf{A}, \mathbf{M}) \subseteq \mathcal{B}(\mathbf{A}, \mathbf{N})$, we have

$$\alpha(z) = \alpha(\gamma \circ z) = \beta(z) = z(b),$$

as γ fixes M. Hence α is an evaluation. ∎

The remainder of this section is devoted to finding examples of non-dualisable algebras that have dualisable term retracts. Our examples come from the variety of p-semilattices, which was defined in the introduction to this chapter. We will need only a few basic p-semilattice facts, all of which follow relatively easily from the definition. A thorough introduction to p-semilattices can be found in O. Frink's foundational paper [32].

Recall that a p-semilattice is boolean if it satisfies the equation $x^{**} \approx x$. As the name suggests, every boolean p-semilattice is term equivalent to a Boolean algebra [32]. Every finite Boolean algebra is strongly dualisable, by the NU Strong Duality Theorem [6, 8]. Thus every finite boolean p-semilattice is strongly dualisable.

We shall finish the characterisation of dualisability for p-semilattices by proving that every finite non-boolean p-semilattice is non-dualisable. Our proof of this result illustrates the power and simplicity of the ghost-element method for establishing non-dualisability. The basic ghost-element method is described in the Ghost Element Theorem, 1.4.6. This method is adapted straight from the definition of the dualisability of an algebra, and is often easy to apply. In addition, some ghost-element proofs can be extended directly to a proof of a much stronger condition than non-dualisability.

Recall that a finite algebra $\underline{\mathbf{M}}$ is inherently non-dualisable if each finite algebra that has $\underline{\mathbf{M}}$ as a subalgebra is non-dualisable. Inherent non-dualisability was introduced by Davey, Idziak, Lampe and McNulty [23]. They used the following theorem to find inherently non-dualisable graph algebras. (Later we will prove a stronger result, Lemma 7.1.3.)

5.2.2 Inherent Non-dualisability Theorem [23, 8] *Let $\underline{\mathbf{M}}$ be a finite algebra and let $f : \omega \to \omega$. Assume that there is a subalgebra $\mathbf{A}$ of $\underline{\mathbf{M}}^S$, for some set S, and an infinite subset A_0 of A such that*

(i) *for every $n \in \omega$ and every congruence θ on $\mathbf{A}$ of index at most n, the equivalence relation $\theta{\restriction}_{A_0}$ has a unique block of size greater than $f(n)$,*

(ii) *the algebra $\mathbf{A}$ does not contain the element g of M^S that is defined by $g(s) := \rho_s(a_s)$, where a_s is any element of the unique block of $\ker(\rho_s){\restriction}_{A_0}$ of size greater than $f(|M|)$.*

Then $\underline{\mathbf{M}}$ is inherently non-dualisable.

We shall apply this theorem in its simplest form, taking the bounding function f to be constant with value 1. The theorem has been used in the literature with a non-constant bounding function [43].

The Inherent Non-dualisability Theorem enables us to complete the characterisation of dualisability for p-semilattices. An alternative method was used in the text by Clark and Davey [8] to prove that every finite subdirectly irreducible p-semilattice is non-dualisable, except for the two-element one.

5.2.3 Theorem *If a finite p-semilattice is non-boolean, then it is inherently non-dualisable.*

Proof Let $\underline{\mathbf{M}} = \langle M; \wedge, {}^*, 0, 1\rangle$ be a finite p-semilattice. Then $\underline{\mathbf{M}}$ satisfies the equations

$$x^* \wedge x \approx 0, \quad x^{**} \wedge x \approx x, \quad x^{***} \approx x^*, \quad 0^* \approx 1 \quad \text{and} \quad 1^* \approx 0.$$

Now assume that $\underline{\mathbf{M}}$ is non-boolean. There is some $a \in M$ such that $a \neq a^{**}$.

We will represent sequences in M^ω using the notation introduced on page 81. Define $\mathbf{A}$ to be the subalgebra of $\underline{\mathbf{M}}^\omega$ generated by the set A_0, where

$$A_0 := \{\, a_n^0 \mid n \in \omega\backslash\{0\}\,\}.$$

We will show that, if θ is a congruence on $\mathbf{A}$ of finite index, then $\theta{\restriction}_{A_0}$ has a unique non-trivial block.

Let θ be a congruence on $\mathbf{A}$ of finite index. Assume that $k, \ell, m, n \in \omega\backslash\{0\}$, with $k \neq \ell$ and $m \neq n$, such that $a_k^0 \equiv_\theta a_\ell^0$ and $a_m^0 \equiv_\theta a_n^0$. Then

$$(a_k^0)^* = (a_k^0)^* \wedge (a_k^0)^* \equiv_\theta (a_k^0)^* \wedge (a_\ell^0)^* = \widehat{a}^*.$$

By symmetry, we also have $(a_m^0)^* \equiv_\theta \widehat{a}^*$. So $(a_k^0)^* \equiv_\theta (a_m^0)^*$, which gives us

$$a_k^0 = (a_k^0)^{**} \wedge a_k^0 \equiv_\theta (a_m^0)^{**} \wedge a_k^0 = a_{mk}^{0\,0}.$$

By symmetry once again, we have $a_m^0 \equiv_\theta a_{km}^{0\,0}$. Thus $a_k^0 \equiv_\theta a_m^0$, whence $\theta{\restriction}_{A_0}$ has at most one non-trivial block. The equivalence relation $\theta{\restriction}_{A_0}$ has at least one non-trivial block, since A_0 is infinite and θ is of finite index.

Now define $g \in M^\omega$ by $g(n) := \rho_n(c_n)$, where c_n is any element of the non-trivial block of $\ker(\rho_n){\restriction}_{A_0}$. Then g is the constant sequence $\widehat{a}$. It remains to prove that $g \notin A$. Using the equations given at the beginning of this proof, it is easy to see that the set $\{0, 1, a, a^*, a^{**}\}$ forms a subalgebra of $\underline{\mathbf{M}}$. Since $a \neq a^{**}$, it also follows that $a \notin \{0, 1, a^*, a^{**}\}$. Define the subset C of M^ω by

$$C := \{\, c \in \{0, 1, a, a^*, a^{**}\}^\omega \mid c(0) \neq a \text{ or } 0 \in c(\omega)\,\}.$$

As 0 is the least element of $\underline{\mathbf{M}}$ and a is meet-irreducible in $\{0, 1, a, a^*, a^{**}\}$, the set C is closed under $\wedge$. Since $b^* \neq a$, for all $b \in \{0, 1, a, a^*, a^{**}\}$, the set C is closed under *. So C is a subuniverse of $\underline{\mathbf{M}}^\omega$. Thus $g \notin A$, as $A_0 \subseteq C$ and $g = \widehat{a} \notin C$. Hence $\underline{\mathbf{M}}$ is inherently non-dualisable, by the Inherent Non-dualisability Theorem, 5.2.2. ∎

Since every finite boolean p-semilattice is strongly dualisable, we now have a characterisation of dualisability for p-semilattices.

5.2.4 Theorem *A finite p-semilattice is dualisable if and only if it is boolean. Moreover, every dualisable p-semilattice is strongly dualisable, and every non-dualisable p-semilattice is inherently non-dualisable.*

We can use the previous theorem to find a plethora of non-dualisable algebras that have dualisable term retracts.

5.2.5 Example *Every finite non-boolean p-semilattice is non-dualisable yet has a non-trivial dualisable p-semilattice as a term retract.*

Proof Let $\underline{\mathbf{P}}$ be a finite p-semilattice. Then $\underline{\mathbf{P}}$ satisfies the equations

$$0^{**} \approx 0, \quad 1^{**} \approx 1, \quad x^{**} \wedge y^{**} \approx (x \wedge y)^{**} \quad \text{and} \quad x^{***} \approx x^{*}.$$

So we can define the homomorphism $\gamma : \underline{\mathbf{P}} \to \underline{\mathbf{P}}$ by $\gamma(a) = a^{**}$. Let $\underline{\mathbf{Q}}$ be the image of γ. For all $a \in P$, we have

$$\gamma(a)^{**} = a^{****} = a^{**} = \gamma(a).$$

This implies that $\underline{\mathbf{Q}}$ is boolean, and therefore dualisable. Moreover, every element of Q is fixed by γ. Thus $\gamma : \underline{\mathbf{P}} \twoheadrightarrow \underline{\mathbf{Q}}$ is a term retraction. If we assume that $\underline{\mathbf{P}}$ is non-boolean, then $\underline{\mathbf{P}}$ must be non-trivial and therefore $\underline{\mathbf{Q}}$ is also non-trivial. ∎

We have seen that the ghost-element method can be very easy to use. We can now illustrate another advantage of this method: ghost-element proofs can often be extended to encompass more examples. Using the following lemma, we will be able to explain one way in which this can happen.

5.2.6 Lemma *Let $\underline{\mathbf{M}}$ be a term reduct of a finite algebra $\underline{\mathbf{M}}^{\sharp}$. Define the quasi-varieties $\mathcal{A} := \mathbb{ISP}(\underline{\mathbf{M}})$ and $\mathcal{A}^{\sharp} := \mathbb{ISP}(\underline{\mathbf{M}}^{\sharp})$. Let $\mathbf{A}$ be a subalgebra of $\underline{\mathbf{M}}^{S}$, for some set S, and assume that $\alpha : \mathcal{A}(\mathbf{A}, \underline{\mathbf{M}}) \to M$ is a brute-force morphism. Define the algebra $\mathbf{B} := \mathbf{sg}_{(\underline{\mathbf{M}}^{\sharp})^{S}}(A)$, and define the map $\beta : \mathcal{A}^{\sharp}(\mathbf{B}, \underline{\mathbf{M}}^{\sharp}) \to M$ by $\beta(x) := \alpha(x{\restriction}_{A})$. Then β is a brute-force morphism, and $g_{\beta} = g_{\alpha}$.*

Proof This result is a simple corollary of the Brute Force Lemma, 1.4.5. The brute-force morphism α has a finite support and is locally an evaluation. It follows easily that β has a finite support and is locally an evaluation. So β is a brute-force morphism. For all $s \in S$, we have

$$g_{\beta}(s) = \beta(\pi_s{\restriction}_{B}) = \alpha(\pi_s{\restriction}_{A}) = g_{\alpha}(s).$$

Thus $g_{\beta} = g_{\alpha}$. ∎

Say we have a ghost-element proof that the algebra $\underline{\mathbf{M}}$ is non-dualisable. Then there is a brute-force morphism $\alpha : \mathcal{A}(\mathbf{A}, \underline{\mathbf{M}}) \to M$, for some set S and $\mathbf{A} \leqslant \underline{\mathbf{M}}^{S}$, such that $g_{\alpha} \notin A$. Now let $\underline{\mathbf{M}}^{\sharp}$ be a finite algebra that has $\underline{\mathbf{M}}$ as a term reduct. To show that $\underline{\mathbf{M}}^\sharp$ is non-dualisable, it is enough to check that the ghost element is not generated, that is, that $g_\alpha \notin \mathrm{sg}_{(\underline{\mathbf{M}}^\sharp)^S}(A)$. This feature of the ghost-element method was used in the development of some of the results from Chapter 3. For example, the proof of Theorem 3.4.4 was adapted from a proof that the single algebra $\langle\{0,1,2\}; 010, 001, 002, 110\rangle$ is non-dualisable.

Ghost-element proofs can be extended not only by adding extra operations, but also by giving weaker conditions on the behaviour of the existing operations. In Chapter 3, the proof of Theorem 3.4.2 was adapted from a proof that the algebra $\langle\{0,1,2\}; 001, 010\rangle$ is non-dualisable. To finish this section, we give an explicit illustration of this method of extending ghost-element proofs.

The following technical lemma, adapted from Theorem 5.2.3, will be used to construct various examples of inherently non-dualisable algebras.

5.2.7 Lemma *Let* $\underline{\mathbf{M}} = \langle M; F \cup \{\wedge\}\rangle$ *be a finite algebra such that* $\wedge$ *is a meet-semilattice operation on* M *and* F *is a set of unary operations on* M*. Assume that there exists* $^* \in F$ *and a pair of distinct elements* $0, a \in M$ *for which*

(i) $a \leqslant a^{**}$, $a^* \leqslant 0^*$ *and* $0^{**} \wedge a = 0$,

(ii) 0 *is the least element of* $\underline{\mathbf{M}}$,

(iii) a *is meet-irreducible in* $\mathrm{sg}_{\underline{\mathbf{M}}}(a)$,

(iv) $a \neq u(b)$, *for all* $u \in F$ *and all* $b \in \mathrm{sg}_{\underline{\mathbf{M}}}(a)$.

Then $\underline{\mathbf{M}}$ *is inherently non-dualisable.*

Proof This proof is basically the same as the proof of Theorem 5.2.3. Define $\mathbf{A}$ to be the subalgebra of $\underline{\mathbf{M}}^\omega$ generated by $A_0 := \{\, a_n^0 \mid n \in \omega\backslash\{0\} \,\}$. Using (i) and (ii), the third paragraph of the proof of Theorem 5.2.3 shows that, for each congruence θ on $\mathbf{A}$ of finite index, the equivalence relation $\theta{\restriction}_{A_0}$ has a unique non-trivial block.

To prove that the algebra $\underline{\mathbf{M}}$ is inherently non-dualisable, using the Inherent Non-dualisability Theorem, 5.2.2, it remains to check that the element $g := \widehat{a}$ of M^ω does not belong to A. Define

$$C := \{\, c \in M^\omega \mid c(0) \in \mathrm{sg}_{\underline{\mathbf{M}}}(a) \text{ and } (c(0) \neq a \text{ or } 0 \in c(\omega)) \,\}.$$

By (ii), (iii) and (iv), the set C forms a subalgebra of $\underline{\mathbf{M}}^\omega$. Since $A_0 \subseteq C$ and $g \notin C$, we have $g \notin A$. ∎

Figure 5.1 gives some examples of semilattices with added unary operations that satisfy the conditions of the previous lemma and are therefore inherently non-dualisable. In contrast, Davey, Jackson and Talukder [24] have proved that a finite semilattice with added algebraic operations must be dualisable. So, for

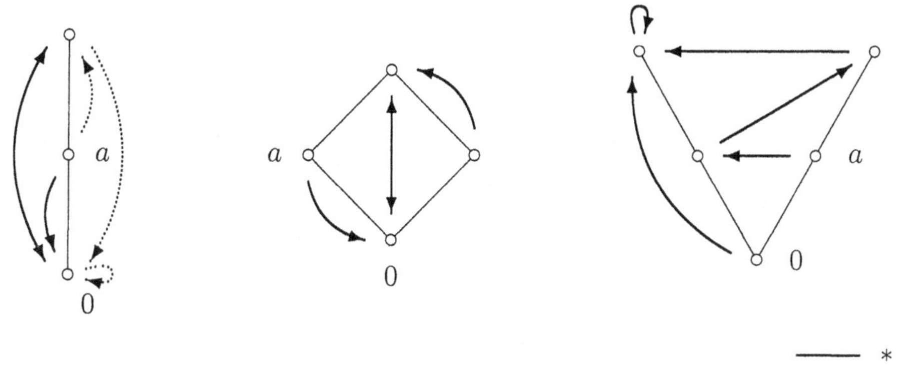

Figure 5.1 Some inherently non-dualisable algebras

example, for each finite p-semilattice $\langle M; \wedge, {}^*, 0, 1\rangle$, the term reduct $\langle M; \wedge, {}^{**}\rangle$ is dualisable.

5.3 Building non-dualisable algebras from dualisable ones

In this section, we find examples to show that non-dualisable algebras can be created from dualisable algebras using natural algebraic constructions.

5.3.1 Definition We shall begin by considering the two unary algebras

$$\underline{\mathbf{P}} := \langle\{0,1,2,3\}; 0011, 0101\rangle \quad \text{and} \quad \underline{\mathbf{Q}} := \langle\{0,1,2\}; 001, 010\rangle,$$

illustrated in Figure 5.2. Both the algebras $\underline{\mathbf{P}}$ and $\underline{\mathbf{Q}}$ are of type $\{u, v\}$, where

$$u^{\underline{\mathbf{P}}} := 0011, \quad v^{\underline{\mathbf{P}}} := 0101, \quad u^{\underline{\mathbf{Q}}} := 001 \quad \text{and} \quad v^{\underline{\mathbf{Q}}} := 010.$$

The algebra $\underline{\mathbf{Q}}$ is a subalgebra of $\underline{\mathbf{P}}$, and so $\underline{\mathbf{Q}} \in \mathbb{ISP}(\underline{\mathbf{P}})$. Note that $\underline{\mathbf{P}}$ and $\underline{\mathbf{Q}}$ do not generate the same quasi-variety, since $\underline{\mathbf{Q}}$ satisfies the quasi-equation $u(x) \approx v(x) \implies x \approx u(x)$ but $\underline{\mathbf{P}}$ does not.

5.3.2 Example *Define the two unary algebras $\underline{\mathbf{P}}$ and $\underline{\mathbf{Q}}$ as in 5.3.1.*

(i) *The dualisable algebra $\underline{\mathbf{P}}$ has a non-dualisable subalgebra $\underline{\mathbf{Q}}$.*

(ii) *The dualisable algebra $\underline{\mathbf{P}}$ and the non-dualisable algebra $\underline{\mathbf{Q}}$ generate the same variety.*

Proof We have already proved that $\underline{\mathbf{Q}}$ is not dualisable; see Theorem 3.0.1. The operations of $\underline{\mathbf{P}}$ are endomorphisms of the lattice $\mathbf{P}_0$ illustrated in Figure 5.2. So $\underline{\mathbf{P}}$ is dualisable, by the Lattice Endomorphism Theorem, 2.1.2.

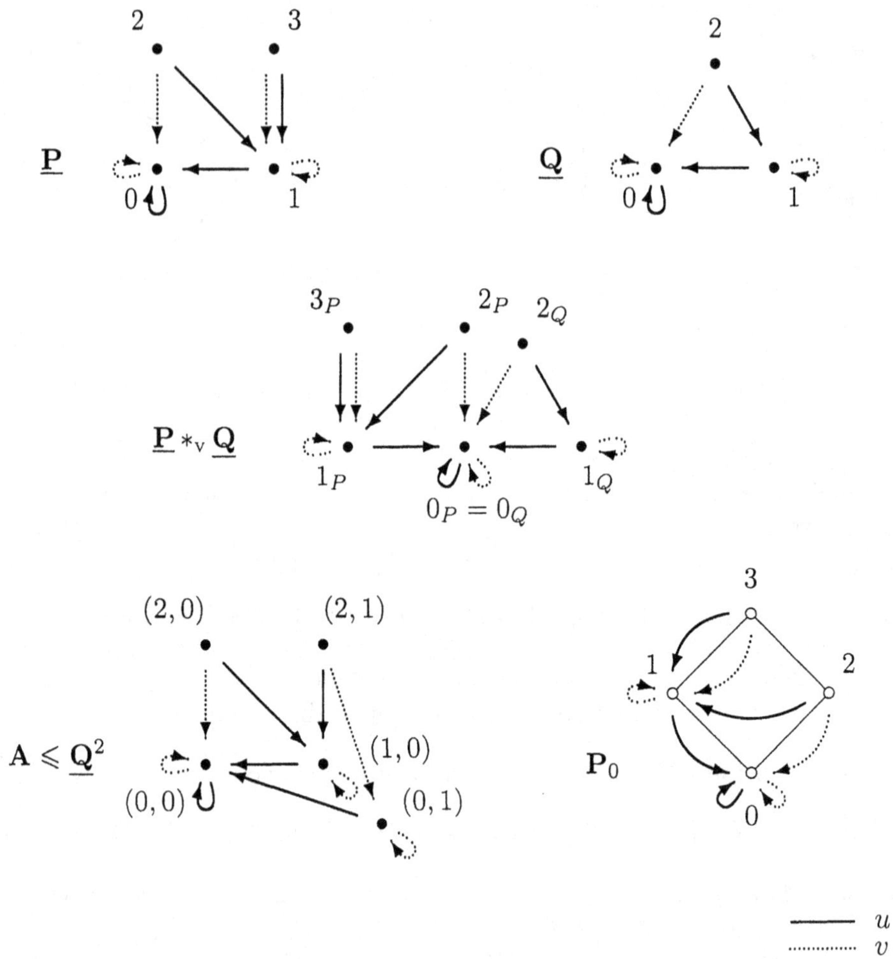

Figure 5.2

It remains to show that $\underline{\mathbf{P}}$ and $\underline{\mathbf{Q}}$ generate the same variety. We know that $\underline{\mathbf{Q}}$ belongs to $\mathrm{Var}(\underline{\mathbf{P}})$. To see that $\underline{\mathbf{P}}$ belongs to $\mathrm{Var}(\underline{\mathbf{Q}})$, let $\mathbf{A}$ be the subalgebra of $\underline{\mathbf{Q}}^2$ drawn in Figure 5.2. There is a surjective homomorphism $x : \mathbf{A} \twoheadrightarrow \underline{\mathbf{P}}$, given by $x\big((a, b)\big) := a + b$, for all $(a, b) \in A$. So $\mathrm{Var}(\underline{\mathbf{P}}) = \mathrm{Var}(\underline{\mathbf{Q}})$. ∎

In general, it is not possible to determine whether or not a finite unary algebra is dualisable simply by studying its abstract monoid of unary term functions. The monoid of unary term functions of an algebra is isomorphic to the monoid of unary term functions of the one-generated free algebra in the variety it generates.

Since $\mathrm{Var}(\underline{\mathbf{P}}) = \mathrm{Var}(\underline{\mathbf{Q}})$, in the previous example, the dualisable algebra $\underline{\mathbf{P}}$ and the non-dualisable algebra $\underline{\mathbf{Q}}$ have isomorphic monoids of unary term functions.

For all finite unary algebras $\mathbf{A}$ and $\mathbf{B}$ of the same type, let $\mathbf{A} *_{\mathrm{v}} \mathbf{B}$ denote the coproduct of $\mathbf{A}$ and $\mathbf{B}$ in the variety $\mathrm{Var}(\mathbf{A}, \mathbf{B})$, and let $\mathbf{A} *_{\mathrm{q}} \mathbf{B}$ denote the coproduct of $\mathbf{A}$ and $\mathbf{B}$ in the quasi-variety $\mathbb{ISP}(\mathbf{A}, \mathbf{B})$. These coproducts must exist, by Lemmas 5.1.2 and 5.1.3.

5.3.3 Example *Define the two unary algebras $\underline{\mathbf{P}}$ and $\underline{\mathbf{Q}}$ as in 5.3.1. Then the non-dualisable algebra $\underline{\mathbf{Q}}$ is a retract of the dualisable algebra $\underline{\mathbf{P}} *_{\mathrm{v}} \underline{\mathbf{Q}}$.*

Proof The only element of $\underline{\mathbf{P}}$ that is the value of a constant term function is 0. So we have $|F_{\mathrm{Var}(\underline{\mathbf{P}})}(0)| = 1$. Since $\mathrm{Var}(\underline{\mathbf{P}}) = \mathrm{Var}(\underline{\mathbf{Q}})$, it follows by Lemma 5.1.3 that $\underline{\mathbf{M}} := \underline{\mathbf{P}} *_{\mathrm{v}} \underline{\mathbf{Q}}$ is as drawn in Figure 5.2. Let $-_P : \underline{\mathbf{P}} \hookrightarrow \underline{\mathbf{M}}$ and $-_Q : \underline{\mathbf{Q}} \hookrightarrow \underline{\mathbf{M}}$ be the natural embeddings. We can define a retraction $x : \underline{\mathbf{M}} \twoheadrightarrow \underline{\mathbf{Q}}$ by

$$x(a_P) = 0, \text{ for all } a \in P, \text{ and } x(b_Q) = b, \text{ for all } b \in Q.$$

Define the homomorphism $y : \underline{\mathbf{M}} \to \underline{\mathbf{P}}$ by

$$y(a_P) = a, \text{ for all } a \in P, \text{ and } y(b_Q) = 0, \text{ for all } b \in Q.$$

Then x and y separate the elements of M. Since $\underline{\mathbf{Q}} \leqslant \underline{\mathbf{P}}$, this tells us that $\underline{\mathbf{M}} \in \mathbb{ISP}(\underline{\mathbf{P}})$, using the $\mathbb{ISP}$ Theorem, 1.1.1. As $\underline{\mathbf{P}}$ embeds into $\underline{\mathbf{M}}$, it follows that $\mathbb{ISP}(\underline{\mathbf{M}}) = \mathbb{ISP}(\underline{\mathbf{P}})$. We know that $\underline{\mathbf{P}}$ is dualisable. So $\underline{\mathbf{M}} = \underline{\mathbf{P}} *_{\mathrm{v}} \underline{\mathbf{Q}}$ is dualisable, by Independence Theorem, 1.4.1. ∎

The previous example is a special case of the following more general result.

5.3.4 Example *Every finite unary algebra with a one-element subalgebra is a retract of a dualisable algebra.*

Proof The proof of this result is almost identical to the proof of the previous example. Consider a finite unary algebra $\underline{\mathbf{M}}$, and assume there is some $m \in M$ that determines a one-element subalgebra of $\underline{\mathbf{M}}$. We know that $\underline{\mathbf{M}}$ is a subalgebra of a dualisable algebra $\underline{\mathbf{N}}$, by Theorem 2.1.4. Construct the new unary algebra $\underline{\mathbf{M}} \cup_m \underline{\mathbf{N}}$ by taking the disjoint union of $\underline{\mathbf{M}}$ and $\underline{\mathbf{N}}$, and identifying $(m, 0)$ in the copy of $\underline{\mathbf{M}}$ with $(m, 1)$ in the copy of $\underline{\mathbf{N}}$. Then $\underline{\mathbf{M}}$ is a retract of $\underline{\mathbf{M}} \cup_m \underline{\mathbf{N}}$. It is straightforward to check that $\underline{\mathbf{M}} \cup_m \underline{\mathbf{N}}$ is separated by homomorphisms into $\underline{\mathbf{N}}$. So the algebras $\underline{\mathbf{M}} \cup_m \underline{\mathbf{N}}$ and $\underline{\mathbf{N}}$ generate the same quasi-variety. Therefore $\underline{\mathbf{M}} \cup_m \underline{\mathbf{N}}$ is dualisable, by the Independence Theorem, 1.4.1. ∎

5.3.5 Definition Now define the unary algebras

$$\underline{\mathbf{K}} := \langle \{0,1,2,3\}; 0010, 0001 \rangle \quad \text{and} \quad \underline{\mathbf{L}} := \langle \{0,1,2\}; 001, 001 \rangle,$$

shown in Figure 5.3. We will prove that both $\underline{\mathbf{K}}$ and $\underline{\mathbf{L}}$ are dualisable, but that the product $\underline{\mathbf{K}} \times \underline{\mathbf{L}}$, the coproducts $\underline{\mathbf{K}} *_{\mathrm{v}} \underline{\mathbf{L}}$ and $\underline{\mathbf{K}} *_{\mathrm{q}} \underline{\mathbf{L}}$, and the disjoint union $\underline{\mathbf{K}} \dot{\cup} \underline{\mathbf{L}}$ are all non-dualisable. The algebras $\underline{\mathbf{K}}$ and $\underline{\mathbf{L}}$ are of type $\{u, v\}$, where

$$u^{\underline{\mathbf{K}}} := 0010, \quad v^{\underline{\mathbf{K}}} := 0001, \quad u^{\underline{\mathbf{L}}} := 001 \quad \text{and} \quad v^{\underline{\mathbf{L}}} := 001.$$

The odd-looking algebra $\underline{\mathbf{L}}$ actually belongs to the variety $\mathrm{Var}(\underline{\mathbf{K}})$: there is a subalgebra $\mathbf{B}$ of $\underline{\mathbf{K}}^2$, drawn in Figure 5.3, that has $\underline{\mathbf{L}}$ as a homomorphic image.

The only element of $\underline{\mathbf{K}}$ that is the value of a constant term function is 0. This implies that $|F_{\mathrm{Var}(\underline{\mathbf{K}})}(0)| = 1$. Using Lemma 5.1.3, it is easy to see that the coproduct $\underline{\mathbf{K}} *_{\mathrm{v}} \underline{\mathbf{L}}$ is as depicted in Figure 5.3. The elements of $\underline{\mathbf{K}} *_{\mathrm{v}} \underline{\mathbf{L}}$ are separated by homomorphisms into $\underline{\mathbf{K}}$ and $\underline{\mathbf{L}}$. So $\underline{\mathbf{K}} *_{\mathrm{v}} \underline{\mathbf{L}} \in \mathbb{ISP}(\underline{\mathbf{K}}, \underline{\mathbf{L}})$. It follows by Lemma 5.1.2 that $\underline{\mathbf{K}} *_{\mathrm{q}} \underline{\mathbf{L}} = \underline{\mathbf{K}} *_{\mathrm{v}} \underline{\mathbf{L}}$.

5.3.6 Lemma *Define the two unary algebras $\underline{\mathbf{K}}$ and $\underline{\mathbf{L}}$ as in 5.3.5. Then both $\underline{\mathbf{K}}$ and $\underline{\mathbf{L}}$ are dualisable.*

Proof The dualisability of $\underline{\mathbf{L}}$ follows easily from the Lattice Endomorphism Theorem, 2.1.2. To prove that $\underline{\mathbf{K}}$ is dualisable, we use Theorem 2.2.9. The element $0 \in K$ is the value of a constant term function of $\underline{\mathbf{K}}$.

Both the fundamental operations of $\underline{\mathbf{K}}$ are endomorphisms of the meet semilattice $\mathbf{K}_0 = \langle K; \wedge_0 \rangle$ drawn in Figure 5.3. So $\wedge_0 : \underline{\mathbf{K}}^2 \to \underline{\mathbf{K}}$ is a binary homomorphism of $\underline{\mathbf{K}}$. We now want to show that $g : K^2 \to K$, given by

$$g(a,b) = \begin{cases} 1 & \text{if } a = 1 \text{ and } b = 0, \\ 2 & \text{if } a = 2 \text{ and } b \neq 2, \\ 3 & \text{if } a = 3 \text{ and } b \neq 3, \\ 0 & \text{otherwise,} \end{cases}$$

is a binary homomorphism of $\underline{\mathbf{K}}$. To do this, let $a, b \in K$. Then

$$\begin{aligned} u(g(a,b)) = 1 &\iff g(a,b) = 2 \\ &\iff a = 2 \ \& \ b \neq 2 \\ &\iff u(a) = 1 \ \& \ u(b) = 0 \\ &\iff g(u(a), u(b)) = 1. \end{aligned}$$

Since $u(a) \in \{0,1\}$, we must have $g(u(a), u(b)) \in \{0,1\}$. As we also have $u(g(a,b)) \in \{0,1\}$, it now follows that $u(g(a,b)) = g(u(a), u(b))$. Thus g preserves u and, by symmetry, it also preserves v. Therefore g is a binary homomorphism of $\underline{\mathbf{K}}$.

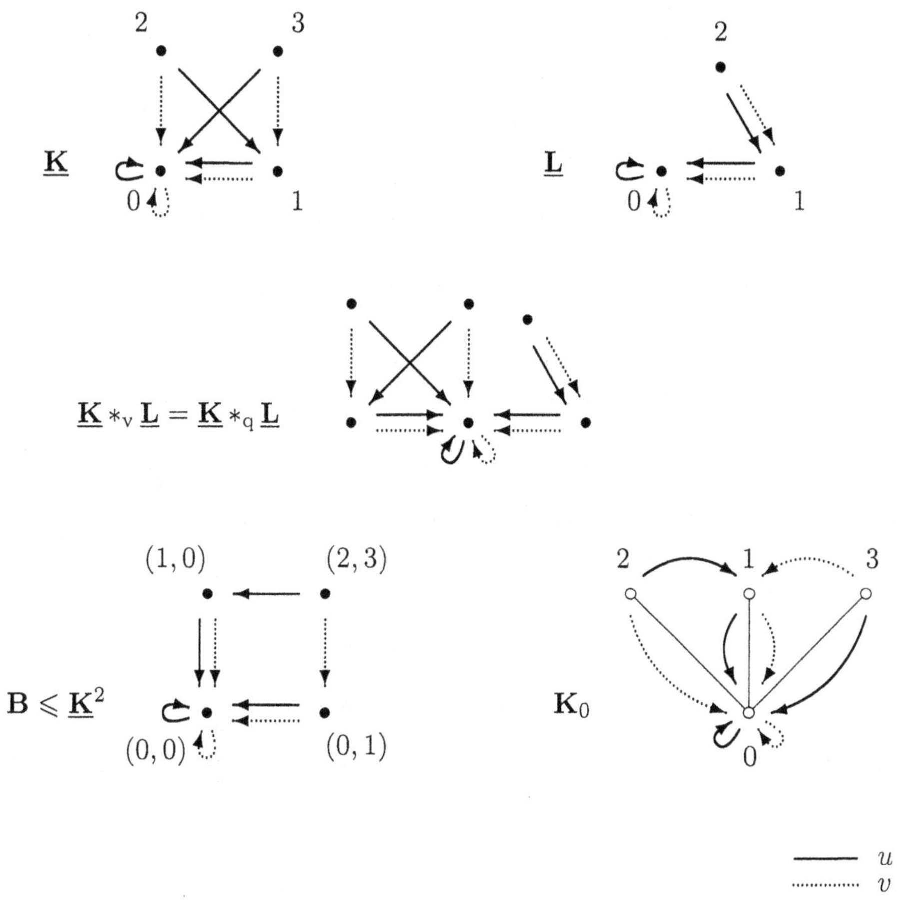

Figure 5.3

Define the set $G := \{\wedge_0, g\}$ of binary homomorphisms of $\underline{\mathbf{K}}$ and the subset $S := \{1, 2, 3\}$ of K. Every element of S is a strong idempotent of $\wedge_0$. Now let $k \in K \backslash \{0\}$. As $\underline{\mathbf{K}}$ satisfies $u(v(x)) \approx v(u(x))$, the operations $u^{\underline{\mathbf{K}}} = 0010$ and $v^{\underline{\mathbf{K}}} = 0001$ are endomorphisms of $\underline{\mathbf{K}}$. So $1 \in S \cap \mathrm{End}(\underline{\mathbf{K}})(k)$. We have

$$g(1, m) = 1 \iff m = 0,$$

for all $m \in K$. Thus G and $S \cap \mathrm{End}(\underline{\mathbf{K}})(k)$ distinguish 0 within K. By Theorem 2.2.9, the algebra $\underline{\mathbf{K}}$ is dualisable. ∎

The non-dualisability of $\underline{\mathbf{K}} \times \underline{\mathbf{L}}$ and $\underline{\mathbf{K}} *_{\mathrm{v}} \underline{\mathbf{L}}$ will follow once we have established that $\underline{\mathbf{K}} \dot{\cup} \underline{\mathbf{L}}$ is non-dualisable.

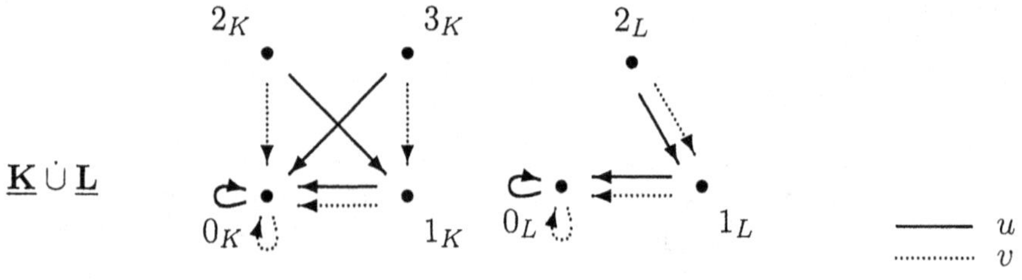

Figure 5.4 A non-dualisable disjoint union of two dualisable algebras

5.3.7 Lemma *Define the two unary algebras* $\underline{\mathbf{K}}$ *and* $\underline{\mathbf{L}}$ *as in 5.3.5. Then the disjoint union* $\underline{\mathbf{K}} \dot{\cup} \underline{\mathbf{L}}$ *is not dualisable.*

Proof Define the algebra $\underline{\mathbf{M}} := \underline{\mathbf{K}} \dot{\cup} \underline{\mathbf{L}}$. Then there are natural embeddings $-_K : \underline{\mathbf{K}} \hookrightarrow \underline{\mathbf{M}}$ and $-_L : \underline{\mathbf{L}} \hookrightarrow \underline{\mathbf{M}}$; see Figure 5.4. We shall prove that $\underline{\mathbf{M}}$ is not dualisable by applying the Non-dualisability Lemma, 3.4.1.

For each $n \in \omega\backslash\{0\}$, define $a_n \in M^\omega$ by

$$a_n(i) = \begin{cases} 1_L & \text{if } i = 0, \\ 1_K & \text{if } i = n, \\ 0_K & \text{otherwise.} \end{cases}$$

For all $m, n \in \omega\backslash\{0\}$ such that $m \neq n$, define $b_{mn} \in M^\omega$ by

$$b_{mn}(i) = \begin{cases} 2_L & \text{if } i = 0, \\ 2_K & \text{if } i = m, \\ 3_K & \text{if } i = n, \\ 0_K & \text{otherwise.} \end{cases}$$

Now define two subsets of M^ω by

$$A_0 := \{\, a_n \mid n \in \omega\backslash\{0\} \,\} \quad \text{and} \quad B := \{\, b_{mn} \mid m, n \in \omega\backslash\{0\} \text{ and } m \neq n \,\}.$$

Let $\mathbf{A}$ denote the subalgebra of $\underline{\mathbf{M}}^\omega$ generated by $A_0 \cup B$.

Let $x : \mathbf{A} \to \underline{\mathbf{M}}$ be a homomorphism. We want to show that $\ker(x{\restriction}_{A_0})$ has a unique non-trivial block. For each $n \in \omega\backslash\{0\}$, we have

$$x(a_n) = x(u(b_{nn+1})) = u(x(b_{nn+1})).$$

Therefore $x(A_0) \subseteq u(M) = \{0_K, 1_K, 0_L, 1_L\}$. Since we want to prove that $\ker(x{\restriction}_{A_0})$ has a unique non-trivial block, we can assume that $x(A_0) \neq \{0_K\}$. So one of the following three cases must apply.

Case 1: $1_K \in x(A_0)$. There is some $m \in \omega\backslash\{0\}$ such that $x(a_m) = 1_K$. Let $n \in \omega\backslash\{0\}$ with $m \neq n$. Then

$$a_m \xleftarrow{\;u\;} b_{mn} \xrightarrow{\;v\;} a_n$$

in $\mathbf{A}$. Under the homomorphism x, this gives us

$$\boxed{1_K} \xleftarrow{\;u\;} 2_K \xrightarrow{\;v\;} 0_K$$

in $\underline{\mathbf{M}}$. So $x(a_n) = 0_K$, and therefore $A_0\backslash\{a_m\}$ is the unique non-trivial block of $\ker(x{\restriction}_{A_0})$.

Case 2: $0_L \in x(A_0)$. There is some $m \in \omega\backslash\{0\}$ for which $x(a_m) = 0_L$. Choose some $n \in \omega\backslash\{0\}$ such that $m \neq n$. Then

$$a_m \xleftarrow{\;u\;} b_{mn} \xrightarrow{\;v\;} a_n \quad \overset{x}{\Longrightarrow} \quad \boxed{0_L} \xleftarrow{\;u\;} 0_L, 1_L \xrightarrow{\;v\;} 0_L.$$

This implies that $x(a_n) = 0_L$. So A_0 is the only block of $\ker(x{\restriction}_{A_0})$.

Case 3: $1_L \in x(A_0)$. There is some $m \in \omega\backslash\{0\}$ such that $x(a_m) = 1_L$. Let $n \in \omega\backslash\{0\}$ with $m \neq n$. Then

$$a_m \xleftarrow{\;u\;} b_{mn} \xrightarrow{\;v\;} a_n \quad \overset{x}{\Longrightarrow} \quad \boxed{1_L} \xleftarrow{\;u\;} 2_L \xrightarrow{\;v\;} 1_L,$$

and therefore $x(a_n) = 1_L$. Thus A_0 is the only block of $\ker(x{\restriction}_{A_0})$.

Now define $g \in M^\omega$ by $g(i) := \rho_i(a_{n_i})$, where a_{n_i} is any element of the unique non-trivial block of $\ker(\rho_i{\restriction}_{A_0})$. Then

$$g(i) = \begin{cases} 1_L & \text{if } i = 0, \\ 0_K & \text{otherwise,} \end{cases}$$

for all $i \in \omega$. To show that $\underline{\mathbf{M}}$ is non-dualisable, it suffices, by the Non-dualisability Lemma, 3.4.1, to prove that $g \notin A$. Define $c \in M^\omega$ by

$$c(i) = \begin{cases} 0_L & \text{if } i = 0, \\ 0_K & \text{otherwise.} \end{cases}$$

We shall show that $C := \{c\} \cup A_0 \cup B$ forms a subalgebra of $\underline{\mathbf{M}}^\omega$. We have $u(c) = c = v(c)$ and, for each $n \in \omega\backslash\{0\}$, we have $u(a_n) = c = v(a_n)$. Lastly, for all $m, n \in \omega\backslash\{0\}$ with $m \neq n$, we know that $u(b_{mn}) = a_m \in A_0$ and $v(b_{mn}) = a_n \in A_0$. So C forms a subalgebra of $\underline{\mathbf{M}}^\omega$. Since $A \subseteq C$, we get $g \notin A$. Thus $\underline{\mathbf{M}}$ is not dualisable. ∎

construction	*preserves non-dualisability?*	*references*
non-trivial subalgebra	✗	5.2.5, 7.1.2, 7.1.6
non-trivial homomorphic image	✗	5.2.5
non-trivial retract	✗	5.2.5
non-trivial term retract	✗	5.2.5, 2.3.2
finite power	✓	[61, 30]
finite product	?	
finite coproduct	?	
one-point extension	?	5.1.13

Table 5.3

5.3.8 Example *Define the two unary algebras* $\underline{\mathbf{K}}$ *and* $\underline{\mathbf{L}}$ *as in 5.3.5. Then both* $\underline{\mathbf{K}}$ *and* $\underline{\mathbf{L}}$ *are dualisable, but the product* $\underline{\mathbf{K}} \times \underline{\mathbf{L}}$*, the coproducts* $\underline{\mathbf{K}} *_{\mathrm{v}} \underline{\mathbf{L}}$ *and* $\underline{\mathbf{K}} *_{\mathrm{q}} \underline{\mathbf{L}}$*, and the disjoint union* $\underline{\mathbf{K}} \dot{\cup} \underline{\mathbf{L}}$ *are all non-dualisable.*

Proof We have just proved that $\underline{\mathbf{K}}$ and $\underline{\mathbf{L}}$ are dualisable and that $\underline{\mathbf{K}} \dot{\cup} \underline{\mathbf{L}}$ is not dualisable. Let $\underline{\mathbf{1}}$ be a one-element algebra of the same type as $\underline{\mathbf{K}}$ and $\underline{\mathbf{L}}$. It is straightforward to check that the disjoint union $\underline{\mathbf{K}} \dot{\cup} \underline{\mathbf{L}}$ is separated by homomorphisms into $(\underline{\mathbf{K}} *_{\mathrm{v}} \underline{\mathbf{L}}) \dot{\cup} \underline{\mathbf{1}}$, and that $(\underline{\mathbf{K}} *_{\mathrm{v}} \underline{\mathbf{L}}) \dot{\cup} \underline{\mathbf{1}}$ is separated by homomorphisms into $\underline{\mathbf{K}} \dot{\cup} \underline{\mathbf{L}}$. So

$$\mathbb{ISP}(\underline{\mathbf{K}} \dot{\cup} \underline{\mathbf{L}}) = \mathbb{ISP}((\underline{\mathbf{K}} *_{\mathrm{v}} \underline{\mathbf{L}}) \dot{\cup} \underline{\mathbf{1}}).$$

Using the Independence Theorem, 1.4.1, and Corollary 5.1.11, it follows that $\underline{\mathbf{K}} *_{\mathrm{v}} \underline{\mathbf{L}} = \underline{\mathbf{K}} *_{\mathrm{q}} \underline{\mathbf{L}}$ must be non-dualisable.

The algebra $\underline{\mathbf{K}} *_{\mathrm{v}} \underline{\mathbf{L}}$ is isomorphic to the subalgebra of $\underline{\mathbf{K}} \times \underline{\mathbf{L}}$ with the underlying set $\big(K \times \{0\}\big) \cup \big(\{0\} \times L\big)$. Therefore $\underline{\mathbf{K}} *_{\mathrm{v}} \underline{\mathbf{L}} \in \mathbb{ISP}(\underline{\mathbf{K}} \times \underline{\mathbf{L}})$. As both $\underline{\mathbf{K}}$ and $\underline{\mathbf{L}}$ are isomorphic to a subalgebra of $\underline{\mathbf{K}} *_{\mathrm{v}} \underline{\mathbf{L}}$, we must have $\underline{\mathbf{K}} \times \underline{\mathbf{L}} \in \mathbb{ISP}(\underline{\mathbf{K}} *_{\mathrm{v}} \underline{\mathbf{L}})$. Thus the product $\underline{\mathbf{K}} \times \underline{\mathbf{L}}$ is not dualisable, by the Independence Theorem. ∎

It is also reasonable to ask how the property of non-dualisability interacts with natural algebraic constructions. We know that non-dualisability is not always preserved by taking non-trivial subalgebras or non-trivial homomorphic images; see Example 5.2.5. The current state of our knowledge about non-dualisability and algebraic constructions is summarised in Table 5.3. This table reveals several open problems: for example, find a pair of non-dualisable algebras whose product is dualisable.

6

Dualisability and clones

We show that, for any natural number n, there is a chain of n unary clones such that the corresponding algebras are alternately dualisable and non-dualisable. We also solve two other clone-related problems in duality theory. We find an example of a non-dualisable algebra that can be obtained by adding a nullary operation to a dualisable algebra, and an example of a non-dualisable entropic algebra.

Of course, adding some extra fundamental operations to a finite algebra can change its dualisability. We gave a dramatic illustration of this in Chapter 3. There we found a chain of six unary clones on a three-element set, where the clones in the chain determine alternately dualisable and non-dualisable algebras (see Table 3.1). In this chapter, we extend this example by showing that there are arbitrarily long alternating chains of unary clones. More precisely, given any $n \in \omega$, there is a finite set M and a chain $F_0 \subseteq \cdots \subseteq F_{n-1}$ of unary clones on M such that, for each $i \in n$, the algebra $\langle M; F_i \rangle$ is dualisable if and only if i is even. There are only finitely many unary clones on any given finite set. So, for unary clones, this result is the best possible.

We will not be tackling the more difficult problem of finding an infinite ascending chain of non-unary clones $F_0 \subseteq F_1 \subseteq F_2 \subseteq \cdots$ such that the corresponding algebras are alternately dualisable and non-dualisable. At the moment, the longest known alternating chains of non-unary clones have length only two. For example, we can build an alternating chain based on the join semilattice $\underline{\mathbf{S}} = \langle \{0,1\}; \vee \rangle$, the implication algebra $\underline{\mathbf{I}} = \langle \{0,1\}; \to \rangle$ and the Boolean algebra $\underline{\mathbf{B}} = \langle \{0,1\}; \vee, \wedge, ', 0, 1 \rangle$. The two-element algebras $\underline{\mathbf{S}}$ and $\underline{\mathbf{B}}$ are early examples of dualisable algebras [29, 63]. The implication algebra $\underline{\mathbf{I}}$ was the first known example of a non-dualisable algebra [29]. Since we have

$a \vee b = (a \to b) \to b$ and $a \to b = a' \vee b$, for all $a, b \in \{0,1\}$, the algebra $\underline{\mathbf{S}}$ is a term reduct of $\underline{\mathbf{I}}$, which in turn is a term reduct of $\underline{\mathbf{B}}$.

Adding a complicated fundamental operation to a finite algebra may alter its dualisability. But can adding something as simple as a single nullary operation have the same effect? It is easy to create a dualisable algebra by adding a nullary operation to a non-dualisable algebra. The two-element implication algebra $\underline{\mathbf{I}} = \langle \{0,1\}; \to \rangle$ is not dualisable. However, the algebra $\langle \{0,1\}; \to, 0 \rangle$ is term equivalent to the two-element Boolean algebra $\underline{\mathbf{B}}$ and is therefore dualisable. There are also examples amongst unary algebras. By Theorem 3.0.1, the unary algebra $\langle \{0,1,2\}; 010, 002, 001, 110 \rangle$ is non-dualisable but the unary algebra $\langle \{0,1,2\}; 010, 002, 001, 110, 222 \rangle$ is dualisable.

It is not so easy to create a non-dualisable algebra by adding a nullary operation to a dualisable algebra. There are large classes of algebras for which dualisability is preserved by adding nullaries: for example, the classes of two-element algebras [16], three-element unary algebras (Theorem 3.0.1), and all finite algebras that generate a congruence-distributive variety [22, 29].

Nevertheless, there are examples of non-dualisable algebras that can be created by adding a nullary operation to a dualisable algebra. Davey and Quackenbush [27] have shown that, for each odd number m, the dihedral group $\underline{\mathbf{D}}_m$ of order $2m$ is dualisable. In an unpublished manuscript [42], P. M. Idziak proved that, for each odd m, the algebra obtained from $\underline{\mathbf{D}}_m$ by adding all the nullaries is not dualisable. In Section 6.3, we shall give another such example. There is a seven-element non-dualisable unary algebra that can be obtained by adding a constant operation to a dualisable unary algebra. These examples are significant because they show that the powerful techniques of tame congruence theory [36], which are unable to detect the addition of a constant to the type of an algebra, are unlikely to yield deep results about dualisability.

We finish this chapter by considering another clone-related question in duality theory: 'Is every finite entropic algebra dualisable?' An algebra is **entropic** if the operations in its clone of term functions preserve one another. Equivalently, the algebra $\mathbf{A}$ is entropic if every fundamental operation f of $\mathbf{A}$ is a homomorphism $f : \mathbf{A}^n \to \mathbf{A}$, where n is the arity of f. Some of the first known examples of dualisable algebras were entropic: for instance, the finite cyclic groups [55, 29], and the two-element semilattice [37]. These examples were extended by the second author [17]: every finite abelian group is entropic and dualisable, and every finite semilattice is entropic and dualisable. Davey, Idziak, Lampe and McNulty [23] proved that a finite graph algebra is dualisable if and only if it is entropic. Similarly, the dualisable finite flat graph algebras are precisely the entropic ones [45].

These results led Lampe, McNulty and Willard [45] to speculate that every finite entropic algebra might be dualisable. Given a finite entropic algebra $\underline{\mathbf{M}} = \langle M; F\rangle$, we know that we can include all the operations in F in the type of a potential dualising structure for $\underline{\mathbf{M}}$. We then aim to find a set R of algebraic relations on $\underline{\mathbf{M}}$ such that $\underset{\sim}{\mathbf{M}} = \langle M; F, R, \mathcal{T}\rangle$ dualises $\underline{\mathbf{M}}$. We shall see that having F as a starting point for $\underset{\sim}{\mathbf{M}}$ is not enough. There are finite entropic algebras that are not dualisable.

In fact, we have already seen an example of a non-dualisable entropic algebra: the six-element unary algebra $\underline{\mathbf{K}} *_{\mathrm{V}} \underline{\mathbf{L}}$ from Example 5.3.8. In this chapter, we exhibit a non-dualisable entropic unary algebra with five elements. This five-element algebra was the first known example of a non-dualisable entropic algebra. More recently, a non-dualisable entropic semigroup has been found by M. Jackson [43].

A unary algebra is entropic if and only if its monoid of unary term functions is commutative. It is straightforward (but tedious) to check that, up to term equivalence and isomorphism, there are eleven entropic three-element unary algebras. It follows from Theorem 3.0.1 that these algebras are all dualisable. It is not known whether every entropic four-element unary algebra is dualisable.

The example in Section 6.3 comes from a paper by the first author [53], and the example in Section 6.4 comes from a paper by both authors [54]. The other results in this chapter have not appeared in print before.

6.1 Distant unions

In this section, we set up one of the tools that we shall use to construct our alternating chains of clones. There is a way to combine two unary algebras together that keeps them even more separated than in their disjoint union. Consider two unary algebras $\mathbf{A}_0 = \langle A_0; F_0\rangle$ and $\mathbf{A}_1 = \langle A_1; F_1\rangle$, not necessarily of the same type. The **distant union** $\mathbf{A}_0 \dot{\uplus} \mathbf{A}_1$ is obtained by putting $\mathbf{A}_0$ and $\mathbf{A}_1$ next to each other with their operations separated:

$$\mathbf{A}_0 \dot{\uplus} \mathbf{A}_1 := \big\langle\, (A_0 \times \{0\}) \cup (A_1 \times \{1\});\ \ (F_0 \times \{0\}) \cup (F_1 \times \{1\})\,\big\rangle,$$

where, for each $i \in \{0,1\}$ and all $u \in F_i$, the unary operation (u,i) of $\mathbf{A}_0 \dot{\uplus} \mathbf{A}_1$ is given by

$$(u,i)\big((a,j)\big) = \begin{cases} (u(a),i) & \text{if } i = j,\\ (a,j) & \text{otherwise,}\end{cases}$$

for each $j \in \{0,1\}$ and all $a \in A_j$. For example, if we define the unary algebras $\underline{\mathbf{P}} = \langle\{0,1,2,3\}; 0011, 0101\rangle$ and $\underline{\mathbf{Q}} = \langle\{0,1,2\}; 001, 010\rangle$ as in 5.3.1, then the distant union $\underline{\mathbf{P}} \dot{\uplus} \underline{\mathbf{Q}}$ is as shown in Figure 6.1.

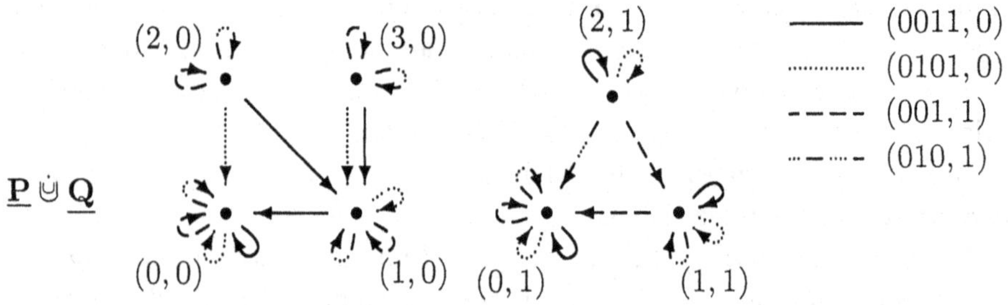

Figure 6.1 A distant union

We want to find conditions under which we can guarantee that the distant union of two dualisable unary algebras is also dualisable. We begin by introducing a new finiteness condition on unary algebras. Let $n \in \omega$, let $\underline{\mathbf{M}}$ be a finite unary algebra and define the quasi-variety $\mathcal{A} := \mathbb{ISP}(\underline{\mathbf{M}})$. We shall say that $\underline{\mathbf{M}}$ is **n-separable** if, for each finite connected algebra $\mathbf{C}$ in $\mathcal{A}$ and each $c \in C$, there is a unary term function τ of $\mathbf{C}$ and a subset X of $\mathcal{A}(\mathbf{C}, \underline{\mathbf{M}})$, with $|X| \leqslant n$, such that

$$\mathop{\&}_{x \in X} x(b) = x(c) \implies \tau(b) = c,$$

for all $b \in C$. The algebra $\underline{\mathbf{M}}$ is **finitely separable** if there is some $n \in \omega$ for which $\underline{\mathbf{M}}$ is n-separable.

We will say that a pair of algebras $\mathbf{A}$ and $\mathbf{B}$ are **term isomorphic** if there is an isomorphic copy $\mathbf{B}'$ of $\mathbf{A}$ such that $\mathbf{B}'$ and $\mathbf{B}$ are term equivalent.

6.1.1 Lemma *Let $\underline{\mathbf{M}}_0$ and $\underline{\mathbf{M}}_1$ be finite unary algebras, not necessarily of the same type. Assume that $\underline{\mathbf{M}}_0$ and $\underline{\mathbf{M}}_1$ are finitely dualisable and finitely separable. Then the distant union $\underline{\mathbf{M}}_0 \dot{\uplus} \underline{\mathbf{M}}_1$ is finitely dualisable.*

Proof Say $\underline{\mathbf{M}}_0$ is of type F_0 and $\underline{\mathbf{M}}_1$ is of type F_1. Define the unary algebra $\underline{\mathbf{N}} := \underline{\mathbf{M}}_0 \dot{\uplus} \underline{\mathbf{M}}_1$ and the quasi-variety $\mathcal{A} := \mathbb{ISP}(\underline{\mathbf{N}})$. For each $i \in \{0, 1\}$, there is some $n_i \in \omega$ such that $\underline{\mathbf{M}}_i$ is n_i-separable. Both the algebras $\underline{\mathbf{M}}_0$ and $\underline{\mathbf{M}}_1$ are finitely dualisable. So we can choose $m \in \omega \backslash \{0\}$ large enough so that $m \geqslant n_0 + n_1$ and, for each $i \in \{0, 1\}$, the set $R_m(\underline{\mathbf{M}}_i)$ of relations yields a duality on $\mathbb{ISP}(\underline{\mathbf{M}}_i)$.

We will show that $R_m(\underline{\mathbf{N}})$ yields a duality on every finite connected algebra in $\mathcal{A}$. It will then follow, by the Petal Duality Lemma, 3.1.6, that $\underline{\mathbf{N}}$ is finitely dualisable. Let $\mathbf{C}$ be a finite connected algebra in $\mathcal{A}$ and let $\alpha : \mathcal{A}(\mathbf{C}, \underline{\mathbf{N}}) \to N$ preserve $R_m(\underline{\mathbf{N}})$. We want to prove that α is an evaluation.

Let $i \in \{0,1\}$. The subset $M_i \times \{i\}$ of N forms a subalgebra $\underline{\mathbf{M}}_i^\sharp$ of $\underline{\mathbf{N}}$. The only non-identity fundamental operations of $\underline{\mathbf{M}}_i^\sharp$ are those of type $F_i \times \{i\}$. So the algebra $\underline{\mathbf{M}}_i^\sharp$ is term isomorphic to $\underline{\mathbf{M}}_i$. The quasi-variety $\boldsymbol{\mathcal{A}}_i := \mathbb{ISP}(\underline{\mathbf{M}}_i^\sharp)$ is contained in $\boldsymbol{\mathcal{A}}$. By Lemma 5.1.1, there exists a smallest congruence θ_i on $\mathbf{C}$ such that $\mathbf{C}/\theta_i \in \boldsymbol{\mathcal{A}}_i$. Let $\eta_i : \mathbf{C} \twoheadrightarrow \mathbf{C}/\theta_i$ be the natural quotient homomorphism.

The map α preserves the unary algebraic relation $M_i \times \{i\}$ on $\underline{\mathbf{N}}$. So we can define the map

$$\alpha_i : \boldsymbol{\mathcal{A}}_i(\mathbf{C}/\theta_i, \underline{\mathbf{M}}_i^\sharp) \to M_i \times \{i\} \quad \text{by} \quad \alpha_i(x) := \alpha(x \circ \eta_i).$$

We know that $R_m(\underline{\mathbf{M}}_i^\sharp)$ yields a duality on $\boldsymbol{\mathcal{A}}_i$. So, to prove that the map α_i is an evaluation, it is enough to show that α_i preserves $R_m(\underline{\mathbf{M}}_i^\sharp)$. Assume that $X \subseteq \boldsymbol{\mathcal{A}}_i(\mathbf{C}/\theta_i, \underline{\mathbf{M}}_i^\sharp)$ with $|X| \leqslant m$. By the Preservation Lemma, 1.4.4, there is some $c \in C$ such that α is given by evaluation at c on $\{\, x \circ \eta_i \mid x \in X \,\}$. For all $x \in X$, we have

$$\alpha_i(x) = \alpha(x \circ \eta_i) = (x \circ \eta_i)(c) = x(\eta_i(c)).$$

So α_i is given by evaluation at $\eta_i(c)$ on X. Using the Preservation Lemma again, it follows that α_i preserves the relations in $R_m(\underline{\mathbf{M}}_i^\sharp)$. Since $R_m(\underline{\mathbf{M}}_i^\sharp)$ yields a duality on $\boldsymbol{\mathcal{A}}_i$, the map α_i must be given by evaluation at some $a_i \in C/\theta_i$.

The term functions of $\underline{\mathbf{M}}_i^\sharp$ can all be built from operation symbols in $F_i \times \{i\}$. Since $\mathbf{C}/\theta_i \in \mathbb{ISP}(\underline{\mathbf{M}}_i^\sharp)$, this implies that the term functions of $\mathbf{C}/\theta_i$ can also be built from $F_i \times \{i\}$. Now, as $\underline{\mathbf{M}}_i^\sharp$ is n_i-separable and $\mathbf{C}/\theta_i$ is connected, there exists a unary term τ_i of type $F_i \times \{i\}$ and a set $X_i \subseteq \boldsymbol{\mathcal{A}}_i(\mathbf{C}/\theta_i, \underline{\mathbf{M}}_i^\sharp)$, with $|X_i| \leqslant n_i$, such that

$$\mathop{\&}_{x \in X_i} x(b) = x(a_i) \implies \tau_i^{\mathbf{C}/\theta_i}(b) = a_i,$$

for all $b \in C/\theta_i$.

As $m \geqslant n_0 + n_1$, the Preservation Lemma tells us that the map α is given by evaluation at some $a \in C$ on the set

$$\big\{\, x \circ \eta_i \;\big|\; i \in \{0,1\} \text{ and } x \in X_i \,\big\}.$$

We will show that α is given by evaluation at $\tau_0^{\mathbf{C}} \circ \tau_1^{\mathbf{C}}(a)$.

Let $z \in \boldsymbol{\mathcal{A}}(\mathbf{C}, \underline{\mathbf{N}})$. As the algebra $\mathbf{C}$ is connected, there is some $j \in \{0,1\}$ with $z(C) \subseteq M_j \times \{j\}$. By the minimality of θ_j, we must have $\theta_j \leqslant \ker(z)$. So there is a homomorphism $z' : \mathbf{C}/\theta_j \to \underline{\mathbf{M}}_j^\sharp$ such that $z = z' \circ \eta_j$. For all $x \in X_j$, we have

$$x(\eta_j(a)) = x \circ \eta_j(a) = \alpha(x \circ \eta_j) = \alpha_j(x) = x(a_j).$$

By construction, this implies that $\tau_j^{\mathbf{C}/\theta_j}(\eta_j(a)) = a_j$. For $k \in \{0,1\}\backslash\{j\}$, the term τ_k is of type $F_k \times \{k\}$, and consequently τ_k interpreted on $\underline{\mathbf{M}}_j^\sharp$ fixes every element of $M_j \times \{j\}$. Therefore

$$\begin{aligned}\alpha(z) &= \alpha(z' \circ \eta_j) = \alpha_j(z') = z'(a_j)\\ &= z'(\tau_j^{\mathbf{C}/\theta_j}(\eta_j(a))) = \tau_j^{\underline{\mathbf{M}}_j^\sharp}(z' \circ \eta_j(a))\\ &= \tau_j^{\underline{\mathbf{M}}_j^\sharp}(z(a)) = \tau_0^{\underline{\mathbf{M}}_j^\sharp} \circ \tau_1^{\underline{\mathbf{M}}_j^\sharp}(z(a)) = z(\tau_0^{\mathbf{C}} \circ \tau_1^{\mathbf{C}}(a)).\end{aligned}$$

Thus α is an evaluation, whence $\underline{\mathbf{N}} = \underline{\mathbf{M}}_0 \dot{\uplus} \underline{\mathbf{M}}_1$ is finitely dualisable. ∎

In the previous chapter, we saw that the disjoint union of a pair of dualisable unary algebras does not have to be dualisable. The above lemma raises a similar problem: 'Is there a pair of dualisable unary algebras whose distant union is non-dualisable?' An answer to this question would tell us whether or not the separation assumption in the previous lemma is really necessary.

The next result will allow us to apply the previous lemma repeatedly.

6.1.2 Lemma *Let $\underline{\mathbf{M}}_0$ and $\underline{\mathbf{M}}_1$ be finite unary algebras, not necessarily of the same type. Assume that $\underline{\mathbf{M}}_0$ and $\underline{\mathbf{M}}_1$ are finitely separable. Then the distant union $\underline{\mathbf{M}}_0 \dot{\uplus} \underline{\mathbf{M}}_1$ is also finitely separable.*

Proof Let F_0 and F_1 be the types of $\underline{\mathbf{M}}_0$ and $\underline{\mathbf{M}}_1$, respectively, and define $\underline{\mathbf{N}} := \underline{\mathbf{M}}_0 \dot{\uplus} \underline{\mathbf{M}}_1$ and $\boldsymbol{\mathcal{A}} := \mathbb{ISP}(\underline{\mathbf{N}})$. For each $i \in \{0,1\}$, the set $M_i \times \{i\}$ forms a subalgebra $\underline{\mathbf{M}}_i^\sharp$ of $\underline{\mathbf{N}}$ that is term isomorphic to $\underline{\mathbf{M}}_i$. So, for each $i \in \{0,1\}$, there is some $n_i \in \omega$ such that $\underline{\mathbf{M}}_i^\sharp$ is n_i-separable. We will show that $\underline{\mathbf{N}}$ is n-separable, where $n := n_0 + n_1$.

Let $\mathbf{C}$ be a finite connected algebra in $\boldsymbol{\mathcal{A}}$, choose some $c \in C$ and let $i \in \{0,1\}$. By Lemma 5.1.1, there exists a smallest congruence θ_i on $\mathbf{C}$ such that $\mathbf{C}/\theta_i \in \mathbb{ISP}(\underline{\mathbf{M}}_i^\sharp)$. Let $\eta_i : \mathbf{C} \twoheadrightarrow \mathbf{C}/\theta_i$ denote the natural quotient map. Since $\underline{\mathbf{M}}_i^\sharp$ is n_i-separable and $\mathbf{C}/\theta_i$ is connected, there is a unary term τ_i of type $F_i \times \{i\}$ and a set $X_i \subseteq \boldsymbol{\mathcal{A}}(\mathbf{C}/\theta_i, \underline{\mathbf{M}}_i^\sharp)$, with $|X_i| \leqslant n_i$, such that

$$\mathop{\&}_{x \in X_i} x(\eta_i(b)) = x(\eta_i(c)) \implies \tau_i^{\mathbf{C}/\theta_i}(\eta_i(b)) = \eta_i(c),$$

for all $b \in C$.

Define the subset Y of $\boldsymbol{\mathcal{A}}(\mathbf{C}, \underline{\mathbf{N}})$ by

$$Y := \{\, x \circ \eta_i \mid i \in \{0,1\} \text{ and } x \in X_i \,\},$$

and define the term τ of type $(F_0 \times \{0\}) \cup (F_1 \times \{1\})$ by $\tau(z) := \tau_0(\tau_1(z))$. Note that $|Y| \leqslant n_0 + n_1 = n$. Now let $b \in C$ and assume that $y(b) = y(c)$, for all $y \in Y$. We want to show that $\tau^{\mathbf{C}}(b) = c$.

Let $i \in \{0, 1\}$. For every $x \in X_i$, we have $x \circ \eta_i \in Y$, and consequently $x(\eta_i(b)) = x(\eta_i(c))$. So $\tau_i^{\mathbf{C}/\theta_i}(\eta_i(b)) = \eta_i(c)$. For $j \in \{0,1\}\backslash\{i\}$, the term τ_j is of type $F_j \times \{j\}$, and so τ_j interpreted on $\underline{\mathbf{M}}_i^\sharp$ fixes each element of $M_i \times \{i\}$. As $\mathbf{C}/\theta_i \in \mathbb{ISP}(\underline{\mathbf{M}}_i^\sharp)$, this implies that $\tau_j^{\mathbf{C}/\theta_i}$ fixes C/θ_i. We now have

$$\begin{aligned}\eta_i\big(\tau^{\mathbf{C}}(b)\big) &= \tau^{\mathbf{C}/\theta_i}\big(\eta_i(b)\big) = \tau_0^{\mathbf{C}/\theta_i} \circ \tau_1^{\mathbf{C}/\theta_i}\big(\eta_i(b)\big)\\ &= \tau_i^{\mathbf{C}/\theta_i}\big(\eta_i(b)\big) = \eta_i(c).\end{aligned}$$

Since the algebra $\mathbf{C}$ is connected, each homomorphism $x : \mathbf{C} \to \underline{\mathbf{N}}$ satisfies $x(C) \subseteq M_0 \times \{0\}$ or $x(C) \subseteq M_1 \times \{1\}$, which implies that $\theta_0 \leqslant \ker(x)$ or $\theta_1 \leqslant \ker(x)$. As $\mathbf{C}$ is separated by homomorphisms into $\underline{\mathbf{N}}$, this tells us that the maps η_0 and η_1 must separate the elements of C. Thus $\tau^{\mathbf{C}}(b) = c$, whence $\underline{\mathbf{N}}$ is n-separable. ∎

Throughout the remainder of this chapter, it will be notationally convenient to use the usual set-theoretic construction of the natural numbers. For all $n \in \omega$, we have $n = \{0, \dots, n-1\}$. Now let M be a set and let $n \in \omega\backslash\{0\}$. For each unary operation $u : M \to M$ and each $i \in n$, define the unary operation $(u, i) : M \times n \to M \times n$ by

$$(u,i)\big((a,j)\big) = \begin{cases}(u(a), i) & \text{if } i = j,\\ (a,j) & \text{otherwise,}\end{cases}$$

for all $a \in M$ and $j \in n$. For each unary algebra $\underline{\mathbf{M}} = \langle M; F\rangle$, we can define the new unary algebra

$$\dot{\biguplus_n} \underline{\mathbf{M}} := \langle M \times n; F \times n\rangle.$$

It is easy to see that $\dot\biguplus_n \underline{\mathbf{M}}$ is term isomorphic to the algebra

$$(\cdots((\underline{\mathbf{M}} \mathbin{\dot\uplus} \underline{\mathbf{M}}) \mathbin{\dot\uplus} \underline{\mathbf{M}}) \cdots \mathbin{\dot\uplus} \underline{\mathbf{M}}) \mathbin{\dot\uplus} \underline{\mathbf{M}},$$

where $\underline{\mathbf{M}}$ appears n times.

The next result follows from Lemmas 6.1.1 and 6.1.2.

6.1.3 Corollary *Let $\underline{\mathbf{M}}$ be a finite unary algebra. Assume that $\underline{\mathbf{M}}$ is finitely dualisable and finitely separable. Then $\dot\biguplus_n \underline{\mathbf{M}}$ is dualisable, for all $n \in \omega\backslash\{0\}$.*

6.2 Alternating chains of clones

In this section, we construct the promised alternating chains of unary clones. Our construction will be based on two particular three-element unary algebras:

the dualisable algebra $\langle\{0,1,2\};001,010,002\rangle$ and its non-dualisable reduct $\langle\{0,1,2\};001,010\rangle$. Using the tools of the previous section, we will be able to use these two small algebras to build bigger algebras.

6.2.1 Lemma *The unary algebra* $\langle\{0,1,2\};001,010,002\rangle$ *is* 2*-separable.*

Proof Define $\underline{\mathbf{M}} := \langle\{0,1,2\};001,010,002\rangle$ and $\boldsymbol{\mathcal{A}} := \mathbb{ISP}(\underline{\mathbf{M}})$. Since the constant map 000 is a term function of $\underline{\mathbf{M}}$, every algebra in $\boldsymbol{\mathcal{A}}$ is connected. Let $\mathbf{A}$ be a non-trivial finite algebra in $\boldsymbol{\mathcal{A}}$. We can assume that $\mathbf{A} \leqslant \underline{\mathbf{M}}^n$, for some $n \in \omega\backslash\{0\}$. The set

$$A_* := A \cap \{0,1\}^n$$

forms a subalgebra $\mathbf{A}_*$ of $\mathbf{A}$. (The algebra $\mathbf{A}_*$ is term equivalent to a 0-pointed set.) Let $a \in A_*\backslash\{\widehat{0}\}$. Then it is straightforward to check that we can define the homomorphism $x_a : \mathbf{A}_* \to \underline{\mathbf{M}}$ by

$$x_a(b) = \begin{cases} 1 & \text{if } b = a, \\ 0 & \text{otherwise.} \end{cases}$$

There is only one element of A_* that does not map to 0 under x_a. So it follows easily from Lemma 4.2.4, (iii) $\Rightarrow$ (i), that the homomorphism $x_a : \mathbf{A}_* \to \underline{\mathbf{M}}$ extends to a homomorphism $\overline{x}_a : \mathbf{A} \to \underline{\mathbf{M}}$.

Now let $c \in A$. To show that $\underline{\mathbf{M}}$ is 2-separable, we need to find a unary term function τ of $\mathbf{A}$ and a subset X of $\boldsymbol{\mathcal{A}}(\mathbf{A}, \underline{\mathbf{M}})$ such that $|X| \leqslant 2$ and, for all $b \in A$, we have

$$\mathop{\&}_{x \in X} x(b) = x(c) \quad \Longrightarrow \quad \tau(b) = c.$$

Case 1: $c = \widehat{0}$. Choose the term function $\tau := 000^{\mathbf{A}}$ of $\mathbf{A}$ and the subset $X := \varnothing$ of $\boldsymbol{\mathcal{A}}(\mathbf{A}, \underline{\mathbf{M}})$. For all $b \in A$, we have $\tau(b) = 000(b) = \widehat{0} = c$.

Case 2: $c \in \{0,1\}^n\backslash\{\widehat{0}\}$. Choose the term function $\tau := 010^{\mathbf{A}}$ and the set $X := \{\overline{x}_c\}$. Let $b \in A$ and assume that $\overline{x}_c(b) = \overline{x}_c(c)$. We have $010(b) \in A_*$, with

$$x_c(010(b)) = 010(\overline{x}_c(b)) = 010(\overline{x}_c(c)) = 010(1) = 1.$$

This implies that $\tau(b) = 010(b) = c$.

Case 3: $c \in \{0,2\}^n\backslash\{\widehat{0}\}$. We have $001(c) \in \{0,1\}^n\backslash\{\widehat{0}\}$. Define $\tau := 002^{\mathbf{A}}$ and $X := \{\overline{x}_{001(c)}\}$. Let $b \in A$ such that $\overline{x}_{001(c)}(b) = \overline{x}_{001(c)}(c)$. Then

$$\begin{aligned} x_{001(c)}(001(b)) &= 001(\overline{x}_{001(c)}(b)) = 001(\overline{x}_{001(c)}(c)) \\ &= x_{001(c)}(001(c)) = 1. \end{aligned}$$

So $001(b) = 001(c)$, which implies that $002(b) = 002(c)$. Therefore we have $\tau(b) = 002(b) = 002(c) = c$.

Case 4: $c \notin \{0,1\}^n \cup \{0,2\}^n$. There are $i, j \in \{0, \dots, n-1\}$ such that $c(i) = 1$ and $c(j) = 2$. So $010(c) \in \{0,1\}^n \backslash \{\widehat{0}\}$ and $001(c) \in \{0,1\}^n \backslash \{\widehat{0}\}$. Choose $\tau := \mathrm{id}_A$ and $X := \{\overline{x}_{001(c)}, \overline{x}_{010(c)}\}$. Let $b \in A$ and assume that $x(b) = x(c)$, for each $x \in X$. Then

$$\begin{aligned} x_{001(c)}(001(b)) &= 001(\overline{x}_{001(c)}(b)) = 001(\overline{x}_{001(c)}(c)) \\ &= x_{001(c)}(001(c)) = 1, \end{aligned}$$

and so $001(b) = 001(c)$. Similarly, we have $x_{010(c)}(010(b)) = 1$, which implies that $010(b) = 010(c)$. Since 001 and 010 separate $\{0,1,2\}$, it follows that $\tau(b) = b = c$. ∎

The following general lemma takes care of the non-dualisable part of our construction.

6.2.2 Lemma *Let $\underline{\mathbf{M}}$ and $\underline{\mathbf{N}}$ be finite algebras, with $\underline{\mathbf{M}}$ a subalgebra of $\underline{\mathbf{N}}$. Assume that, for every $\mathbf{A} \in \mathbb{ISP}(\underline{\mathbf{M}})$ and every homomorphism $x : \mathbf{A} \to \underline{\mathbf{N}}$, we have $x(A) \subseteq M$ or $|x(A)| = 1$. If $\underline{\mathbf{M}}$ is non-dualisable, then $\underline{\mathbf{N}}$ is also non-dualisable.*

Proof We shall prove the contrapositive of the claim. Assume that $\underline{\mathbf{N}}$ is dualisable. Let $\mathbf{A}$ belong to the quasi-variety $\mathcal{A} := \mathbb{ISP}(\underline{\mathbf{M}})$ and consider a brute-force morphism $\alpha : \mathcal{A}(\mathbf{A}, \underline{\mathbf{M}}) \to M$. To prove that $\underline{\mathbf{M}}$ is dualisable, it suffices to show that α is an evaluation.

Since $\underline{\mathbf{M}}$ is a subalgebra of $\underline{\mathbf{N}}$, the algebra $\mathbf{A}$ is a member of the quasi-variety $\mathcal{B} := \mathbb{ISP}(\underline{\mathbf{N}})$. Choose some $c \in A$ and define $\beta : \mathcal{B}(\mathbf{A}, \underline{\mathbf{N}}) \to N$ by

$$\beta(x) = \begin{cases} \alpha(x) & \text{if } x(A) \subseteq M, \\ x(c) & \text{otherwise.} \end{cases}$$

As α is a brute-force morphism, the Brute Force Lemma, 1.4.5, tells us that α has a finite support and is locally an evaluation. We shall prove that β also has these two properties, and it will then follow that β is a brute-force morphism.

The map α has a finite non-empty support S. To check that S is also a support for β, let $x, y \in \mathcal{B}(\mathbf{A}, \underline{\mathbf{N}})$ such that $x{\restriction}_S = y{\restriction}_S$. First assume that $x(S) = y(S) \subseteq M$. Then $x(A) \subseteq M$ and $y(A) \subseteq M$. Since S is a support for α, we have

$$\beta(x) = \alpha(x) = \alpha(y) = \beta(y).$$

Now assume that $x(S) = y(S) \nsubseteq M$. We must have $|x(A)| = 1 = |y(A)|$ and $x(A) = y(A)$. So

$$\beta(x) = x(c) = y(c) = \beta(y).$$

Therefore S is a finite support for β.

To see that β is locally an evaluation, let X be a finite subset of $\mathcal{B}(\mathbf{A}, \underline{\mathbf{N}})$. Since α is locally an evaluation, there exists some $a \in A$ such that α is given by evaluation at a on the set $\{ x \in X \mid x(A) \subseteq M \}$. Now let $x \in X$. If $x(A) \subseteq M$, then $\beta(x) = \alpha(x) = x(a)$. Otherwise, we have $|x(A)| = 1$ and therefore $\beta(x) = x(c) = x(a)$. So β is locally an evaluation.

We have shown that β is a brute-force morphism. Since $\underline{\mathbf{N}}$ is dualisable, the map β is given by evaluation at some $b \in A$. We can finish the proof by showing that α is also given by evaluation at b. Let $z \in \mathcal{A}(\mathbf{A}, \underline{\mathbf{M}})$. Then $z \in \mathcal{B}(\mathbf{A}, \underline{\mathbf{N}})$ and $\alpha(z) = \beta(z) = z(b)$. Thus α is an evaluation, whence $\underline{\mathbf{M}}$ is dualisable. ∎

We are now ready to prove the main result of this chapter.

6.2.3 Theorem *Let $n \in \omega \backslash \{0\}$. There is chain $F_0 \subseteq \cdots \subseteq F_{n-1}$ of unary clones on a finite set M such that, for every $i \in n$, the algebra $\langle M; F_i \rangle$ is dualisable if and only if i is even.*

Proof Let $n \in \omega \backslash \{0\}$ and define the set $M := \{0, 1, 2\}$. We will construct a chain

$$F_0^* \subseteq F_0 \subseteq \cdots \subseteq F_{n-1}^* \subseteq F_{n-1}$$

of sets of unary operations on the set $M \times n$. We will then prove that, for each $i \in n$, the algebra $\underline{\mathbf{N}}_i^* := \langle M \times n; F_i^* \rangle$ is non-dualisable and the algebra $\underline{\mathbf{N}}_i := \langle M \times n; F_i \rangle$ is dualisable. The claim will then follow.

Recall from the previous section that, for each unary operation $u : M \to M$ and each $k \in n$, we define the unary operation $(u, k) : M \times n \to M \times n$ by

$$(u, k)((a, \ell)) = \begin{cases} (u(a), k) & \text{if } k = \ell, \\ (a, \ell) & \text{otherwise,} \end{cases}$$

for all $a \in M$ and $\ell \in n$. Define the three-element unary algebra

$$\underline{\mathbf{M}} = \langle \{0, 1, 2\}; F \rangle, \quad \text{where } F := \{001, 010, 002\}.$$

Now, for each $i \in n$, define

$$F_i^* := (F \times i) \cup \{(001, i), (010, i)\} \quad \text{and} \quad F_i := F \times (i + 1).$$

Note that $F_i = F_i^* \cup \{(002, i)\}$, for each $i \in n$, and that the algebra $\underline{\mathbf{N}}_{n-1}$ determined by the largest set of operations F_{n-1} is equal to the distant union

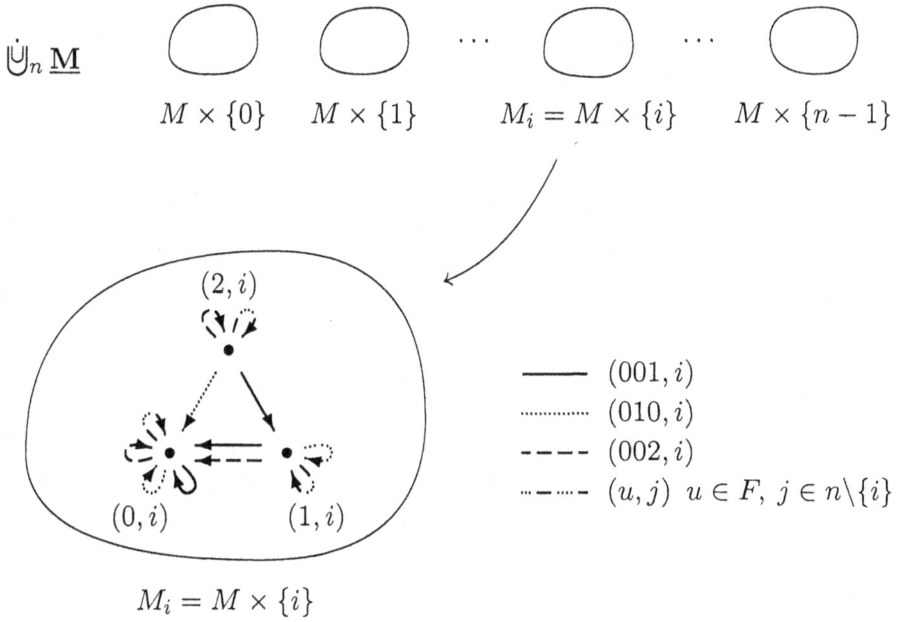

Figure 6.2 The structure of the distant union $\dot{\biguplus}_n \underline{\mathbf{M}}$

$\dot{\biguplus}_n \underline{\mathbf{M}} = \langle M \times n; F \times n \rangle$. The top picture in Figure 6.2 gives an idea of the general shape of the algebra $\dot{\biguplus}_n \underline{\mathbf{M}}$. The bottom picture is a close-up of one of the components of $\dot{\biguplus}_n \underline{\mathbf{M}}$.

Choose some $i \in n$. First we will show that $\underline{\mathbf{N}}_i^*$ is non-dualisable. Let $\underline{\mathbf{M}}_i^*$ denote the subalgebra of $\underline{\mathbf{N}}_i^*$ on the set $M_i := M \times \{i\}$. The only non-identity fundamental operations of $\underline{\mathbf{M}}_i^*$ are $(001, i)^{\underline{\mathbf{M}}_i^*}$ and $(010, i)^{\underline{\mathbf{M}}_i^*}$. So $\underline{\mathbf{M}}_i^*$ is term isomorphic to $\underline{\mathbf{Q}} := \langle \{0, 1, 2\}; 001, 010 \rangle$. We already know that the algebra $\underline{\mathbf{Q}}$ is non-dualisable, by Theorem 3.0.1, and so $\underline{\mathbf{M}}_i^*$ is non-dualisable. We will be using Lemma 6.2.2 to show that $\underline{\mathbf{N}}_i^*$ is also non-dualisable.

Let $\mathbf{A} \in \mathbb{ISP}(\underline{\mathbf{M}}_i^*)$ and let $x : \mathbf{A} \to \underline{\mathbf{N}}_i^*$ be a homomorphism. Assume there is some $a \in A$ with $x(a) \notin M_i$. The operation $(000, i) = (001, i) \circ (001, i)$ is a term function of $\underline{\mathbf{N}}_i^*$. The operation $(000, i)$ is constant on M_i and fixes every element of $(M \times n) \backslash M_i$. As $\mathbf{A} \in \mathbb{ISP}(\underline{\mathbf{M}}_i^*)$, the operation $(000, i)^{\mathbf{A}}$ on $\mathbf{A}$ is constant. So, for all $b \in A$, we have

$$\begin{aligned}(000, i)(x(b)) &= x((000, i)(b)) = x((000, i)(a)) \\ &= (000, i)(x(a)) = x(a) \notin M_i,\end{aligned}$$

and hence $x(b) \notin M_i$ with $x(b) = (000, i)(x(b)) = x(a)$. So $|x(A)| = 1$. As $\underline{\mathbf{M}}_i^*$ is non-dualisable, this implies that $\underline{\mathbf{N}}_i^*$ is non-dualisable, by Lemma 6.2.2.

It remains to show that $\underline{\mathbf{N}}_i$ is a dualisable algebra. The set of operations of $\underline{\mathbf{N}}_i$ is $F_i = F \times (i+1)$. So the subalgebra of $\underline{\mathbf{N}}_i$ on the set $M \times (i+1)$ is the distant union $\dot{\biguplus}_{i+1} \underline{\mathbf{M}}$. The algebra $\underline{\mathbf{M}} = \langle\{0,1,2\}; 001, 010, 002\rangle$ is finitely dualisable; see Theorem 3.3.9. Since $\underline{\mathbf{M}}$ is 2-separable, by Lemma 6.2.1, it follows using Corollary 6.1.3 that $\dot{\biguplus}_{i+1} \underline{\mathbf{M}}$ is dualisable. The elements of the set $(M \times n) \backslash (M \times (i+1))$ are fixed by each operation in $F_i = F \times (i+1)$. So $\underline{\mathbf{N}}_i$ can be obtained from $\dot{\biguplus}_{i+1} \underline{\mathbf{M}}$ by taking $3(n-(i+1))$ successive one-point extensions. Thus Corollary 5.1.11 tells us that $\underline{\mathbf{N}}_i$ is dualisable. ∎

6.3 Adding constants

In this section, we begin an investigation into the effect that adding a nullary operation to an algebra has on its dualisability. Since dualisability is preserved under term equivalence, adding a nullary operation is equivalent to adding any corresponding constant operation.

In the introduction to this chapter, we saw examples of dualisable algebras that can be obtained by adding a nullary operation to a non-dualisable algebra. Here, we will see that it is possible to obtain a non-dualisable algebra by adding a nullary operation to a dualisable algebra. First, we shall give a condition under which adding a nullary does not destroy dualisability.

Let $\mathbf{A} = \langle A; F\rangle$ be a unary algebra and let $a \in A$. Then the element a is said to be **isolated in A** if, for all $u \in F$ and $b \in A$, we have

$$u(b) = a \iff b = a.$$

So a is isolated in $\mathbf{A}$ if and only if $\{a\}$ is the underlying set of a connected component of $\mathbf{A}$.

6.3.1 Lemma *Let $\underline{\mathbf{M}}$ be a finite unary algebra. Assume that m is an isolated element of $\underline{\mathbf{M}}$, and define $\underline{\mathbf{M}}_m$ to be the algebra obtained from $\underline{\mathbf{M}}$ by adding the nullary operation with value m. If $\underline{\mathbf{M}}$ is dualisable, then $\underline{\mathbf{M}}_m$ is dualisable.*

Proof Define the two quasi-varieties $\mathcal{A} := \mathbb{ISP}(\underline{\mathbf{M}})$ and $\mathcal{A}_m := \mathbb{ISP}(\underline{\mathbf{M}}_m)$. Assume that $\underline{\mathbf{M}}$ is dualisable. We want to show that $\underline{\mathbf{M}}_m$ is dualisable, so let $\mathbf{A} \in \mathcal{A}_m$ and let $\alpha : \mathcal{A}_m(\mathbf{A}, \underline{\mathbf{M}}_m) \to M$ be a brute-force morphism. By the Brute Force Lemma, 1.4.5, the map α has a finite support and is locally an evaluation. To prove that $\underline{\mathbf{M}}_m$ is dualisable, it suffices to prove that α is an evaluation.

Case 1: α is constant. Since m is an isolated element of $\underline{\mathbf{M}}$, the set $\{m\}$ is a subuniverse of $\underline{\mathbf{M}}_m$. So there is a constant homomorphism $\underline{m} : \mathbf{A} \to \underline{\mathbf{M}}_m$ with value m. The brute-force morphism α must preserve the unary algebraic relation $\{m\}$ on $\underline{\mathbf{M}}_m$. So $\alpha(\underline{m}) = m$, and therefore m is the value of the constant map α. We have $m^{\mathbf{A}} \in A$ and, for all $w \in \mathcal{A}_m(\mathbf{A}, \underline{\mathbf{M}}_m)$, we have $\alpha(w) = m = w(m^{\mathbf{A}})$. Thus α is an evaluation.

Case 2: α is not constant. Let $\mathbf{A}^\flat$ denote the reduct of $\mathbf{A}$ obtained by removing the nullary operation with value $m^{\mathbf{A}}$. Then $\mathbf{A}^\flat \in \mathcal{A}$. Since m is isolated in $\underline{\mathbf{M}}$, the element $m^{\mathbf{A}}$ is isolated in $\mathbf{A}^\flat$. Thus, for each homomorphism $x : \mathbf{A}^\flat \to \underline{\mathbf{M}}$, we can define the homomorphism

$$x_m : \mathbf{A} \to \underline{\mathbf{M}}_m \quad \text{by} \quad x_m := x\restriction_{A\setminus\{m^{\mathbf{A}}\}} \cup \underline{m}\restriction_{\{m^{\mathbf{A}}\}}.$$

Now we can define the map

$$\alpha^\flat : \mathcal{A}(\mathbf{A}^\flat, \underline{\mathbf{M}}) \to M \quad \text{by} \quad \alpha^\flat(x) := \alpha(x_m).$$

We want to check that $\alpha^\flat$ is a brute-force morphism.

The brute-force morphism α has a finite support, and it is easy to check that $\alpha^\flat$ has the same finite support. To see that $\alpha^\flat$ is locally an evaluation, let X be a finite subset of $\mathcal{A}(\mathbf{A}^\flat, \underline{\mathbf{M}})$. We are assuming that α is not constant, so there are $y, z \in \mathcal{A}_m(\mathbf{A}, \underline{\mathbf{M}}_m)$ with $\alpha(y) \neq \alpha(z)$. We can now define

$$X_m := \{\, x_m \mid x \in X \,\} \cup \{y, z\} \subseteq \mathcal{A}_m(\mathbf{A}, \underline{\mathbf{M}}_m).$$

Since α is locally an evaluation, there is some $a \in A$ such that α is given by evaluation at a on X_m. As

$$y(a) = \alpha(y) \neq \alpha(z) = z(a),$$

we must have $a \neq m^{\mathbf{A}}$. So, for all $x \in X$, we get

$$\alpha^\flat(x) = \alpha(x_m) = x_m(a) = x(a).$$

Thus $\alpha^\flat$ is locally an evaluation. Hence $\alpha^\flat$ is a brute-force morphism, by the Brute Force Lemma.

As $\underline{\mathbf{M}}$ is dualisable, we know that $\alpha^\flat$ is given by evaluation at some $b \in A$. For all $w \in \mathcal{A}_m(\mathbf{A}, \underline{\mathbf{M}}_m)$, we have $w \in \mathcal{A}(\mathbf{A}^\flat, \underline{\mathbf{M}})$ with $w_m = w$, and so

$$\alpha(w) = \alpha(w_m) = \alpha^\flat(w) = w(b).$$

Hence α is an evaluation. ∎

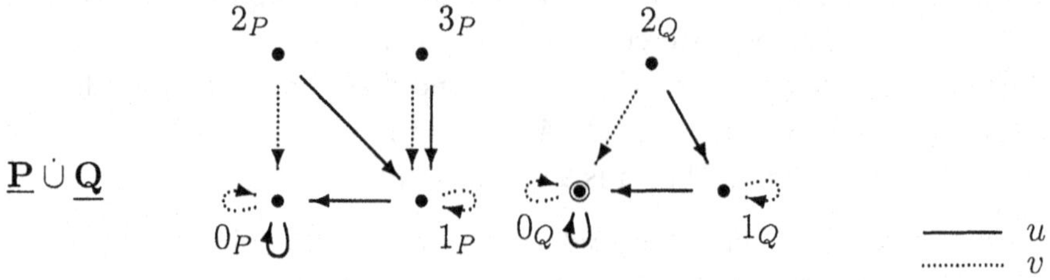

Figure 6.3 Adding a constant can destroy dualisability

We can combine the previous lemma with Corollary 5.1.11. Fix a finite unary algebra $\underline{\mathbf{M}}$, and let $\underline{\mathbf{1}}$ be a one-element algebra of the same type. The one-point extension of $\underline{\mathbf{M}}$ is the disjoint union $\underline{\mathbf{M}} \dot\cup \underline{\mathbf{1}}$. Now let $0'$ denote the element of $\underline{\mathbf{M}} \dot\cup \underline{\mathbf{1}}$ corresponding to the unique element 0 of $\underline{\mathbf{1}}$. The **pointed one-point extension of** $\underline{\mathbf{M}}$ is obtained from $\underline{\mathbf{M}} \dot\cup \underline{\mathbf{1}}$ by adding the nullary operation with value $0'$. Since $0'$ is isolated in $\underline{\mathbf{M}} \dot\cup \underline{\mathbf{1}}$, we obtain the following result.

6.3.2 Corollary *Let $\underline{\mathbf{M}}$ be a finite unary algebra. If $\underline{\mathbf{M}}$ is dualisable, then the pointed one-point extension of $\underline{\mathbf{M}}$ is also dualisable.*

The previous lemma raises several questions. Are there weaker conditions on unary algebras under which adding a nullary will preserve dualisability? What about conditions on non-unary algebras?

We next show that it is possible to create a non-dualisable algebra by adding a nullary operation to a dualisable algebra. We will be building our example using our favourite non-dualisable unary algebra $\underline{\mathbf{Q}} = \langle\{0,1,2\}; 001, 010\rangle$ and its dualisable extension $\underline{\mathbf{P}} = \langle\{0,1,2,3\}; 0011, \overline{0101}\rangle$; see Example 5.3.2.

6.3.3 Example *There is a seven-element non-dualisable unary algebra that can be obtained by adding a constant operation to a dualisable unary algebra.*

Proof Define the unary algebras $\underline{\mathbf{P}}$ and $\underline{\mathbf{Q}}$ as in 5.3.1, and define the disjoint union $\underline{\mathbf{M}} := \underline{\mathbf{P}} \dot\cup \underline{\mathbf{Q}}$. There are natural embeddings

$$-_P : \underline{\mathbf{P}} \hookrightarrow \underline{\mathbf{M}} \quad \text{and} \quad -_Q : \underline{\mathbf{Q}} \hookrightarrow \underline{\mathbf{M}};$$

see Figure 6.3. Define $\underline{\mathbf{M}}^\sharp$ to be the unary algebra obtained from $\underline{\mathbf{M}}$ by adding the constant operation $\underline{0_Q} : M \to M$ with value 0_Q.

The algebra $\underline{\mathbf{P}}$ is dualisable, by the Lattice Endomorphism Theorem, 2.1.2; see Example 5.3.2. As $\underline{\mathbf{Q}} \leqslant \underline{\mathbf{P}}$, we know that $\underline{\mathbf{M}} = \underline{\mathbf{P}} \dot\cup \underline{\mathbf{Q}}$ is dualisable, by Theorem 5.1.10. We will show that $\underline{\mathbf{M}}^\sharp$ is not dualisable, using Lemma 6.2.2.

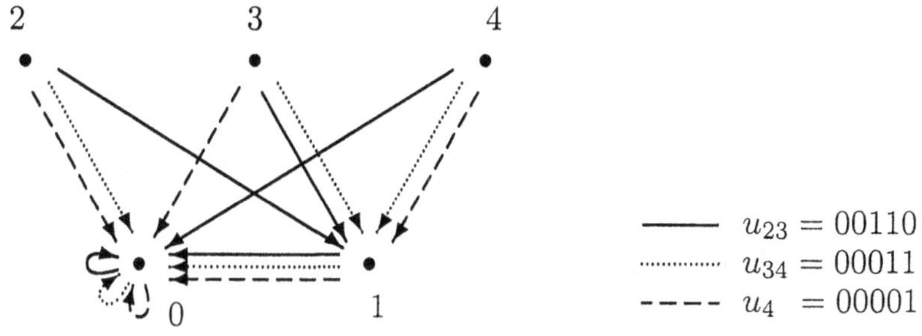

Figure 6.4 A non-dualisable entropic algebra

Define $\underline{\mathbf{Q}}^\sharp$ to be the subalgebra of $\underline{\mathbf{M}}^\sharp$ with universe $Q_Q = \{0_Q, 1_Q, 2_Q\}$. Then $\underline{\mathbf{Q}}^\sharp$ is term isomorphic to $\underline{\mathbf{Q}}$, since the term functions $\underline{0}_Q$ and u^2 of $\underline{\mathbf{M}}^\sharp$ agree on Q_Q. The algebra $\underline{\mathbf{Q}} = \langle\{0,1,2\}; 001, 010\rangle$ is non-dualisable, by Theorem 3.0.1. So we know that $\underline{\mathbf{Q}}^\sharp$ is non-dualisable.

Now let $\mathbf{A} \in \mathbb{ISP}(\underline{\mathbf{Q}}^\sharp)$ and let $x : \mathbf{A} \to \underline{\mathbf{M}}^\sharp$ be a homomorphism. The term function u^2 of $\mathbf{A}$ is constant with value $0_Q^{\mathbf{A}}$. For all $a \in A$, we have

$$u^2(x(a)) = x(u^2(a)) = x(0_Q^{\mathbf{A}}) = 0_Q.$$

So $x(A) \subseteq Q_Q$. It follows that $\underline{\mathbf{M}}^\sharp$ is non-dualisable, by Lemma 6.2.2. ∎

6.4 A non-dualisable entropic algebra

We finish this chapter by showing that the five-element unary algebra drawn in Figure 6.4 is entropic but not dualisable.

6.4.1 Example *The unary algebra* $\langle\{0,1,2,3,4\}; 00110, 00011, 00001\rangle$ *is entropic but not dualisable.*

Proof Define $M := \{0,1,2,3,4\}$ and, for each subset S of M, define the operation $u_S : M \to M$ by

$$u_S(m) = \begin{cases} 1 & \text{if } m \in S, \\ 0 & \text{otherwise.} \end{cases}$$

We will be using the Non-dualisability Lemma, 3.4.1, to prove that

$$\underline{\mathbf{M}} := \langle\{0,1,2,3,4\}; u_{23}, u_{34}, u_4\rangle$$

is not dualisable. The algebra $\underline{\mathbf{M}}$ is entropic, since $u_S \circ u_T = u_\varnothing$, for all $S, T \subseteq M\backslash\{0,1\}$.

Define two subsets of M^ω by

$$A_0 := \{\, 0^{11}_{0n} \mid n \in \omega\backslash\{0\} \,\}$$

and

$$B := \{\, 0^{3\,2\,4}_{0mn} \mid m, n \in \omega\backslash\{0\} \text{ and } m \neq n \,\}.$$

Define $\mathbf{A}$ to be the subalgebra of $\underline{\mathbf{M}}^\omega$ generated by the set $A_0 \cup B$. Now let $x : \mathbf{A} \to \underline{\mathbf{M}}$ be a homomorphism. We will prove that there is a unique non-trivial block of $\ker(x{\restriction}_{A_0})$.

The constant map $\underline{0} : M \to M$, with value 0, is a term function of $\underline{\mathbf{M}}$. So $x(\widehat{0}) = 0$. Now let $n \in \omega\backslash\{0\}$. Then

$$\widehat{0} \xleftarrow{u_{23}} 0^{11}_{0n} \xrightarrow{u_4} \widehat{0}$$

in $\mathbf{A}$. Applying the homomorphism x, this gives us

$$0 \xleftarrow{u_{23}} x(0^{11}_{0n}) \xrightarrow{u_4} 0$$

in $\underline{\mathbf{M}}$, and consequently $x(0^{11}_{0n}) \in \{0, 1\}$. Therefore $x(A_0) \subseteq \{0, 1\}$.

We are trying to show that $\ker(x{\restriction}_{A_0})$ has only one non-trivial block, so we can now assume that there exist $m, n \in \omega\backslash\{0\}$ such that $x(0^{11}_{0m}) = 0$ and $x(0^{11}_{0n}) = 1$. Let $k \in \omega\backslash\{0\}$ with $k \neq n$. We will prove that $x(0^{11}_{0k}) = 0$, and it will then follow that $A_0\backslash\{0^{11}_{0n}\}$ is the unique non-trivial block of $\ker(x{\restriction}_{A_0})$. We have

$$\begin{array}{ccccccccc} & & 0^{11}_{0n} & & & & & & \\ & & \uparrow u_{34} & & & & & & \\ 0^{11}_{0m} & \xleftarrow{u_{23}} & 0^{3\,2\,4}_{0mn} & \xrightarrow{u_4} & 0^{1}_{n} & \xleftarrow{u_4} & 0^{324}_{0kn} & \xrightarrow{u_{23}} & 0^{11}_{0k} \end{array}$$

in $\mathbf{A}$. So, under x, we get

$$\begin{array}{ccccccccc} & & \boxed{1} & & & & & & \\ & & \uparrow u_{34} & & & & & & \\ \boxed{0} & \xleftarrow{u_{23}} & 4 & \xrightarrow{u_4} & 1 & \xleftarrow{u_4} & 4 & \xrightarrow{u_{23}} & 0 \end{array}$$

in $\underline{\mathbf{M}}$. Thus $x(0^{11}_{0k}) = 0$, as required.

Now define $g \in M^\omega$ by $g(i) := \rho_i(a_i)$, where a_i belongs to the unique non-trivial block of $\ker(\rho_i{\restriction}_{A_0})$. Then $g = 0^{1}_{0}$. The set

$$C := A_0 \cup B \cup \{\, a \in M^\omega \mid a(0) = 0 \,\}$$

forms a subalgebra of $\underline{\mathbf{M}}^\omega$. Since $A \subseteq C$ and $g \notin C$, it follows that $g \notin A$. So $\underline{\mathbf{M}}$ is not dualisable, by the Non-dualisability Lemma, 3.4.1. ∎

7

Inherent dualisability

Beginning from any finite unary algebra with at least two fundamental operations, there is an infinite ascending chain of finite algebras that are alternately dualisable and non-dualisable. We will obtain this result while characterising the finite algebras that can be embedded into a non-dualisable algebra.

There is a natural way to strengthen the definition of non-dualisability. A finite algebra $\underline{\mathbf{M}}$ is called inherently non-dualisable if every finite algebra that has $\underline{\mathbf{M}}$ as a subalgebra is non-dualisable. Many of the algebras that are known to be non-dualisable are also inherently non-dualisable. For example, each non-dualisable two-element algebra is inherently non-dualisable, and each non-dualisable graph algebra is inherently non-dualisable [23]. Similarly, all the non-dualisable p-semilattices are inherently non-dualisable, by Theorem 5.2.4. In contrast, there are no inherently non-dualisable unary algebras at all: every finite unary algebra can be embedded into a dualisable algebra, by Theorem 2.1.4.

In this chapter, we shall consider the corresponding notion of inherent dualisability. We will say that a finite algebra $\underline{\mathbf{M}}$ is **inherently dualisable** if each finite algebra that has $\underline{\mathbf{M}}$ as a subalgebra is dualisable. We have already met algebras that are inherently dualisable simply by virtue of their type. Recall that a unar is a unary algebra with only one fundamental operation. Every finite unar is inherently dualisable, since all finite unars are dualisable, by Theorem 3.5.1.

We shall say that a type F is **small** if

- each operation symbol in F is either nullary or unary, and
- there is at most one unary operation symbol in F.

One of the primary aims of this chapter is to prove that a finite algebra is inherently dualisable if and only if it is of small type.

In Section 7.1, we show that every algebra that is not of small type is not inherently dualisable. In other words, we will prove that every finite algebra not of small type can be embedded into a non-dualisable algebra. We already know that every finite unary algebra can be embedded into a dualisable algebra, by Theorem 2.1.4. So we obtain an interesting corollary: for any finite unary algebra $\underline{\mathbf{M}}$ with at least two fundamental operations, there exists an infinite chain $\underline{\mathbf{M}} \leqslant \underline{\mathbf{M}}_0 \leqslant \underline{\mathbf{M}}_1 \leqslant \cdots$ such that $\underline{\mathbf{M}}_{2i}$ is dualisable and $\underline{\mathbf{M}}_{2i+1}$ is non-dualisable, for all $i \in \omega$.

In Section 7.2, we will prove that every finite algebra of small type is dualisable. It will then follow that every finite algebra of small type is inherently dualisable. Each algebra of small type is term equivalent to a unary algebra with at most one non-constant operation. In discussion, we shall refer to such a unary algebra as a **unar with added constants**. We saw in Theorem 3.5.1 that it is quite easy to prove that every finite unar is dualisable, using the binary-homomorphisms methods of Chapter 2. Unfortunately, this approach does not seem to work for unars with added constants. (Indeed, we saw in Chapter 6 that adding constant operations can sometimes destroy dualisability altogether.)

We will use a quite different style of proof to establish the dualisability of a large class of unary algebras, a class that includes all finite unars and all finite unars with added constants. A unary algebra $\underline{\mathbf{M}}$ is said to be **linear** if, for all unary term functions u and v of $\underline{\mathbf{M}}$, there exists a unary term function w of $\underline{\mathbf{M}}$ such that

$$u = w \circ v \quad \text{or} \quad v = w \circ u.$$

We shall prove that every finite linear unary algebra is dualisable. The class of linear unary algebras has been studied in other contexts. M. A. Valeriote [64] showed that a finite unary algebra (of finite type) generates a variety with a decidable first-order theory if and only if it is linear. J. Sichler [62] has shown that a finite unary algebra generates a group-universal variety if and only if it is not linear.

We finish this chapter by proving that all finite linear unary algebras are strongly dualisable. This generalises J. Hyndman's result that all finite unars are strongly dualisable [38]. We prove that all finite linear unary algebras have enough algebraic operations, which lifts dualisability up to strong dualisability. The concept of enough algebraic operations, defined in Chapter 1, is developed and extended in the appendix.

The first two sections of this chapter arose from a paper by the first author [53], and the last section is new.

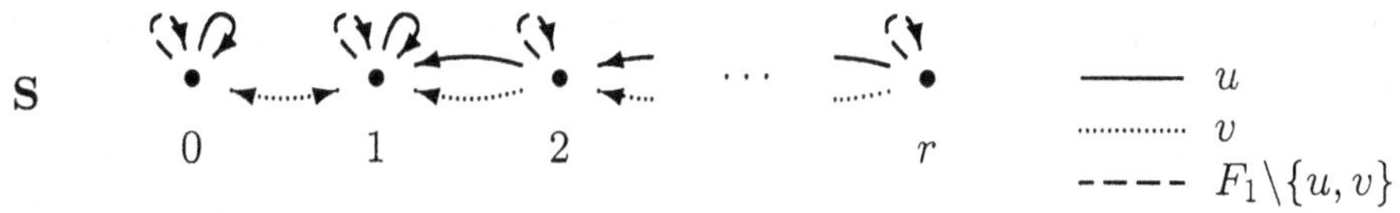

Figure 7.1

7.1 Embeddings into non-dualisable algebras

In this section, we show that every finite algebra that is not of small type can be embedded into a non-dualisable algebra. The following lemma deals with the unary case.

7.1.1 Lemma *Let F be a type such that*

(i) *each operation symbol in F is either nullary or unary, and*

(ii) *there are at least two unary operation symbols in F.*

Then every finite algebra of type F can be embedded into a non-dualisable algebra.

Proof Let F_1 denote the set of all unary operation symbols in F, and choose distinct symbols $u, v \in F_1$. Let $\underline{\mathbf{M}} = \langle M; F^{\underline{\mathbf{M}}}\rangle$ be a finite algebra of type F and assume that $M \cap \omega = \varnothing$. Define $r := |M| + 1$ and $S := \{0, \dots, r\}$. We want to define an extension $\underline{\mathbf{N}} = \langle M \cup S; F^{\underline{\mathbf{N}}}\rangle$ of $\underline{\mathbf{M}}$. To do this, we just need to say how the unary operations in $F^{\underline{\mathbf{N}}}$ act on S. For the two chosen symbols $u, v \in F_1$, set

$$u^{\underline{\mathbf{N}}}(s) = \begin{cases} 0 & \text{if } s = 0, \\ 1 & \text{if } s = 1, \\ s-1 & \text{otherwise,} \end{cases} \quad \text{and} \quad v^{\underline{\mathbf{N}}}(s) = \begin{cases} 1 & \text{if } s = 0, \\ 0 & \text{if } s = 1, \\ s-1 & \text{otherwise,} \end{cases}$$

for all $s \in S$. For every other symbol $w \in F_1\backslash\{u, v\}$, set $w^{\underline{\mathbf{N}}}{\restriction}_S = \mathrm{id}_S$. Now let $\underline{\mathbf{N}}_\flat$ denote the reduct of $\underline{\mathbf{N}}$ with type F_1. Then S is the underlying set of a subalgebra $\mathbf{S}$ of $\underline{\mathbf{N}}_\flat$. The algebra $\mathbf{S}$ is illustrated in Figure 7.1.

We will use the Non-dualisability Lemma, 3.4.1, to prove that $\underline{\mathbf{N}}$ is not dualisable. Define two subsets of S^ω by

$$A_0 := \{\, 1_n^0 \mid n \in \omega \,\} \quad \text{and} \quad A_1 := \{\, r_{mn}^{0\,1} \mid m, n \in \omega \text{ with } m \neq n \,\}.$$

Let $\mathbf{A}$ denote the subalgebra of $\underline{\mathbf{N}}^\omega$ generated by A_1, and let $\mathbf{A}^*$ denote the subalgebra of $(\underline{\mathbf{N}}_\flat)^\omega$ generated by A_1.

The unary algebra $\mathbf{A}^*$ is connected. To see this, let $k, \ell, m, n \in \omega$ such that $k \neq \ell$ and $m \neq n$. We shall show that there is a path between $r^{01}_{k\ell}$ and $r^{0\,1}_{mn}$ in the graph $G(\mathbf{A}^*)$. As A_1 generates $\mathbf{A}^*$, it will then follow that $\mathbf{A}^*$ is connected. First assume that $k \neq m$. Then

$$r^{01}_{k\ell} \xrightarrow{u^{r-1}} 1^0_k \xleftarrow{u^{r-1}} r^{0\,1}_{km} \xrightarrow{v} (r-1)^{1\,0}_{km} \xrightarrow{u^{r-2}} 1^0_m \xleftarrow{u^{r-1}} r^{0\,1}_{mn}$$

in $\mathbf{A}^*$. Next assume that $k = m$. We have

$$r^{01}_{k\ell} \xrightarrow{u^{r-1}} 1^0_k \xleftarrow{u^{r-1}} r^{0\,1}_{mn}$$

in $\mathbf{A}^*$. There is a path in $G(\mathbf{A}^*)$ between $r^{01}_{k\ell}$ and $r^{0\,1}_{mn}$. So $\mathbf{A}^*$ is connected.

Now let $x : \mathbf{A} \to \underline{\mathbf{N}}$ be a homomorphism. For every $n \in \omega$, we have $1^0_n = u^{r-1}(r^{0\;\;1}_{n\,n+1})$. Therefore $A_0 \subseteq A^* \subseteq A$. We want to show that there is a unique non-trivial block of $\ker(x{\restriction}_{A_0})$. Since the algebra $\mathbf{A}^*$ is connected, we know that $x(A^*) \subseteq M$ or $x(A^*) \subseteq S$.

Case 1: $x(A^*) \subseteq M$. Let $m, n \in \omega$ such that $m \neq n$. We shall prove that $x(1^0_m) = x(1^0_n)$. It will then follow that $\ker(x{\restriction}_{A_0})$ has only one block. First, let $k \in \omega \backslash \{m, n\}$ and let $\ell \in \{m, n\}$. Then

$$r^{01}_{k\ell} \xrightarrow{u} (r-1)^{01}_{k\ell} \xrightarrow{u} \cdots \xrightarrow{u} 2^{01}_{k\ell} \xrightarrow{u} 1^0_k \circlearrowleft u$$

in $\mathbf{A}^*$. As $|x(A^*)| \leqslant |M| = r-1$, the map x must collapse two of the above elements of A^*. Since x preserves u, it follows that x collapses $2^{01}_{k\ell}$ and 1^0_k. So

$$x(2^{0\,1}_{km}) = x(1^0_k) = x(2^{01}_{kn}).$$

We now have

$$x(1^0_m) = x(v(2^{0\,1}_{km})) = v(x(2^{0\,1}_{km})) = v(x(2^{01}_{kn})) = x(v(2^{01}_{kn})) = x(1^0_n).$$

Thus $\ker(x{\restriction}_{A_0})$ has only one block.

Case 2: $x(A^*) \subseteq S$. For all $n \in \omega$, we have $u(x(1^0_n)) = x(u(1^0_n)) = x(1^0_n)$. Therefore $x(A_0) \subseteq \{0, 1\}$. Since we are trying to prove that $\ker(x{\restriction}_{A_0})$ has a unique non-trivial block, we can assume that $x(A_0) \neq \{1\}$. There is some $n \in \omega$ such that $x(1^0_n) = 0$. Let $m \in \omega$ with $m \neq n$. Then

$$1^0_n \xleftarrow{u} 2^{0\,1}_{nm} \xrightarrow{v} 1^0_m \qquad \overset{x}{\Longrightarrow} \qquad \boxed{0} \xleftarrow{u} 0 \xrightarrow{v} 1,$$

and therefore $x(1^0_m) = 1$. Thus $A_0 \backslash \{1^0_n\}$ is the unique non-trivial block of $\ker(x{\restriction}_{A_0})$.

Now define $g \in N^\omega$ by $g(i) := \rho_i(a_i)$, where a_i is any element of the unique non-trivial block of $\ker(\rho_i{\restriction}_{A_0})$. Then g is the constant sequence $\widehat{1}$. We can define a subuniverse B of $\underline{\mathbf{N}}^\omega$ by

$$B := \{\, \widehat{m} \mid m \in M \,\} \cup \{\, a \in S^\omega \mid \{0,1\} \subseteq a(\omega) \,\}.$$

Since $A_1 \subseteq B$, it follows that $A \subseteq B$. Thus $\widehat{1} \notin A$, whence $\underline{\mathbf{N}}$ is non-dualisable, by the Non-dualisability Lemma, 3.4.1. ∎

The next result highlights the complexity of dualisability for unary algebras. It follows from the previous lemma and Theorem 2.1.4.

7.1.2 Theorem *Let $\underline{\mathbf{M}} = \langle M; F^{\underline{\mathbf{M}}}\rangle$ be a finite unary algebra with $|F| \geqslant 2$. There is an infinite chain $\underline{\mathbf{M}} \leqslant \underline{\mathbf{M}}_0 \leqslant \underline{\mathbf{M}}_1 \leqslant \cdots$ of finite algebras such that $\underline{\mathbf{M}}_{2i}$ is dualisable and $\underline{\mathbf{M}}_{2i+1}$ is non-dualisable, for all $i \in \omega$.*

It remains to show that each finite non-unary algebra can be embedded into a non-dualisable algebra. We will be building our proof around the two-element implication algebra $\underline{\mathbf{I}} = \langle \{0,1\}; \to\rangle$, where the binary operation $\to$ is given as follows.

$\to$	0	1
0	1	1
1	0	1

Despite its innocent appearance, the algebra $\underline{\mathbf{I}}$ is very badly behaved. Davey, Idziak, Lampe and McNulty [23] proved that the algebra $\underline{\mathbf{I}}$ is inherently non-dualisable. We shall prove that $\underline{\mathbf{I}}$ satisfies an even stronger version of non-dualisability.

Let $\mathbf{A}$ be any algebra. An algebra $\mathbf{B}$ is called a **subreduct of A** if there is a term reduct $\mathbf{A}_\flat$ of $\mathbf{A}$ such that $\mathbf{B} \leqslant \mathbf{A}_\flat$. We will say that a finite algebra $\underline{\mathbf{M}}$ is **contagiously non-dualisable** if each finite algebra $\underline{\mathbf{N}}$ that satisfies the following two conditions is non-dualisable:

(i) $\underline{\mathbf{M}}$ is a subreduct of $\underline{\mathbf{N}}$;

(ii) for each $k \in \omega$ and each term function $\tau : N^k \to N$ of $\underline{\mathbf{N}}$, we have

  (a) $\tau(M^k) \subseteq N\backslash M$, or

  (b) $\tau(M^k) \subseteq M$ and $\tau{\restriction}_{M^k} : M^k \to M$ is a term function of $\underline{\mathbf{M}}$.

Every contagiously non-dualisable algebra is inherently non-dualisable.

The following lemma provides a method for showing that an algebra is contagiously non-dualisable, based on a result due to Davey, Idziak, Lampe and McNulty [23, 8]; see the Inherent Non-dualisability Theorem, 5.2.2.

7.1.3 Lemma *Let $\underline{\mathbf{M}}$ be a finite algebra and let $f : \omega \to \omega$. Assume that there is a subalgebra $\mathbf{A}$ of $\underline{\mathbf{M}}^S$, for some set S, and an infinite subset A_0 of A such that*

(i) *for every $n \in \omega$ and every congruence θ on $\mathbf{A}$ of index at most n, the equivalence relation $\theta{\restriction}_{A_0}$ has a unique block of size greater than $f(n)$,*

(ii) *the algebra $\mathbf{A}$ does not contain the element g of M^S that is defined by $g(s) := \rho_s(a_s)$, where a_s is any element of the unique block of $\ker(\rho_s){\restriction}_{A_0}$ of size greater than $f(|M|)$.*

Then $\underline{\mathbf{M}}$ is contagiously non-dualisable.

Proof Let $\underline{\mathbf{N}}$ be a finite algebra such that $\underline{\mathbf{M}}$ is a subreduct of $\underline{\mathbf{N}}$ and, for each $k \in \omega$, every term function $\tau : N^k \to N$ of $\underline{\mathbf{N}}$ satisfies

(a) $\tau(M^k) \subseteq N \backslash M$, or

(b) $\tau(M^k) \subseteq M$ and $\tau{\restriction}_{M^k} : M^k \to M$ is a term function of $\underline{\mathbf{M}}$.

We will prove that the algebra $\underline{\mathbf{N}}$ is non-dualisable, using the Non-dualisability Lemma, 3.4.1, with the bound $n := f(|N|)$. As $A \subseteq M^S \subseteq N^S$, we can define $\mathbf{A}^+$ to be the subalgebra of $\underline{\mathbf{N}}^S$ generated by A. We have $A_0 \subseteq A \subseteq A^+$, and so A_0 is an infinite subset of A^+.

Let $x : \mathbf{A}^+ \to \underline{\mathbf{N}}$ be a homomorphism. We want to show that the equivalence relation $\ker(x{\restriction}_{A_0})$ has a unique block of size greater than n. Since $\underline{\mathbf{M}}$ is a subreduct of $\underline{\mathbf{N}}$, the algebra $\mathbf{A}$ is a subreduct of the algebra $\mathbf{A}^+$. So $\ker(x{\restriction}_A)$ is a congruence on $\mathbf{A}$ of index at most $|N|$. Thus, by assumption, there is a unique block of the equivalence relation $\ker(x{\restriction}_A){\restriction}_{A_0} = \ker(x{\restriction}_{A_0})$ that has size greater than $f(|N|) = n$.

For each $s \in S$, let $\rho_s^+ : \mathbf{A}^+ \to \underline{\mathbf{N}}$ be the projection $\pi_s{\restriction}_{A^+}$. Now define $g^+ \in N^S$ by $g^+(s) := \rho_s^+(a_s)$, where a_s is any element of the unique block of $\ker(\rho_s^+{\restriction}_{A_0})$ of size greater than n. It remains to show that g^+ is a ghost element of $\mathbf{A}^+$, that is, to show that $g^+ \notin A^+$.

We are assuming that A_0 is infinite. For all $s \in S$, the congruence $\ker(\rho_s)$ on $\mathbf{A}$ is of finite index, and so the equivalence relation $\ker(\rho_s){\restriction}_{A_0} = \ker(\rho_s^+{\restriction}_{A_0})$ has a unique infinite block. It follows that $g^+ = g \in M^S \backslash A$. We now want to show that $A^+ \subseteq A \cup (N \backslash M)^S$. To this end, let $\tau : N^k \to N$ be a term function of $\underline{\mathbf{N}}$, for some $k \in \omega$, and let $a_0, \dots, a_{k-1} \in A$. If $\tau(M^k) \subseteq N \backslash M$, then $\tau(a_0, \dots, a_{k-1}) \in (N \backslash M)^S$. If $\tau(M^k) \subseteq M$ and $\tau{\restriction}_{M^k} : M^k \to M$ is a term function of $\underline{\mathbf{M}}$, then $\tau(a_0, \dots, a_{k-1}) \in A$. Thus $A^+ \subseteq A \cup (N \backslash M)^S$. Since $g^+ \in M^S \backslash A$, this implies that $g^+ \notin A^+$. We can now use the Non-dualisability Lemma to conclude that $\underline{\mathbf{N}}$ is non-dualisable. Hence $\underline{\mathbf{M}}$ is contagiously non-dualisable. ∎

The proof of the lemma below comes from the proof used by Davey, Idziak, Lampe and McNulty to show that the two-element implication algebra $\underline{\mathbf{I}}$ is inherently non-dualisable [23, Lemma 5].

7.1.4 Lemma *The two-element implication algebra is contagiously non-dualisable.*

Proof Let $\underline{\mathbf{I}} = \langle\{0,1\};\rightarrow\rangle$ be the two-element implication algebra. We will use Lemma 7.1.3 to show that $\underline{\mathbf{I}}$ is contagiously non-dualisable. Our bounding function will be the constant map $\underline{1}: \omega \to \omega$. Define $\mathbf{A}$ to be the subalgebra of $\underline{\mathbf{I}}^\omega$ generated by the set

$$A_0 := \{\, 0^1_n \mid n \in \omega \}.$$

Let θ be a congruence on $\mathbf{A}$ of finite index. Assume we have $0^1_k \equiv_\theta 0^1_\ell$ and $0^1_m \equiv_\theta 0^1_n$, for some $k, \ell, m, n \in \omega$ with $k \neq \ell$ and $m \neq n$. To prove that $\theta{\restriction}_{A_0}$ has a unique non-trivial block, it suffices to show that $0^1_k \equiv_\theta 0^1_m$.

Using the definition of the operation $\rightarrow$ from page 183, we have

$$(0^1_k \to 0^1_\ell) \to 0^1_m = 1^0_k \to 0^1_m = 0^{1\,1}_{km}$$

and

$$(0^1_k \to 0^1_k) \to 0^1_m = \widehat{1} \to 0^1_m = 0^1_m$$

in $\mathbf{A}$. This gives us

$$0^1_m = (0^1_k \to 0^1_k) \to 0^1_m \equiv_\theta (0^1_k \to 0^1_\ell) \to 0^1_m = 0^{1\,1}_{km}.$$

By symmetry, we also have $0^1_k \equiv_\theta 0^{1\,1}_{mk}$. Thus $0^1_k \equiv_\theta 0^1_m$, and so $\theta{\restriction}_{A_0}$ has a unique non-trivial block.

Now define g in I^ω by $g(s) := \rho_s(a_s)$, where a_s belongs to the unique non-trivial block of $\ker(\rho_s){\restriction}_{A_0}$. Then g is the constant sequence $\widehat{0}$. The algebra $\underline{\mathbf{I}}$ satisfies $a \to b = 0 \implies b = 0$, for all $a, b \in I$, and so it follows that $g = \widehat{0} \notin A$. Thus $\underline{\mathbf{I}}$ is contagiously non-dualisable. ∎

We are now ready to show that every finite non-unary algebra can be embedded into a non-dualisable algebra.

7.1.5 Lemma *Let $\underline{\mathbf{M}}$ be a finite algebra that has an n-ary fundamental operation, for some $n \in \omega\backslash\{0,1\}$. Then $\underline{\mathbf{M}}$ can be embedded into a non-dualisable algebra.*

Proof Assume that $0, 1 \notin M$. Let F denote the type of $\underline{\mathbf{M}}$ and choose an operation $h \in F$ such that $k := \text{arity}(h) \geqslant 2$. We want to define an extension $\underline{\mathbf{N}} = \langle M \cup \{0,1\}; F^{\underline{\mathbf{N}}}\rangle$ of $\underline{\mathbf{M}}$. Let $\rightarrow$ denote the usual implication operation on $\{0,1\}$. We define the operation $h^{\underline{\mathbf{N}}} : N^k \to N$ by

$$h^{\underline{\mathbf{N}}}(a_0,\dots,a_{k-1}) = \begin{cases} h^{\underline{\mathbf{M}}}(a_0,\dots,a_{k-1}) & \text{if } a_0,\dots,a_{k-1} \in M, \\ a_0 \to a_1 & \text{if } a_0,\dots,a_{k-1} \in \{0,1\}, \\ a_0 & \text{otherwise.} \end{cases}$$

For each $f \in F \backslash \{h\}$ such that $\ell := \operatorname{arity}(f) \neq 0$, we shall define the operation $f^{\underline{\mathbf{N}}} : N^\ell \to N$ by

$$f^{\underline{\mathbf{N}}}(a_0,\dots,a_{\ell-1}) = \begin{cases} f^{\underline{\mathbf{M}}}(a_0,\dots,a_{\ell-1}) & \text{if } a_0,\dots,a_{\ell-1} \in M, \\ a_0 & \text{otherwise.} \end{cases}$$

The nullary operations of $\underline{\mathbf{N}}$ have the same values as the corresponding nullary operations of $\underline{\mathbf{M}}$.

Define the term function $\twoheadrightarrow : N^2 \to N$ of $\underline{\mathbf{N}}$ by $a \twoheadrightarrow b := h^{\underline{\mathbf{N}}}(a,b,\dots,b)$. Then $\twoheadrightarrow$ agrees with the implication operation $\to$ on $I := \{0,1\}$. So the two-element implication algebra $\underline{\mathbf{I}}$ is a subreduct of $\underline{\mathbf{N}}$.

We already know that $\underline{\mathbf{I}}$ is contagiously non-dualisable. So let $n \in \omega$. To see that $\underline{\mathbf{N}}$ is non-dualisable, it is enough to show that, for each n-ary term τ of type F, one of the following conditions is satisfied:

(a) $\tau^{\underline{\mathbf{N}}}(I^n) \subseteq M$;

(b) $\tau^{\underline{\mathbf{N}}}(I^n) \subseteq I$ and $\tau^{\underline{\mathbf{N}}}{\restriction}_{I^n} : I^n \to I$ is a term function of $\underline{\mathbf{I}}$.

For each nullary term τ of type F, the value of $\tau^{\underline{\mathbf{N}}}$ belongs to M. So we can assume that $n \neq 0$. We will argue by induction. Every variable, viewed as an n-ary term of type F, satisfies (b) and every nullary operation symbol in F, viewed as an n-ary term of type F, satisfies (a).

Now let $f \in F$ with $\ell := \operatorname{arity}(f) \neq 0$. Assume that $\tau_0,\dots,\tau_{\ell-1}$ are n-ary terms of type F, each of which satisfies (a) or (b). We want to show that the term $f(\tau_0,\dots,\tau_{\ell-1})$ satisfies (a) or (b).

Case 1: τ_i satisfies (a), for all $i \in \{0,\dots,\ell-1\}$. For all $a \in I^n$, we have

$$\begin{aligned} f(\tau_0,\dots,\tau_{\ell-1})^{\underline{\mathbf{N}}}(a) &= f^{\underline{\mathbf{N}}}(\tau_0^{\underline{\mathbf{N}}}(a),\dots,\tau_{\ell-1}^{\underline{\mathbf{N}}}(a)) \\ &= f^{\underline{\mathbf{M}}}(\tau_0^{\underline{\mathbf{N}}}(a),\dots,\tau_{\ell-1}^{\underline{\mathbf{N}}}(a)) \in M, \end{aligned}$$

as $\underline{\mathbf{M}} \leqslant \underline{\mathbf{N}}$. So $f(\tau_0,\dots,\tau_{\ell-1})$ satisfies (a).

Case 2: τ_i satisfies (b), for all $i \in \{0,\dots,\ell-1\}$. First assume that $f \neq h$. Then, for all $a \in I^n$, we have

$$f(\tau_0,\dots,\tau_{\ell-1})^{\underline{\mathbf{N}}}(a) = f^{\underline{\mathbf{N}}}(\tau_0^{\underline{\mathbf{N}}}(a),\dots,\tau_{\ell-1}^{\underline{\mathbf{N}}}(a)) = \tau_0^{\underline{\mathbf{N}}}(a).$$

So $f(\tau_0,\dots,\tau_{\ell-1})$ satisfies (b).

Next assume that $f = h$. Then, for every $a \in I^n$, we get

$$f(\tau_0, \dots, \tau_{\ell-1})^{\underline{\mathbf{N}}}(a) = h^{\underline{\mathbf{N}}}(\tau_0^{\underline{\mathbf{N}}}(a), \dots, \tau_{k-1}^{\underline{\mathbf{N}}}(a))$$
$$= \tau_0^{\underline{\mathbf{N}}}(a) \to \tau_1^{\underline{\mathbf{N}}}(a) = (\tau_0^{\underline{\mathbf{N}}}{\restriction_{I^n}} \to \tau_1^{\underline{\mathbf{N}}}{\restriction_{I^n}})(a).$$

So $f(\tau_0, \dots, \tau_{\ell-1})$ satisfies (b).

Case 3: there are $i, j \in \{0, \dots, \ell - 1\}$ such that τ_i satisfies (a) and τ_j satisfies (b). In this case, we have

$$f(\tau_0, \dots, \tau_{\ell-1})^{\underline{\mathbf{N}}}(a) = f^{\underline{\mathbf{N}}}(\tau_0^{\underline{\mathbf{N}}}(a), \dots, \tau_{\ell-1}^{\underline{\mathbf{N}}}(a)) = \tau_0^{\underline{\mathbf{N}}}(a),$$

for all $a \in I^n$. So $f(\tau_0, \dots, \tau_{\ell-1})$ satisfies (a) or (b), as τ_0 satisfies (a) or (b).

It now follows by induction that each n-ary term of type F satisfies (a) or (b). Hence $\underline{\mathbf{N}}$ is non-dualisable. ∎

The next theorem follows from Lemmas 7.1.1 and 7.1.5.

7.1.6 Theorem *Each finite algebra that is not of small type can be embedded into a non-dualisable algebra.*

Our proof for the unary case (Lemma 7.1.1) was slightly more subtle than our proof for the non-unary case (Lemma 7.1.5). To extend a finite non-unary algebra to make it non-dualisable, we simply attached the two-element implication algebra to it. But the way we extended a finite unary algebra to make it non-dualisable depended on its size. As the next lemma shows, there is no finite unary algebra that is as badly behaved as the two-element implication algebra.

7.1.7 Lemma *There does not exist a finite unary algebra $\underline{\mathbf{N}}$ such that, for each finite algebra $\underline{\mathbf{M}}$ of the same type as $\underline{\mathbf{N}}$, the algebra $\underline{\mathbf{M}} \,\dot\cup\, \underline{\mathbf{N}}$ is non-dualisable.*

Proof Suppose a finite unary algebra $\underline{\mathbf{N}}$ exists in opposition to the lemma. By Theorem 2.1.4, we know that $\underline{\mathbf{N}}$ is a subalgebra of a dualisable algebra $\underline{\mathbf{N}}^+$. As $\underline{\mathbf{N}} \in \mathbb{ISP}(\underline{\mathbf{N}}^+)$, the algebra $\underline{\mathbf{N}}^+ \,\dot\cup\, \underline{\mathbf{N}}$ is dualisable, by Theorem 5.1.10. But $\underline{\mathbf{N}}^+ \,\dot\cup\, \underline{\mathbf{N}}$ must be non-dualisable, by our supposition, giving a contradiction. ∎

7.2 Linear unary algebras are dualisable

This section is devoted to proving that every finite algebra of small type is dualisable. In fact, we will show that every finite linear unary algebra is dualisable. (Recall that the unary algebra $\underline{\mathbf{M}}$ is linear provided that, for all unary term functions u and v of $\underline{\mathbf{M}}$, there exists a unary term function w of $\underline{\mathbf{M}}$ such that $u = w \circ v$ or $v = w \circ u$.) Each algebra of small type is term equivalent to a unary algebra with at most one non-constant operation. Our first lemma shows that these unary algebras are all linear.

7.2.1 Lemma *Every unary algebra with at most one non-constant operation is linear. In particular, every algebra of small type is term equivalent to a linear unary algebra.*

Proof Let $\underline{\mathbf{M}}$ be a unary algebra with at most one non-constant operation, and let u and v be unary term functions of $\underline{\mathbf{M}}$. If u is constant, then $u = u \circ v$. So we can assume that neither u nor v is constant. There is a unique non-constant operation $w : M \to M$ of $\underline{\mathbf{M}}$. So there must exist $m, n \in \omega$ such that $u = w^m$ and $v = w^n$. Assume $m \leqslant n$. Then w^{n-m} is a unary term function of $\underline{\mathbf{M}}$, and

$$v = w^n = w^{n-m} \circ w^m = w^{n-m} \circ u.$$

Thus $\underline{\mathbf{M}}$ is linear. ∎

The linearity of $\underline{\mathbf{M}}$ implies that the algebras in $\mathbb{ISP}(\underline{\mathbf{M}})$ have a very simple structure. To see this, we will use ordered sets to capture the overall structure of unary algebras. Let $\mathbf{A}$ be a unary algebra. Recall that, for each $a \in A$, the subuniverse of $\mathbf{A}$ generated by a is denoted by $\mathrm{sg}_{\mathbf{A}}(a)$. Now define

$$\mathrm{Sub}_1(\mathbf{A}) := \{\, \mathrm{sg}_{\mathbf{A}}(a) \mid a \in A \,\} \cup \{\varnothing\},$$

and let $\mathbf{Sub_1}(\mathbf{A})$ denote the ordered set consisting of $\mathrm{Sub}_1(\mathbf{A})$ under set inclusion. We are including $\varnothing$ in $\mathrm{Sub}_1(\mathbf{A})$ to ensure that $\mathbf{Sub_1}(\mathbf{A})$ has a minimum element.

7.2.2 Example Define $\mathbf{A}$ to be the unary algebra shown in Figure 7.2. We will show how the structure of $\mathbf{A}$ is reflected in the ordered set $\mathbf{Sub_1}(\mathbf{A})$. Define the equivalence relation $\approx$ on A by

$$a \approx b \iff \mathrm{sg}_{\mathbf{A}}(a) = \mathrm{sg}_{\mathbf{A}}(b).$$

There is a natural order $\preccurlyeq$ on $A/{\approx}$, given by

$$a/{\approx} \preccurlyeq b/{\approx} \iff \mathrm{sg}_{\mathbf{A}}(a) \subseteq \mathrm{sg}_{\mathbf{A}}(b).$$

The ordered set $\widetilde{\mathbf{A}} = \langle A/{\approx}; \preccurlyeq \rangle$, drawn in Figure 7.2, captures the overall structure of $\mathbf{A}$. In particular, for each subset B of A, we have

$$\mathrm{sg}_{\mathbf{A}}(B) = \bigcup \{\, a/{\approx} \mid a/{\approx} \preccurlyeq b/{\approx} \text{ for some } b \in B \,\}.$$

The ordered set $\mathbf{Sub_1}(\mathbf{A})$ is isomorphic to $\mathbf{1} \oplus \widetilde{\mathbf{A}}$.

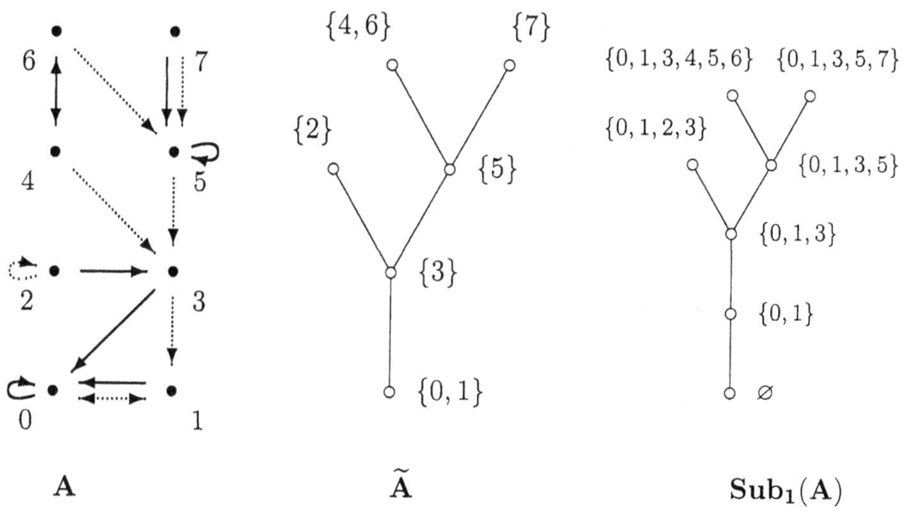

Figure 7.2 The structure of a unary algebra

For any ordered set $\mathbf{S} = \langle S; \leqslant \rangle$ and $s \in S$, we define the ordered set $\downarrow_{\mathbf{S}}(s)$ to be the set

$$\downarrow_{\mathbf{S}}(s) := \{ t \in S \mid t \leqslant s \}$$

equipped with the order induced from $\mathbf{S}$; the ordered set $\uparrow_{\mathbf{S}}(s)$ is defined dually. We say that an ordered set $\mathbf{S}$ is a **tree** if it is connected and the ordered set $\downarrow_{\mathbf{S}}(s)$ is a chain, for all $s \in S$.

For each algebra $\underline{\mathbf{M}}$, define $\mathbf{F}_{\underline{\mathbf{M}}}(1)$ to be the one-generated free algebra in the class $\mathbb{ISP}(\underline{\mathbf{M}})$, taking the universe of $\mathbf{F}_{\underline{\mathbf{M}}}(1)$ to be the set $F_{\underline{\mathbf{M}}}(1)$ of all unary term functions of $\underline{\mathbf{M}}$. Our next lemma demonstrates the origin of the name 'linear'.

7.2.3 Lemma *Let $\underline{\mathbf{M}}$ be a unary algebra. The following are equivalent:*

(i) *the algebra $\underline{\mathbf{M}}$ is linear;*

(ii) *the ordered set $\mathbf{Sub_1}(\mathbf{F}_{\underline{\mathbf{M}}}(1))$ is a chain;*

(iii) *the ordered set $\mathbf{Sub_1}(\mathbf{A})$ is a tree, for all $\mathbf{A} \in \mathbb{ISP}(\underline{\mathbf{M}})$.*

Proof The ordered set $\mathbf{Sub_1}(\mathbf{F}_{\underline{\mathbf{M}}}(1))$ has maximum element $F_{\underline{\mathbf{M}}}(1)$. So (iii) implies (ii). To show that (ii) implies (i), assume that $\mathbf{Sub_1}(\mathbf{F}_{\underline{\mathbf{M}}}(1))$ is a chain and let u and v be unary term functions of $\underline{\mathbf{M}}$. Then we have

$$\mathrm{sg}_{\mathbf{F}_{\underline{\mathbf{M}}}(1)}(u) \subseteq \mathrm{sg}_{\mathbf{F}_{\underline{\mathbf{M}}}(1)}(v) \quad \text{or} \quad \mathrm{sg}_{\mathbf{F}_{\underline{\mathbf{M}}}(1)}(v) \subseteq \mathrm{sg}_{\mathbf{F}_{\underline{\mathbf{M}}}(1)}(u).$$

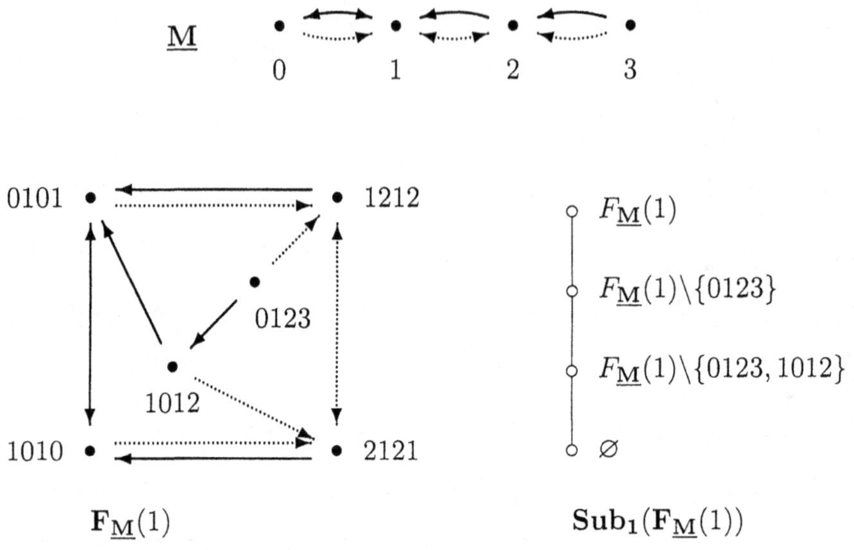

Figure 7.3 A linear unary algebra: $\underline{\mathbf{M}} = \langle \{0, 1, 2, 3\}; 1012, 1212 \rangle$

This implies that $u \in \mathrm{sg}_{\mathbf{F}_{\underline{\mathbf{M}}}(1)}(v)$ or $v \in \mathrm{sg}_{\mathbf{F}_{\underline{\mathbf{M}}}(1)}(u)$. So there is some unary term function w of $\underline{\mathbf{M}}$ such that $u = w^{\mathbf{F}_{\underline{\mathbf{M}}}(1)}(v)$ or $v = w^{\mathbf{F}_{\underline{\mathbf{M}}}(1)}(u)$. This gives us $u = w \circ v$ or $v = w \circ u$, whence $\underline{\mathbf{M}}$ is linear.

To prove that (i) implies (iii), assume that $\underline{\mathbf{M}}$ is linear and let $\mathbf{A} \in \mathbb{ISP}(\underline{\mathbf{M}})$. The ordered set $\mathbf{Sub_1}(\mathbf{A})$ is connected, since it has $\varnothing$ as a bottom element. Choose $a, b, c \in A$ such that $\mathrm{sg}_{\mathbf{A}}(a) \supseteq \mathrm{sg}_{\mathbf{A}}(b)$ and $\mathrm{sg}_{\mathbf{A}}(a) \supseteq \mathrm{sg}_{\mathbf{A}}(c)$. There are unary term functions u and v of $\underline{\mathbf{M}}$ such that $b = u^{\mathbf{A}}(a)$ and $c = v^{\mathbf{A}}(a)$. As $\underline{\mathbf{M}}$ is linear, there exists a unary term function w of $\underline{\mathbf{M}}$ such that $u = w \circ v$ or $v = w \circ u$. Since $\mathbf{A} \in \mathbb{ISP}(\underline{\mathbf{M}})$, we have

$$b = u^{\mathbf{A}}(a) = (w^{\mathbf{A}} \circ v^{\mathbf{A}})(a) = w^{\mathbf{A}}(c)$$

or

$$c = v^{\mathbf{A}}(a) = (w^{\mathbf{A}} \circ u^{\mathbf{A}})(a) = w^{\mathbf{A}}(b).$$

Thus $\mathrm{sg}_{\mathbf{A}}(b) \subseteq \mathrm{sg}_{\mathbf{A}}(c)$ or $\mathrm{sg}_{\mathbf{A}}(c) \subseteq \mathrm{sg}_{\mathbf{A}}(b)$. So $\mathbf{Sub_1}(\mathbf{A})$ is a tree. ∎

There are linear unary algebras with more than one non-constant operation. Using Figure 7.3 and Lemma 7.2.3, it is easy to check that, for example, the unary algebra $\langle \{0, 1, 2, 3\}; 1012, 1212 \rangle$ is linear.

We want to show that there is not too much variety amongst the algebras in $\mathbb{ISP}(\underline{\mathbf{M}})$ whenever $\underline{\mathbf{M}}$ is a finite linear unary algebra. Let $\mathbf{A}$ and $\mathbf{B}$ be

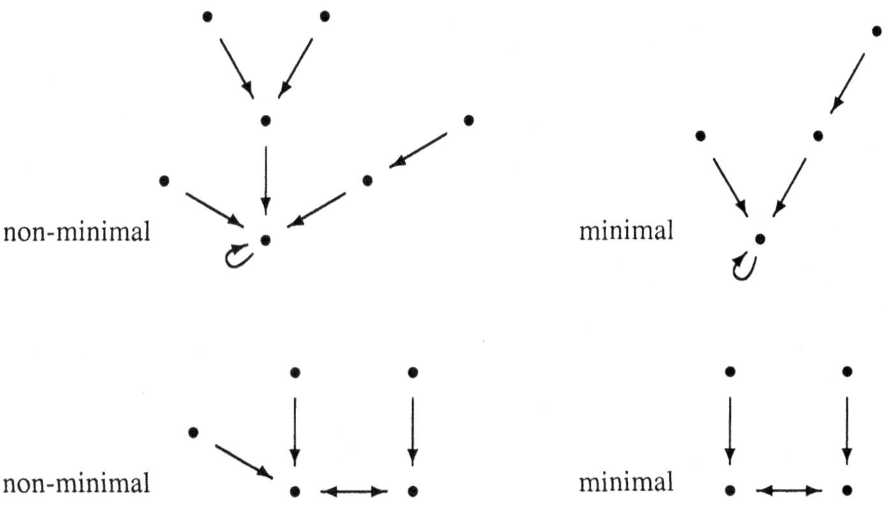

Figure 7.4 Minimal and non-minimal unars

algebras. We say that a retraction $\varphi : \mathbf{A} \twoheadrightarrow \mathbf{B}$ is **unbiased** if there is a set $\{ \psi_i : \mathbf{B} \hookrightarrow \mathbf{A} \mid i \in I \}$ of jointly surjective coretractions for φ. An unbiased retraction $\varphi : \mathbf{A} \twoheadrightarrow \mathbf{B}$ does not destroy too much of the structure of $\mathbf{A}$. We will say that the algebra $\mathbf{A}$ is **minimal** if, for each algebra $\mathbf{C}$, every unbiased retraction $\varphi : \mathbf{A} \twoheadrightarrow \mathbf{C}$ is an isomorphism.

In some sense, a minimal algebra has no repeated structure. Figure 7.4 gives some examples of minimal and non-minimal unars. There is an unbiased retraction from each non-minimal unar in Figure 7.4 onto the minimal unar to its right. It is easy to check that a composition of unbiased retractions is again an unbiased retraction. So, for each finite algebra $\mathbf{A}$, there is an unbiased retraction $\varphi : \mathbf{A} \twoheadrightarrow \mathbf{B}$, for some minimal algebra $\mathbf{B}$. We will be proving that, if $\underline{\mathbf{M}}$ is linear, then, up to isomorphism, there are only finitely many minimal algebras in $\mathbb{ISP}(\underline{\mathbf{M}})$.

7.2.4 Example We already know enough to verify that, for the unary algebra $\underline{\mathbf{R}} = \langle\{0, 1, 2\}; 121, 010\rangle$ studied in Chapter 1, there are only finitely many minimal algebras in $\mathbb{ISP}(\underline{\mathbf{R}})$. Using Lemma 7.2.3, it is straightforward to show that $\underline{\mathbf{R}}$ is linear. The petals of $\mathbb{ISP}(\underline{\mathbf{R}})$ were described in Lemma 1.2.2. So it is easy to check that the minimal petals of $\mathbb{ISP}(\underline{\mathbf{R}})$ are all shown in Figure 3.4. Every algebra in $\mathbb{ISP}(\underline{\mathbf{R}})$ is the coproduct of its petals. It follows that every minimal algebra of $\mathbb{ISP}(\underline{\mathbf{R}})$ is a coproduct of non-isomorphic minimal petals of $\mathbb{ISP}(\underline{\mathbf{R}})$. Therefore $\mathbb{ISP}(\underline{\mathbf{R}})$ has finitely many minimal algebras.

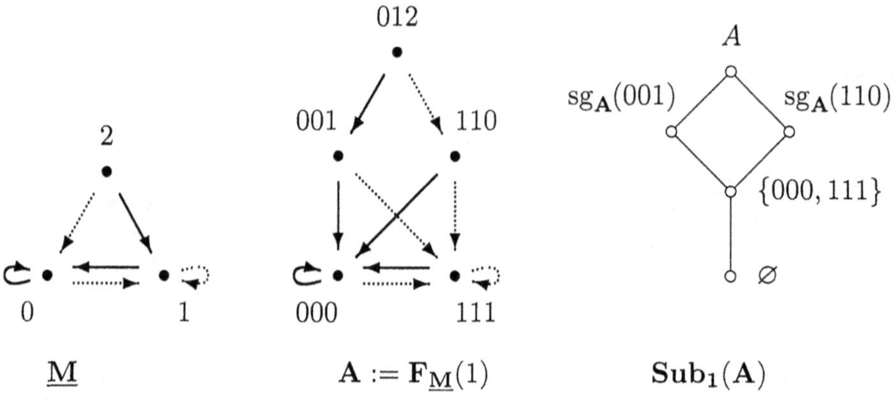

Figure 7.5 A non-linear one-kernel algebra: $\underline{\mathbf{M}} = \langle\{0,1,2\}; 001, 110\rangle$

In Chapter 3, we viewed the petals in Figure 3.4 as a gentle basis for $\mathbb{ISP}(\underline{\mathbf{R}})$. In general, if $\underline{\mathbf{M}}$ is a one-kernel algebra, then any gentle basis for $\mathbb{ISP}(\underline{\mathbf{M}})$ must include all the finite minimal petals of $\mathbb{ISP}(\underline{\mathbf{M}})$. This is because every gentle surjection is an unbiased retraction, by Lemma 3.2.4. There are certainly linear unary algebras that have more than one kernel: for example, the unar $\langle\{0,1,2,3\}; 0012\rangle$. The one-kernel three-element unary algebra $\langle\{0,1,2\}; 001, 110\rangle$ is not linear; see Figure 7.5.

In our proof that each finite linear unary algebra $\underline{\mathbf{M}}$ has only finitely many minimal algebras in its quasi-variety $\mathbb{ISP}(\underline{\mathbf{M}})$, we will be inducting on the complexity of the algebras in $\mathbb{ISP}(\underline{\mathbf{M}})$. Before we can define of our measure of complexity, we need the following easy lemma. For each ordered set $\mathbf{S}$, an element s of S is called a **node of S** if s is comparable with every other element of S.

7.2.5 Lemma *Let $\underline{\mathbf{M}}$ be a finite unary algebra and let $\mathbf{A} \in \mathbb{ISP}(\underline{\mathbf{M}})$.*

(i) *The ordered set $\mathbf{Sub_1}(\mathbf{A})$ has height at most $|F_{\underline{\mathbf{M}}}(1)|$.*

(ii) *The ordered set $\mathbf{Sub_1}(\mathbf{A})$ has a greatest node.*

Proof For each $S \in \mathrm{Sub}_1(\mathbf{A})$, we have

$$\left|{\downarrow}_{\mathbf{Sub_1(A)}}(S)\right| \leqslant |S| + 1 \leqslant \left|F_{\underline{\mathbf{M}}}(1)\right| + 1.$$

So the ordered set $\mathbf{Sub_1}(\mathbf{A})$ has height at most $|F_{\underline{\mathbf{M}}}(1)|$.

The minimum element $\varnothing$ of $\mathbf{Sub_1}(\mathbf{A})$ is a node of $\mathbf{Sub_1}(\mathbf{A})$. The nodes of any ordered set always form a chain inside the ordered set. Since $\mathbf{Sub_1}(\mathbf{A})$ has finite height, it follows that $\mathbf{Sub_1}(\mathbf{A})$ has a greatest node. ∎

Now let $\underline{\mathbf{M}}$ be a finite unary algebra and let $\mathbf{A} \in \mathbb{ISP}(\underline{\mathbf{M}})$. By the previous lemma, we know that $\mathbf{Sub_1}(\mathbf{A})$ has finite height and a greatest node. So we can define the **depth of A** to be the height of the ordered set $\uparrow_{\mathbf{Sub_1}(\mathbf{A})}(Y)$, where Y is the greatest node of $\mathbf{Sub_1}(\mathbf{A})$. For example, the algebra $\mathbf{A}$ in Figure 7.2 has depth 2.

7.2.6 Remark The concept of depth defined above is most useful within a quasi-variety that is generated by a linear unary algebra. Assume that $\underline{\mathbf{M}}$ is a finite linear unary algebra. Then we know that $\mathbf{Sub_1}(\mathbf{A})$ is a tree of finite height, for all $\mathbf{A} \in \mathbb{ISP}(\underline{\mathbf{M}})$, using Lemmas 7.2.3 and 7.2.5. Within a tree of finite height, the nodes form a covering chain from the bottom of the tree to the greatest node. (As an example, in Figure 7.2 the greatest node of $\mathbf{Sub_1}(\mathbf{A})$ is $\{0, 1, 3\}$.) So, in a sense, the depth of an algebra $\mathbf{A} \in \mathbb{ISP}(\underline{\mathbf{M}})$ gives us the height of the 'non-trivial' part of the ordered set $\mathbf{Sub_1}(\mathbf{A})$. Thus depth is a loose measure of complexity in $\mathbb{ISP}(\underline{\mathbf{M}})$.

Before launching into our proof that the quasi-variety generated by a finite linear unary algebra has a finite number of minimal algebras, we prove four useful lemmas.

We shall be invoking the following technical lemma many times during the rest of this chapter. Figure 7.6 illustrates this lemma in a particular instance for the algebra $\mathbf{A}$ considered in Example 7.2.2. It would be helpful to keep Figure 7.6 in mind throughout the rest of this chapter.

7.2.7 Lemma *Let* $\mathbf{A}$ *be a unary algebra such that the ordered set* $\mathbf{Sub_1}(\mathbf{A})$ *is a tree of finite height. Let* Z *belong to* $\mathrm{Sub_1}(\mathbf{A})$ *and let* $\mathfrak{U}_Z$ *denote the set of all upper covers of* Z *in* $\mathbf{Sub_1}(\mathbf{A})$. *For every* $S \in \mathfrak{U}_Z$, *define the sets* $S^{\vartriangle} := \{\, a \in A \mid S \subseteq \mathrm{sg}_{\mathbf{A}}(a) \,\}$ *and* $S^{\blacktriangle} := \mathrm{sg}_{\mathbf{A}}(S^{\vartriangle})$. *Then*

(i) $A \backslash S^{\vartriangle}$ *is a subuniverse of* $\mathbf{A}$, *for each* $S \in \mathfrak{U}_Z$,

(ii) $S^{\blacktriangle} = Z \cup S^{\vartriangle}$ *and* $S^{\blacktriangle} \cap (A \backslash S^{\vartriangle}) = Z$, *for each* $S \in \mathfrak{U}_Z$,

(iii) $S^{\vartriangle} \cap T^{\vartriangle} = \varnothing$ *and* $S^{\blacktriangle} \cap T^{\blacktriangle} = Z$, *for all* $S, T \in \mathfrak{U}_Z$ *with* $S \neq T$,

(iv) $A = \bigcup\{\, S^{\blacktriangle} \mid S \in \mathfrak{U}_Z \,\}$, *if* Z *is a node of* $\mathbf{Sub_1}(\mathbf{A})$ *and* $\mathfrak{U}_Z \neq \varnothing$.

Proof Let $S \in \mathfrak{U}_Z$. To prove (i), let $a \in A \backslash S^{\vartriangle}$ and let $b \in \mathrm{sg}_{\mathbf{A}}(a)$. Then we have $S \nsubseteq \mathrm{sg}_{\mathbf{A}}(a)$ and $\mathrm{sg}_{\mathbf{A}}(b) \subseteq \mathrm{sg}_{\mathbf{A}}(a)$, which implies that $S \nsubseteq \mathrm{sg}_{\mathbf{A}}(b)$. So $b \in A \backslash S^{\vartriangle}$, and therefore (i) holds.

To prove $\mathrm{sg}_{\mathbf{A}}(S^{\vartriangle}) = Z \cup S^{\vartriangle}$ for (ii), first let $a \in S^{\vartriangle}$ and let $b \in \mathrm{sg}_{\mathbf{A}}(a)$. Then $S \subseteq \mathrm{sg}_{\mathbf{A}}(a)$ and $\mathrm{sg}_{\mathbf{A}}(b) \subseteq \mathrm{sg}_{\mathbf{A}}(a)$. So $S \subseteq \mathrm{sg}_{\mathbf{A}}(b)$ or $\mathrm{sg}_{\mathbf{A}}(b) \subset S$, since $\mathbf{Sub_1}(\mathbf{A})$ is a tree. As Z is the lower cover of S in the tree $\mathbf{Sub_1}(\mathbf{A})$, this implies that $b \in S^{\vartriangle}$ or $b \in Z$. Therefore we have $\mathrm{sg}_{\mathbf{A}}(S^{\vartriangle}) \subseteq Z \cup S^{\vartriangle}$.

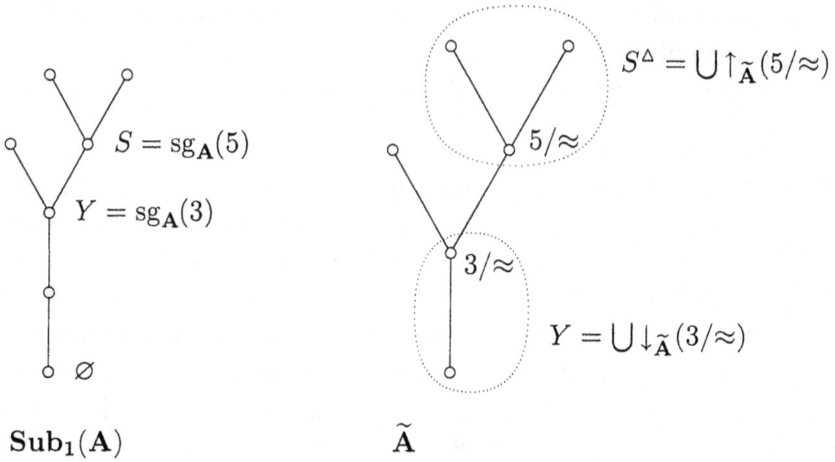

Figure 7.6 An illustration of Lemma 7.2.7, with Z the greatest node Y

Clearly, $S^{\Delta} \subseteq \mathrm{sg}_{\mathbf{A}}(S^{\Delta})$. So it remains to show that we have $Z \subseteq \mathrm{sg}_{\mathbf{A}}(S^{\Delta})$. As $Z \subset S$, the set S is non-empty. So there is some $c \in A$ such that $S = \mathrm{sg}_{\mathbf{A}}(c)$. We have $c \in S^{\Delta}$ and $Z \subseteq S = \mathrm{sg}_{\mathbf{A}}(c)$. This gives us $Z \subseteq \mathrm{sg}_{\mathbf{A}}(S^{\Delta})$, and therefore $S^{\blacktriangle} = \mathrm{sg}_{\mathbf{A}}(S^{\Delta}) = Z \cup S^{\Delta}$.

Since $Z \subset S$, we must have $Z \cap S^{\Delta} = \varnothing$. So we get

$$S^{\blacktriangle} \cap (A \backslash S^{\Delta}) = (Z \cup S^{\Delta}) \cap (A \backslash S^{\Delta}) = Z \backslash S^{\Delta} = Z.$$

Thus (ii) holds.

To prove (iii), let $S, T \in \mathcal{U}_Z$ such that $S \neq T$. First let $a \in S^{\Delta}$. Then $S \subseteq \mathrm{sg}_{\mathbf{A}}(a)$. Since S and T are non-comparable in the tree $\mathbf{Sub_1}(\mathbf{A})$, this implies that $T \nsubseteq \mathrm{sg}_{\mathbf{A}}(a)$. So $a \notin T^{\Delta}$, giving $S^{\Delta} \cap T^{\Delta} = \varnothing$. Using (ii), it follows that

$$S^{\blacktriangle} \cap T^{\blacktriangle} = (Z \cup S^{\Delta}) \cap (Z \cup T^{\Delta}) = Z \cup (S^{\Delta} \cap T^{\Delta}) = Z.$$

So (iii) holds.

Finally, for (iv), assume that Z is a node of $\mathbf{Sub_1}(\mathbf{A})$ and that $\mathcal{U}_Z \neq \varnothing$. Let $a \in A$. We must have $\mathrm{sg}_{\mathbf{A}}(a) \subseteq Z$ or $Z \subset \mathrm{sg}_{\mathbf{A}}(a)$. Assume that $Z \subset \mathrm{sg}_{\mathbf{A}}(a)$. As $\mathbf{Sub_1}(\mathbf{A})$ has finite height, we have $S \subseteq \mathrm{sg}_{\mathbf{A}}(a)$, for some $S \in \mathcal{U}_Z$. So $a \in S^{\Delta}$. It now follows that

$$A = Z \cup \bigcup \{\, S^{\Delta} \mid S \in \mathcal{U}_Z \,\} = \bigcup \{\, S^{\blacktriangle} \mid S \in \mathcal{U}_Z \,\},$$

using (ii). Hence (iv) holds. ∎

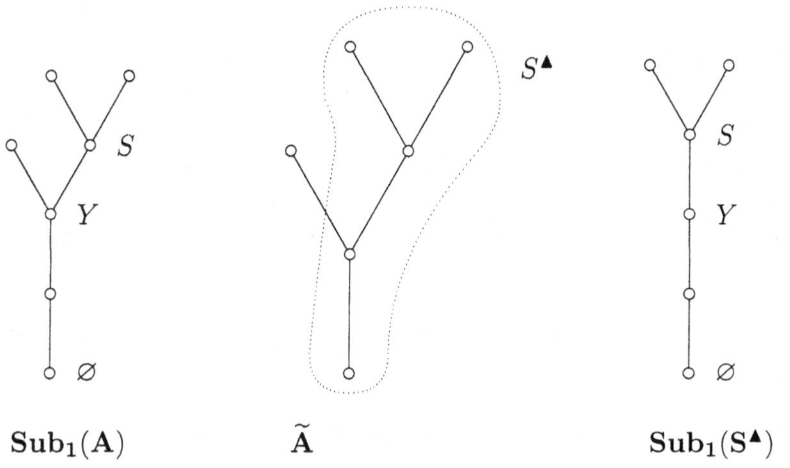

Figure 7.7 An illustration of Lemma 7.2.9

The next lemma helps with the initial step in our induction on the depth of unary algebras.

7.2.8 Lemma *Let* $\underline{\mathbf{M}}$ *be a finite unary algebra and let* $\mathbf{A} \in \mathbb{ISP}(\underline{\mathbf{M}})$. *Then* $\mathbf{A}$ *has depth* 0 *if and only if* $\mathbf{A}$ *is one-generated.*

Proof First assume that $\mathbf{A}$ has depth 0. Then the greatest node Y of $\mathbf{Sub_1}(\mathbf{A})$ is a maximal element of $\mathbf{Sub_1}(\mathbf{A})$. As Y is a node, it must be the maximum element of $\mathbf{Sub_1}(\mathbf{A})$. So $\mathbf{A}$ is a one-generated algebra.

Now assume that $\mathbf{A}$ is one-generated. Then A is the maximum element of $\mathbf{Sub_1}(\mathbf{A})$. So A is a node of $\mathbf{Sub_1}(\mathbf{A})$, and therefore $\mathbf{A}$ has depth 0. ∎

The next lemma helps with the inductive step. An illustration of this lemma is given in Figure 7.7. It may be helpful to look back on this picture during the rather dense proofs of Lemmas 7.2.11 and 7.3.1.

7.2.9 Lemma *Let* $\underline{\mathbf{M}}$ *be a finite linear unary algebra. Let* $\mathbf{A} \in \mathbb{ISP}(\underline{\mathbf{M}})$ *and let* Y *be the greatest node of* $\mathbf{Sub_1}(\mathbf{A})$. *Assume* $S \in \mathrm{Sub_1}(\mathbf{A})$ *with* $Y \subset S$. *Define the algebra* $\mathbf{S}^{\blacktriangle} := \mathbf{sg_A}(S^{\vartriangle})$, *where* $S^{\vartriangle} := \{\, a \in A \mid S \subseteq \mathrm{sg}_{\mathbf{A}}(a) \,\}$.

(i) *Both* S *and* Y *are nodes of* $\mathbf{Sub_1}(\mathbf{S}^{\blacktriangle})$.

(ii) *The depth of* $\mathbf{S}^{\blacktriangle}$ *is strictly less than the depth of* $\mathbf{A}$.

Proof By Lemma 7.2.3, the ordered set $\mathbf{Sub_1}(\mathbf{A})$ is a tree. We have

$$\mathrm{Sub_1}(\mathbf{S}^{\blacktriangle}) = \{\, T \in \mathrm{Sub_1}(\mathbf{A}) \bigm| T \subseteq S^{\blacktriangle} \,\},$$

and so $Y, S \in \mathrm{Sub}_1(\mathbf{S}^\blacktriangle)$. Let $b \in S^\blacktriangle$. Then $b \in \mathrm{sg}_\mathbf{A}(a)$, for some $a \in A$ with $S \subseteq \mathrm{sg}_\mathbf{A}(a)$. We know that $\mathrm{sg}_\mathbf{A}(b) \subseteq \mathrm{sg}_\mathbf{A}(a)$ and $Y \subseteq S \subseteq \mathrm{sg}_\mathbf{A}(a)$. Since $\mathbf{Sub_1}(\mathbf{A})$ is a tree, it follows that $\mathrm{sg}_\mathbf{A}(b) \subseteq S$ or $S \subseteq \mathrm{sg}_\mathbf{A}(b)$, and it also follows that $\mathrm{sg}_\mathbf{A}(b) \subseteq Y$ or $Y \subseteq \mathrm{sg}_\mathbf{A}(b)$. Therefore S and Y are nodes of $\mathbf{Sub_1}(\mathbf{S}^\blacktriangle)$.

As S is above Y in $\mathbf{Sub_1}(\mathbf{A})$, the height of the ordered set ${\uparrow}_{\mathbf{Sub_1}(\mathbf{S}^\blacktriangle)}(S)$ is less than the height of the ordered set ${\uparrow}_{\mathbf{Sub_1}(\mathbf{A})}(Y)$. We have shown that S is a node of $\mathbf{Sub_1}(\mathbf{S}^\blacktriangle)$. So the depth of $\mathbf{S}^\blacktriangle$ is strictly less than the depth of $\mathbf{A}$. ■

We shall prove one last lemma before we embark on our proof that the quasi-variety generated by a finite linear unary algebra only has a finite number of minimal algebras.

7.2.10 Lemma *Let* $\mathbf{A}$ *be a locally finite unary algebra, and let* $\mathbf{B}$ *and* $\mathbf{C}$ *be subalgebras of* $\mathbf{A}$.

(i) *Assume that* $\varphi : \mathbf{A} \twoheadrightarrow \mathbf{B}$ *is an unbiased retraction and that* S *is a node of* $\mathbf{Sub_1}(\mathbf{A})$. *Then* $S \subseteq B$, *with* $\varphi(S) = S$ *and* $\varphi^{-1}(S) = S$.

(ii) *Assume that* $\varphi : \mathbf{C} \to \mathbf{B}$ *is one-to-one on each one-generated subalgebra of* $\mathbf{C}$ *and that* S *is a node of* $\mathbf{Sub_1}(\mathbf{A})$ *with* $S \subseteq C$. *Then* $S \subseteq B$, *with* $\varphi(S) = S$ *and* $\varphi^{-1}(S) = S$.

Proof We begin by proving that every unbiased retraction is one-to-one on each one-generated subalgebra of its domain. Then (i) will follow from (ii), as a special case.

Assume $\varphi : \mathbf{A} \twoheadrightarrow \mathbf{B}$ is an unbiased retraction, and let $a \in A$. Since φ is unbiased, there is a coretraction $\psi : \mathbf{B} \hookrightarrow \mathbf{A}$ for φ with $a \in \psi(B)$. We have $\varphi \circ \psi = \mathrm{id}_B$, and therefore φ is one-to-one on $\psi(B) \supseteq \mathrm{sg}_\mathbf{A}(a)$.

We now prove (ii). Assume that $\varphi : \mathbf{C} \to \mathbf{B}$ is one-to-one on every one-generated subalgebra of $\mathbf{C}$. Let S be a node of $\mathbf{Sub_1}(\mathbf{A})$ with $S \subseteq C$. Then

$$\varphi(S) \in \mathrm{Sub}_1(\mathbf{B}) \subseteq \mathrm{Sub}_1(\mathbf{A}).$$

As S is a node of $\mathbf{Sub_1}(\mathbf{A})$, this implies that $\varphi(S) \subseteq S$ or $S \subset \varphi(S)$. Since $\mathbf{A}$ is locally finite, the set S is finite. So it follows that $\varphi(S) \subseteq S$. The map φ is one-to-one on the one-generated subalgebra S of $\mathbf{C}$. So $\varphi(S) = S$, which implies that $S \subseteq B$ and that $S \subseteq \varphi^{-1}(S)$.

To check that $\varphi^{-1}(S) \subseteq S$, let $c \in \varphi^{-1}(S) \subseteq C$. Then $\varphi(\mathrm{sg}_\mathbf{A}(c)) \subseteq S$. The map φ is one-to-one on $\mathrm{sg}_\mathbf{A}(c)$, and therefore $|\mathrm{sg}_\mathbf{A}(c)| \leqslant |S|$. Since S is a node of $\mathbf{Sub_1}(\mathbf{A})$, we have $\mathrm{sg}_\mathbf{A}(c) \subseteq S$ or $S \subset \mathrm{sg}_\mathbf{A}(c)$. So it follows that $c \in \mathrm{sg}_\mathbf{A}(c) \subseteq S$. We have now shown that $\varphi^{-1}(S) = S$. ■

The following result is a major step towards proving that every finite linear unary algebra is dualisable.

7.2.11 Lemma *Let* $\underline{\mathbf{M}}$ *be a finite linear unary algebra. Up to isomorphism, there are finitely many minimal algebras in* $\mathbb{ISP}(\underline{\mathbf{M}})$*, all of which are finite.*

Proof Define the quasi-variety $\mathcal{A} := \mathbb{ISP}(\underline{\mathbf{M}})$. We will prove, by induction, that there is a finite upper bound on the sizes of the minimal algebras in $\mathcal{A}$. Every algebra in $\mathcal{A}$ of depth 0 is one-generated, by Lemma 7.2.8. Therefore $|F_{\underline{\mathbf{M}}}(1)|$ is an upper bound on the sizes of the algebras in $\mathcal{A}$ of depth 0.

Now let $n \in \omega$. Assume that there is a finite upper bound k on the sizes of the minimal algebras in $\mathcal{A}$ of depth at most n. Let $\mathcal{M}$ consist of exactly one isomorphic copy of each minimal algebra in $\mathcal{A}$ of depth at most n. Since the quasi-variety $\mathcal{A}$ is locally finite, it follows that $\mathcal{M}$ is a finite set. Now assume that $\mathbf{A}$ is a minimal algebra in $\mathcal{A}$ of depth $n+1$. We shall bound the size of $\mathbf{A}$ by proving that $|A| \leqslant k^2|\mathcal{M}|$.

By Lemmas 7.2.3 and 7.2.5, the ordered set $\mathbf{Sub_1}(\mathbf{A})$ is a tree of finite height. Let Y denote the greatest node of $\mathbf{Sub_1}(\mathbf{A})$ and let $\mathcal{U}_Y$ denote the set of all upper covers of Y. Since $\mathbf{A}$ has depth $n+1$, the set $\mathcal{U}_Y$ is non-empty. For every $S \in \mathcal{U}_Y$, define the set

$$S^{\vartriangle} := \{\, a \in A \mid S \subseteq \mathrm{sg}_{\mathbf{A}}(a)\,\}$$

and define the algebra $\mathbf{S}^{\blacktriangle} := \mathbf{sg}_{\mathbf{A}}(S^{\vartriangle})$.

Claim 1 For all $S \in \mathcal{U}_Y$, the algebra $\mathbf{S}^{\blacktriangle}$ is minimal and has depth at most n.

Let $S \in \mathcal{U}_Y$. Since $\mathbf{A}$ has depth $n+1$, the algebra $\mathbf{S}^{\blacktriangle}$ has depth at most n, by Lemma 7.2.9(ii). To see that $\mathbf{S}^{\blacktriangle}$ is minimal, let $\varphi : \mathbf{S}^{\blacktriangle} \twoheadrightarrow \mathbf{B}$ be an unbiased retraction. We want to show that φ is an isomorphism. Without loss of generality, we can assume that $\mathbf{B}$ is a subalgebra of $\mathbf{S}^{\blacktriangle}$ and that $\varphi{\restriction}_B = \mathrm{id}_B$. Using Lemma 7.2.9(i), we know that S is a node of $\mathbf{Sub_1}(\mathbf{S}^{\blacktriangle})$. It follows by Lemma 7.2.10(i) that $S \subseteq B$, with $\varphi(S) = S$ and $\varphi^{-1}(S) = S$.

We will be applying the results in Lemma 7.2.7 many times throughout this proof, with $Z := Y$. By 7.2.7(i), the set $B^+ := B \cup (A\backslash S^{\vartriangle})$ forms a subalgebra of $\mathbf{A}$. In order to use the minimality of $\mathbf{A}$, we want to define an unbiased retraction

$$\varphi^+ : \mathbf{A} \twoheadrightarrow \mathbf{B}^+ \ \text{ by } \ \varphi^+ := \varphi \cup \mathrm{id}_{A\backslash S^{\vartriangle}}.$$

Since $S^{\vartriangle} \subseteq S^{\blacktriangle}$, we have $A = S^{\blacktriangle} \cup (A\backslash S^{\vartriangle})$. We know that $A\backslash S^{\vartriangle}$ is a subuniverse of $\mathbf{A}$, by 7.2.7(i), and that $S^{\blacktriangle} \cap (A\backslash S^{\vartriangle}) = Y \subseteq S$, by 7.2.7(ii). Since $S \subseteq B$, we must have $\varphi{\restriction}_S = \mathrm{id}_S$. It follows that φ^+ is a well-defined homomorphism. We have $\varphi^+{\restriction}_{B^+} = \mathrm{id}_{B^+}$, and thus φ^+ is a retraction.

We now want to show that φ^+ is unbiased. The inclusion map $\iota : \mathbf{B}^+ \hookrightarrow \mathbf{A}$ is a coretraction for φ^+ with image B^+. So choose some $a \in A\backslash B^+$. Then $a \in S^\vartriangle \subseteq S^\blacktriangle$. Since φ is unbiased, there is a coretraction $\psi : \mathbf{B} \hookrightarrow \mathbf{S}^\blacktriangle$ for φ such that $a \in \psi(B)$. Since $S \subseteq B$, we have $\varphi \circ \psi\restriction_S = \mathrm{id}_S$. As $\varphi^{-1}(S) = S$ and $\varphi\restriction_S = \mathrm{id}_S$, this implies that $\psi\restriction_S = \mathrm{id}_S$. We know that

$$B \cap (A\backslash S^\vartriangle) \subseteq S^\blacktriangle \cap (A\backslash S^\vartriangle) \subseteq S.$$

So we can define $\psi^+ : \mathbf{B}^+ \to \mathbf{A}$ by $\psi^+ := \psi \cup \mathrm{id}_{A\backslash S^\vartriangle}$. As

$$\varphi^+ \circ \psi^+ = (\varphi \circ \psi) \cup \mathrm{id}_{A\backslash S^\vartriangle} = \mathrm{id}_{B^+},$$

the homomorphism ψ^+ is a coretraction for φ^+ with $a \in \psi(B) \subseteq \psi^+(B^+)$. Thus $\varphi^+ : \mathbf{A} \twoheadrightarrow \mathbf{B}^+$ is an unbiased retraction. The algebra $\mathbf{A}$ is minimal, and therefore φ^+ is an isomorphism. This implies that φ is an isomorphism, whence $\mathbf{S}^\blacktriangle$ is minimal.

Claim 2 There do not exist $S, T \in \mathcal{U}_Y$, with $S \neq T$, for which there is an isomorphism $\varphi : \mathbf{S}^\blacktriangle \hookrightarrow\!\!\!\!\rightarrow \mathbf{T}^\blacktriangle$ such that $\varphi\restriction_Y = \mathrm{id}_Y$.
Let $S, T \in \mathcal{U}_Y$ with $S \neq T$. Suppose there is an isomorphism $\varphi : \mathbf{S}^\blacktriangle \hookrightarrow\!\!\!\!\rightarrow \mathbf{T}^\blacktriangle$ such that $\varphi\restriction_Y = \mathrm{id}_Y$. Since $A\backslash S^\vartriangle$ is a subuniverse of $\mathbf{A}$, by 7.2.7(i), and $S^\blacktriangle \cap (A\backslash S^\vartriangle) = Y$, by 7.2.7(ii), we can define the homomorphism

$$\psi : \mathbf{A} \to \mathbf{A} \ \text{ by } \ \psi := \varphi \cup \mathrm{id}_{A\backslash S^\vartriangle}.$$

By symmetry, we can define $\xi : \mathbf{A} \to \mathbf{A}$ by $\xi := \varphi^{-1} \cup \mathrm{id}_{A\backslash T^\vartriangle}$.

The set $A^* := A\backslash S^\vartriangle$ is a subuniverse of $\mathbf{A}$. We must have $S^\vartriangle \cap T^\vartriangle = \varnothing$, by 7.2.7(iii), and therefore

$$\varphi(S^\blacktriangle) = T^\blacktriangle = Y \cup T^\vartriangle \subseteq A\backslash S^\vartriangle = A^*,$$

using 7.2.7(ii). It follows that $\psi : \mathbf{A} \twoheadrightarrow \mathbf{A}^*$ is a retraction with $\psi\restriction_{A^*} = \mathrm{id}_{A^*}$.

The inclusion homomorphism $\iota : \mathbf{A}^* \hookrightarrow \mathbf{A}$ is a coretraction for ψ with image $A^* = A\backslash S^\vartriangle$. The homomorphism $\xi\restriction_{A^*} : \mathbf{A}^* \to \mathbf{A}$ is a coretraction for ψ with image $\xi(A^*) \supseteq S^\blacktriangle$. Thus $\psi : \mathbf{A} \twoheadrightarrow \mathbf{A}^*$ is an unbiased retraction. But ψ is not one-to-one, as

$$\psi(S^\vartriangle) \subseteq A^* = A\backslash S^\vartriangle = \psi(A\backslash S^\vartriangle).$$

This is a contradiction, since $\mathbf{A}$ is minimal.

We can now prove that $|A| \leqslant k^2|\mathcal{M}|$. Using 7.2.7(iv), we have

$$A = \bigcup\{\, S^\blacktriangle \mid S \in \mathcal{U}_Y \,\}.$$

By Claim 1, for all $S \in \mathcal{U}_Y$, there must be a minimal algebra $\mathbf{B} \in \mathcal{M}$ such that $\mathbf{S}^{\blacktriangle} \cong \mathbf{B}$. So $|S^{\blacktriangle}| \leqslant k$, for each $S \in \mathcal{U}_Y$. We need to bound the size of $\mathcal{U}_Y$.

Let $\mathbf{B} \in \mathcal{M}$ and assume that $S, T \in \mathcal{U}_Y$, with $S \neq T$, such that there are isomorphisms $\varphi_S : \mathbf{S}^{\blacktriangle} \hookrightarrow\!\!\!\!\rightarrow \mathbf{B}$ and $\varphi_T : \mathbf{T}^{\blacktriangle} \hookrightarrow\!\!\!\!\rightarrow \mathbf{B}$. Then there is an isomorphism $\varphi_T^{-1} \circ \varphi_S : \mathbf{S}^{\blacktriangle} \hookrightarrow\!\!\!\!\rightarrow \mathbf{T}^{\blacktriangle}$. We have $Y \subseteq S \subseteq S^{\blacktriangle}$ and $Y \subseteq T \subseteq T^{\blacktriangle}$. By Claim 2, we must have $\varphi_T^{-1} \circ \varphi_S{\restriction}_Y \neq \mathrm{id}_Y$. So $\varphi_S{\restriction}_Y \neq \varphi_T{\restriction}_Y$, and therefore $Y \neq \varnothing$. Thus $\varphi_S{\restriction}_Y : \mathbf{Y} \hookrightarrow \mathbf{B}$ and $\varphi_T{\restriction}_Y : \mathbf{Y} \hookrightarrow \mathbf{B}$ are distinct embeddings. There are at most $|B|$ ways to embed the one-generated algebra $\mathbf{Y}$ into $\mathbf{B}$. So

$$\big|\{\, S \in \mathcal{U}_Y \mid \mathbf{S}^{\blacktriangle} \cong \mathbf{B} \,\}\big| \leqslant |B| \leqslant k.$$

It follows that $|\mathcal{U}_Y| \leqslant k|\mathcal{M}|$. We have shown that $|S^{\blacktriangle}| \leqslant k$, for each $S \in \mathcal{U}_Y$, and so

$$|A| = \Big|\bigcup\{\, S^{\blacktriangle} \mid S \in \mathcal{U}_Y \,\}\Big| \leqslant \sum\{\, |S^{\blacktriangle}| \mid S \in \mathcal{U}_Y \,\} \leqslant k^2|\mathcal{M}|.$$

Thus there is a finite upper bound on the sizes of the minimal algebras in $\mathcal{A}$ of depth at most $n + 1$.

We know that $\mathbf{Sub_1}(\mathbf{A})$ has height at most $|F_{\underline{\mathbf{M}}}(1)|$, for all $\mathbf{A} \in \mathcal{A}$, by Lemma 7.2.5. So algebras in $\mathcal{A}$ can have depth at most $|F_{\underline{\mathbf{M}}}(1)|$. It now follows by induction that there is a finite upper bound on the sizes of the minimal algebras in $\mathcal{A}$. Hence, up to isomorphism, there are finitely many minimal algebras in $\mathcal{A}$, all of which are finite. ∎

We can now show that a finite unary algebra is dualisable if it is linear.

7.2.12 Theorem *Every finite linear unary algebra is dualisable.*

Proof Let $\underline{\mathbf{M}}$ be a finite linear unary algebra and define $\mathcal{A} := \mathbb{ISP}(\underline{\mathbf{M}})$. Let $\mathcal{B}$ consist of exactly one isomorphic copy of each minimal algebra in $\mathcal{A}$. Then $\mathcal{B}$ is a finite set of finite algebras, by Lemma 7.2.11. So we can choose $n \in \omega$ large enough so that $n \geqslant 4$ and $n \geqslant |\mathcal{A}(\mathbf{B}, \underline{\mathbf{M}})|$, for all $\mathbf{B} \in \mathcal{B}$. Define an alter ego of $\underline{\mathbf{M}}$ by $\underset{\sim}{\mathbf{M}} := \langle M; R_n, \mathcal{T}\rangle$. Let $\mathbf{A}$ be a finite algebra in $\mathcal{A}$ and let $\alpha : \mathrm{D}(\mathbf{A}) \to \underset{\sim}{\mathbf{M}}$ be a morphism. By the Duality Compactness Theorem, 1.4.2, it will follow that $\underset{\sim}{\mathbf{M}}$ yields a duality on $\mathcal{A}$ once we have proven that α is an evaluation.

There is an unbiased retraction $\varphi : \mathbf{A} \twoheadrightarrow \mathbf{B}$, for some minimal algebra $\mathbf{B}$ in $\mathcal{B}$. Let $\{\, \psi_i : \mathbf{B} \hookrightarrow \mathbf{A} \mid i \in I \,\}$ be a set of jointly surjective coretractions for φ. We want to show that $\psi_i(B)$ is a support for α, for some $i \in I$. To do this, we can assume that α is not constant. We will find some $S \in \mathrm{Sub}_1(\mathbf{A})$ such that S is a support for α. For each $S \in \mathrm{Sub}_1(\mathbf{A})$, define

$$S^{\vartriangle} := \big\{\, a \in A \;\big|\; S \subseteq \mathrm{sg}_{\mathbf{A}}(a) \,\big\}.$$

During this proof, we shall say that a set $S \in \mathrm{Sub}_1(\mathbf{A})$ is a **hold for** α if there exist $y, z \in \mathcal{A}(\mathbf{A}, \underline{\mathbf{M}})$ such that $\alpha(y) \neq \alpha(z)$ and $A \backslash S^{\Delta} \subseteq \mathrm{eq}(y, z)$.

By Lemma 7.2.3, the ordered set $\mathbf{Sub_1}(\mathbf{A})$ is a tree with minimum element $\varnothing$. We have $\varnothing^{\Delta} = A$ and we are assuming that α is not constant. So $\varnothing$ is a hold for α. Since $\mathbf{Sub_1}(\mathbf{A})$ is finite, there is some $S \in \mathrm{Sub}_1(\mathbf{A})$ such that S is a hold for α and there is no upper cover of S in $\mathbf{Sub_1}(\mathbf{A})$ that is also a hold for α. There are homomorphisms $y, z \in \mathcal{A}(\mathbf{A}, \underline{\mathbf{M}})$ with $\alpha(y) \neq \alpha(z)$ and $A \backslash S^{\Delta} \subseteq \mathrm{eq}(y, z)$.

To see that S is a support for α, let $w, x \in \mathcal{A}(\mathbf{A}, \underline{\mathbf{M}})$ such that $w{\restriction}_S = x{\restriction}_S$. Let $T_1, \dots, T_k$ be the upper covers of S in $\mathbf{Sub_1}(\mathbf{A})$, where $k \in \omega$. We now define a sequence $w_0, \dots, w_k$ of homomorphisms in $\mathcal{A}(\mathbf{A}, \underline{\mathbf{M}})$. First define $w_0 := w$. Then, for each $i \in \{0, \dots, k-1\}$, we can define

$$w_{i+1} := w_i{\restriction}_{A \backslash T_{i+1}^{\Delta}} \cup x{\restriction}_{T_{i+1}^{\Delta}},$$

by Lemma 7.2.7(i), (ii). Since

$$S^{\Delta} \subseteq S \cup T_1^{\Delta} \cup \dots \cup T_k^{\Delta},$$

it follows that $w_k{\restriction}_{S^{\Delta}} = x{\restriction}_{S^{\Delta}}$.

The map $\alpha : \mathcal{A}(\mathbf{A}, \underline{\mathbf{M}}) \to M$ preserves the relations in R_n. By the Preservation Lemma, 1.4.4, there is some $a \in A$ such that α is given by evaluation at a on $\{w_k, x, y, z\}$. We must have $a \in S^{\Delta}$, as

$$y(a) = \alpha(y) \neq \alpha(z) = z(a)$$

and $A \backslash S^{\Delta} \subseteq \mathrm{eq}(y, z)$. Since $w_k{\restriction}_{S^{\Delta}} = x{\restriction}_{S^{\Delta}}$, this implies that

$$\alpha(w_k) = w_k(a) = x(a) = \alpha(x).$$

For each $i \in \{0, \dots, k-1\}$, we have $\alpha(w_i) = \alpha(w_{i+1})$, as T_{i+1} is not a hold for α and $A \backslash T_{i+1}^{\Delta} \subseteq \mathrm{eq}(w_i, w_{i+1})$. Therefore

$$\alpha(w) = \alpha(w_0) = \cdots = \alpha(w_k) = \alpha(x),$$

whence S is a support for α. Since $S \in \mathrm{Sub}_1(\mathbf{A})$ and the maps in the set $\{\psi_i : \mathbf{B} \hookrightarrow \mathbf{A} \mid i \in I\}$ are jointly surjective, we must have $S \subseteq \psi_j(B)$, for some $j \in I$.

We have shown that there is some $j \in I$ for which $\psi_j(B)$ is a support for α. By the Preservation Lemma, the map α is given by evaluation at some $b \in A$ on the set $\{ w \circ \varphi \mid w \in \mathcal{A}(\mathbf{B}, \underline{\mathbf{M}}) \}$. We will show that α is given by evaluation at $\psi_j \circ \varphi(b) \in A$. Let $x \in \mathcal{A}(\mathbf{A}, \underline{\mathbf{M}})$. As $\varphi \circ \psi_j = \mathrm{id}_B$, we have

$$x{\restriction}_{\psi_j(B)} = x \circ \psi_j \circ \varphi{\restriction}_{\psi_j(B)}.$$

Therefore

$$\alpha(x) = \alpha(x \circ \psi_j \circ \varphi) = x \circ \psi_j \circ \varphi(b) = x(\psi_j \circ \varphi(b)).$$

Thus α is an evaluation. ∎

The main result of this chapter follows immediately from Lemma 7.2.1 and Theorems 7.2.12 and 7.1.6.

7.2.13 Theorem *A finite algebra is inherently dualisable if and only if it is of small type.*

7.3 Linear unary algebras are strongly dualisable

Building on the results of the previous section, we conclude this chapter by proving that all finite linear unary algebras are strongly dualisable. By the EAO Theorem, 1.5.4, it suffices to prove that all finite linear unary algebras have enough algebraic operations. It will be easy to accomplish this once we have proved the following technical lemma. The proof of this lemma is an induction on the depth of unary algebras, as defined in the previous section.

A homomorphism $\varphi : \mathbf{A} \to \mathbf{B}$ is called a **subretraction** if $\mathbf{B} \leqslant \mathbf{A}$ and $\varphi{\restriction}_B = \mathrm{id}_B$. Every subretraction is a retraction, with the inclusion map as a coretraction.

7.3.1 Lemma *Let $\underline{\mathbf{M}}$ be a finite linear unary algebra. Then there is a function $f : \omega \to \omega$ for which the following condition holds:*

for all finite algebras $\mathbf{A} \leqslant \mathbf{B} \leqslant \mathbf{C}$ in $\mathbb{ISP}(\underline{\mathbf{M}})$, there exists a subretraction $\varphi : \mathbf{C} \twoheadrightarrow \mathbf{D}$, with $\mathbf{A} \leqslant \mathbf{D} \leqslant \mathbf{C}$, such that $\varphi(B) \subseteq B$ and $|D| \leqslant f(|A|)$.

Proof Define $\mathcal{A} := \mathbb{ISP}(\underline{\mathbf{M}})$. For each $k \in \omega$, let $\mathcal{A}_k$ consist of exactly one isomorphic copy of each algebra in $\mathcal{A}$ of size at most k. Since $\mathcal{A}$ is locally finite, the set $\mathcal{A}_k$ is finite, for each $k \in \omega$.

For each $n \in \omega$ and each function $f : \omega \to \omega$, define the condition $\mathcal{C}(n, f)$ as follows.

$\mathcal{C}(n, f)$ *For all finite algebras $\mathbf{C}$ in $\mathcal{A}$ of depth at most n and all $A, B \subseteq C$, there is a subretraction $\varphi : \mathbf{C} \twoheadrightarrow \mathbf{D}$, for some $\mathbf{D} \leqslant \mathbf{C}$, such that*

(i) *φ is one-to-one on every one-generated subalgebra of $\mathbf{C}$,*
(ii) *$A \subseteq D$ and $\varphi(B) \subseteq B$,*
(iii) *$|D| \leqslant f(|A|)$.*

If we can find some $f : \omega \to \omega$ such that $\mathcal{C}(n, f)$ holds, for all $n \in \omega$, then we will have proved the lemma.

We will first define a sequence $f_0, f_1, f_2, \ldots$ of order-preserving functions on ω such that condition $\mathcal{C}(n, f_n)$ holds, for each $n \in \omega$. We will define these functions inductively.

We begin by defining the constant function $f_0 : \omega \to \omega$ by

$$f_0(k) := |F_{\underline{\mathbf{M}}}(1)|,$$

where $\mathbf{F}_{\underline{\mathbf{M}}}(1)$ is the one-generated free algebra in $\mathcal{A}$. To see that $\mathcal{C}(0, f_0)$ holds, choose any $\mathbf{C}$ in $\mathcal{A}$ of depth 0 and let $A, B \subseteq C$. Then $\mathrm{id}_C : \mathbf{C} \twoheadrightarrow \mathbf{C}$ is a subretraction. The map id_C is one-to-one, with $A \subseteq C$ and $\mathrm{id}_C(B) \subseteq B$. As $\mathbf{C}$ has depth 0, we know that $\mathbf{C}$ is one-generated, by Lemma 7.2.8. Therefore $|C| \leqslant |F_{\underline{\mathbf{M}}}(1)| = f_0(|A|)$. Thus condition $\mathcal{C}(0, f_0)$ is satisfied.

Now let $n \in \omega$ and assume that we have already defined an order-preserving function $f_n : \omega \to \omega$ such that $\mathcal{C}(n, f_n)$ is satisfied. Define $f_{n+1} : \omega \to \omega$ by

$$f_{n+1}(k) := f_n(k) \cdot \big(k + |F_{\underline{\mathbf{M}}}(1)| \cdot |\mathcal{A}_{f_n(k)}| \cdot 2^{f_n(k)}\big).$$

Then f_{n+1} is order-preserving, since f_n is order-preserving.

We wish to show that condition $\mathcal{C}(n+1, f_{n+1})$ holds. To this end, let $\mathbf{C}$ be a finite algebra in $\mathcal{A}$ of depth at most $n+1$, and let $A, B \subseteq C$. Since $\mathcal{C}(n, f_n)$ holds and $f_n \leqslant f_{n+1}$, we need only consider the case where $\mathbf{C}$ has depth $n+1$.

By Lemma 7.2.3, the finite ordered set $\mathbf{Sub_1}(\mathbf{C})$ is a tree. As in the proof of Lemma 7.2.11, we let Y denote the greatest node of $\mathbf{Sub_1}(\mathbf{C})$ and let $\mathcal{U}_Y$ denote the set of all upper covers of Y. Since $\mathbf{C}$ has depth $n+1$, the set $\mathcal{U}_Y$ is non-empty. For every $S \in \mathcal{U}_Y$, define

$$S^{\vartriangle} := \{\, c \in C \mid S \subseteq \mathrm{sg}_{\mathbf{C}}(c) \,\}$$

and $\mathbf{S}^{\blacktriangle} := \mathbf{sg}_{\mathbf{C}}(S^{\vartriangle})$. Then $Y \subseteq S \subseteq S^{\blacktriangle}$, for all $S \in \mathcal{U}_Y$.

Consider some $S \in \mathcal{U}_Y$. By Lemma 7.2.9(ii), the algebra $\mathbf{S}^{\blacktriangle}$ has depth at most n. So, as we are assuming that $\mathcal{C}(n, f_n)$ holds, there is a subretraction $\varphi_S : \mathbf{S}^{\blacktriangle} \twoheadrightarrow \mathbf{D}_S$, where $\mathbf{D}_S \leqslant \mathbf{S}^{\blacktriangle} \leqslant \mathbf{C}$, such that

(i)$_S$ φ_S is one-to-one on every one-generated subalgebra of $\mathbf{S}^{\blacktriangle}$,

(ii)$_S$ $A \cap S^{\blacktriangle} \subseteq D_S$ and $\varphi_S(B \cap S^{\blacktriangle}) \subseteq B \cap S^{\blacktriangle}$,

(iii)$_S$ $|D_S| \leqslant f_n(|A \cap S^{\blacktriangle}|)$.

We know that Y is a node of $\mathbf{Sub_1}(\mathbf{S}^{\blacktriangle})$, using Lemma 7.2.9(i). So it follows that $Y \subseteq D_S$, by Lemma 7.2.10(ii). Thus $\varphi_S{\restriction}_Y = \mathrm{id}_Y$, as φ_S is a subretraction.

We want to 'combine' the subretractions in $\{\, \varphi_S : \mathbf{S}^{\blacktriangle} \twoheadrightarrow \mathbf{D}_S \mid S \in \mathcal{U}_Y \,\}$ to form a subretraction $\varphi : \mathbf{C} \twoheadrightarrow \mathbf{D}$, for some $\mathbf{D} \leqslant \mathbf{C}$. But we will need to bound the size of the algebra $\mathbf{D}$ by some function of the size of A.

Define the set

$$\mathcal{S}_A := \{ S \in \mathcal{U}_Y \mid A \cap S^{\vartriangle} \neq \varnothing \}.$$

Define the equivalence relation $\equiv$ on $\mathcal{U}_Y \backslash \mathcal{S}_A$ by $S \equiv T$ if and only if there is an isomorphism from $\mathbf{D}_S$ onto $\mathbf{D}_T$ that fixes each element of Y and maps $D_S \cap B$ onto $D_T \cap B$. Choose a transversal $\mathcal{T}$ of the blocks of $\equiv$.

For each pair $(S,T) \in \equiv$ with $S \neq T$, fix an isomorphism $\eta_{ST} : \mathbf{D}_S \hookrightarrow\!\!\!\!\rightarrow \mathbf{D}_T$ such that

$$\eta_{ST}{\restriction}_Y = \mathrm{id}_Y \quad \text{and} \quad \eta_{ST}(D_S \cap B) = D_T \cap B.$$

For each $S \in \mathcal{U}_Y \backslash \mathcal{S}_A$, we define $\eta_{SS} := \mathrm{id}_{D_S}$.

Now we can define the subuniverse D of $\mathbf{C}$ by

$$D := \bigcup \{ D_S \mid S \in \mathcal{S}_A \} \cup \bigcup \{ D_T \mid T \in \mathcal{T} \}.$$

Since $Y \subseteq D_S$, for all $S \in \mathcal{U}_Y$, we must have $Y \subseteq D$. We want to define the subretraction $\varphi : \mathbf{C} \twoheadrightarrow \mathbf{D}$ by

$$\varphi := \bigcup \{ \varphi_S \mid S \in \mathcal{S}_A \} \cup \bigcup \{ \eta_{ST} \circ \varphi_S \mid T \in \mathcal{T} \text{ and } S \in T/{\equiv} \}.$$

During the four claims that follow, we first show that φ is a well-defined subretraction, and then use φ to show that condition $\mathcal{C}(n+1, f_{n+1})$ holds.

Claim 1 The relation φ is a well-defined subretraction.

The domain of φ is

$$\bigcup \{ S^{\blacktriangle} \mid S \in \mathcal{S}_A \} \cup \bigcup \{ S^{\blacktriangle} \mid T \in \mathcal{T} \text{ and } S \in T/{\equiv} \}$$
$$= \bigcup \{ S^{\blacktriangle} \mid S \in \mathcal{U}_Y \} = C,$$

by 7.2.7(iv). The range of φ is D. We now want to show that φ is well defined. For all $S, T \in \mathcal{U}_Y$ with $S \neq T$, it follows from 7.2.7(iii) that $S^{\blacktriangle} \cap T^{\blacktriangle} = Y$. We know that $\varphi_S{\restriction}_Y = \mathrm{id}_Y$, for all $S \in \mathcal{U}_Y$, and that $\eta_{ST}{\restriction}_Y = \mathrm{id}_Y$, for all $(S,T) \in \equiv$. Thus φ is a well-defined surjection.

For each $S \in \mathcal{U}_Y$, the map φ_S is a subretraction, and so $\varphi_S{\restriction}_{D_S} = \mathrm{id}_{D_S}$. For each $T \in \mathcal{T}$, we have $\eta_{TT} \circ \varphi_T{\restriction}_{D_T} = \mathrm{id}_{D_T}$. Thus φ fixes each element of its range D, which implies that φ is a subretraction.

Claim 2 The map φ is one-to-one on all one-generated subalgebras of $\mathbf{C}$.

For all $S \in \mathcal{U}_Y$, the map φ_S is one-to-one on every one-generated subalgebra of $\mathbf{S}^{\blacktriangle}$, by (i)$_S$. The map η_{ST} is an isomorphism, for all $(S,T) \in \equiv$, by construction. So φ is one-to-one on all one-generated subalgebras of $\mathbf{C}$.

Claim 3 We have $A \subseteq D$ and $\varphi(B) \subseteq B$.

To see that $A \subseteq D$, let $a \in A$. Since $A \subseteq C = \bigcup\{ S^{\blacktriangle} \mid S \in \mathcal{U}_Y \}$, there is some $S \in \mathcal{U}_Y$ such that $a \in S^{\blacktriangle}$. By 7.2.7(ii), we get $a \in S^{\blacktriangle} = Y \cup S^{\vartriangle}$. Since we know that $Y \subseteq D$, we can assume that $a \in S^{\vartriangle}$. Then $A \cap S^{\vartriangle} \neq \varnothing$, and so $S \in \mathcal{S}_A$. Thus $a \in A \cap S^{\blacktriangle} \subseteq D_S \subseteq D$, by (ii)$_S$.

We now want to show that $\varphi(B) \subseteq B$. For all $S \in \mathcal{U}_Y$, we have

$$\varphi_S(B \cap S^{\blacktriangle}) \subseteq B \cap S^{\blacktriangle} \subseteq B,$$

again by (ii)$_S$. This gives us

$$\eta_{ST} \circ \varphi_S(B \cap S^{\blacktriangle}) \subseteq \eta_{ST}(D_S \cap B) = D_T \cap B \subseteq B,$$

for all $(S, T) \in \equiv$. It follows that $\varphi(B) \subseteq B$.

Claim 4 We have $|D| \leqslant f_{n+1}(|A|)$.

For each $S \in \mathcal{U}_Y$, we know that $|D_S| \leqslant f_n(|A \cap S^{\blacktriangle}|)$, by (iii)$_S$. Since f_n is order-preserving, this implies that $|D_S| \leqslant f_n(|A|)$, for all $S \in \mathcal{U}_Y$.

To bound the size of the algebra $\mathbf{D}$, we need to bound the sizes of the set $\mathcal{S}_A$ and the transversal $\mathcal{T}$. Using 7.2.7(ii)–(iv), the set $\{ S^{\vartriangle} \mid S \in \mathcal{U}_Y \}$ forms a partition of $C \backslash Y$. This gives us

$$|\mathcal{S}_A| = \big|\{ S \in \mathcal{U}_Y \mid A \cap S^{\vartriangle} \neq \varnothing \}\big| \leqslant |A|.$$

We will now bound the size of $\mathcal{T}$. For all $T \in \mathcal{T}$, we have $|D_T| \leqslant f_n(|A|)$, and so $\mathbf{D}_T$ is isomorphic to a member of $\mathcal{A}_{f_n(|A|)}$. Now let $\mathbf{K} \in \mathcal{A}_{f_n(|A|)}$ and assume that there are isomorphisms $\psi_{T_1} : \mathbf{D}_{T_1} \hookrightarrow\!\!\!\rightarrow \mathbf{K}$ and $\psi_{T_2} : \mathbf{D}_{T_2} \hookrightarrow\!\!\!\rightarrow \mathbf{K}$, for some $T_1, T_2 \in \mathcal{T}$ with $T_1 \neq T_2$. Then $\psi_{T_2}^{-1} \circ \psi_{T_1} : \mathbf{D}_{T_1} \hookrightarrow\!\!\!\rightarrow \mathbf{D}_{T_2}$ is an isomorphism.

We know that $Y \subseteq D_{T_1} \subseteq C$ and that Y is a node of $\mathbf{Sub_1(C)}$. So, by Lemma 7.2.10(ii), we must have $\psi_{T_2}^{-1} \circ \psi_{T_1}(Y) = Y$. Now we can define the subuniverse Y_K of $\mathbf{K}$ by $Y_K := \psi_{T_1}(Y) = \psi_{T_2}(Y)$.

Since $T_1, T_2 \in \mathcal{T}$ with $T_1 \neq T_2$, we must have $T_1 \not\equiv T_2$. This tells us that

$$\psi_{T_2}^{-1} \circ \psi_{T_1}{\restriction}_Y \neq \mathrm{id}_Y \quad \text{or} \quad \psi_{T_2}^{-1} \circ \psi_{T_1}(D_{T_1} \cap B) \neq D_{T_2} \cap B.$$

We will consider these two cases separately.

Case 1: $\psi_{T_2}^{-1} \circ \psi_{T_1}{\restriction}_Y \neq \mathrm{id}_Y$. The two isomorphisms $\psi_{T_1}{\restriction}_Y : \mathbf{Y} \hookrightarrow\!\!\!\rightarrow \mathbf{Y}_K$ and $\psi_{T_2}{\restriction}_Y : \mathbf{Y} \hookrightarrow\!\!\!\rightarrow \mathbf{Y}_K$ must be distinct. Since both $\mathbf{Y}$ and $\mathbf{Y}_K$ are one-generated algebras in $\boldsymbol{\mathcal{A}}$, the number of different isomorphisms from $\mathbf{Y}$ to $\mathbf{Y}_K$ is at most $|Y_K| \leqslant |F_{\underline{\mathbf{M}}}(1)|$.

Case 2: $\psi_{T_2}^{-1} \circ \psi_{T_1}(D_{T_1} \cap B) \neq D_{T_2} \cap B$. This implies that $\psi_{T_1}(D_{T_1} \cap B)$ and $\psi_{T_2}(D_{T_2} \cap B)$ are distinct subsets of K. Since $|K| \leqslant f_n(|A|)$, the number of distinct subsets of K is at most $2^{f_n(|A|)}$.

It now follows that there can be at most $|F_{\underline{\mathbf{M}}}(1)| \cdot 2^{f_n(|A|)}$ distinct members T of $\mathcal{T}$ such that $\mathbf{D}_T$ is isomorphic to $\mathbf{K}$. Since $\mathbf{K}$ was chosen from $\mathcal{A}_{f_n(|A|)}$, this implies that

$$|\mathcal{T}| \leqslant \left|\mathcal{A}_{f_n(|A|)}\right| \cdot \left|F_{\underline{\mathbf{M}}}(1)\right| \cdot 2^{f_n(|A|)}.$$

Finally, we have

$$\begin{aligned} |D| &\leqslant \sum \{\, |D_S| \mid S \in \mathcal{S}_A \,\} + \sum \{\, |D_T| \mid T \in \mathcal{T} \,\} \\ &\leqslant f_n(|A|) \cdot |\mathcal{S}_A| + f_n(|A|) \cdot |\mathcal{T}| \\ &\leqslant f_n(|A|) \cdot |A| + f_n(|A|) \cdot \left(\left|\mathcal{A}_{f_n(|A|)}\right| \cdot \left|F_{\underline{\mathbf{M}}}(1)\right| \cdot 2^{f_n(|A|)}\right) \\ &= f_{n+1}(|A|). \end{aligned}$$

So $|D| \leqslant f_{n+1}(|A|)$, as required.

Via Claims 1 to 4, we have established that $\mathcal{C}(n+1, f_{n+1})$ holds. Hence there is a sequence $f_0, f_1, f_2, \ldots$ of order-preserving functions on ω such that condition $\mathcal{C}(n, f_n)$ holds, for each $n \in \omega$. We know that algebras in $\mathcal{A}$ can have depth at most $|F_{\underline{\mathbf{M}}}(1)|$, by Lemma 7.2.5. Thus we can define $f := f_{|F_{\underline{\mathbf{M}}}(1)|}$, and condition $\mathcal{C}(n, f)$ holds, for all $n \in \omega$. Hence f satisfies the condition in the statement of the lemma. ∎

The following general result links the technical condition in the previous lemma to enough algebraic operations.

7.3.2 Lemma *Let $\underline{\mathbf{M}}$ be a finite algebra. Assume that there exists a function $f : \omega \to \omega$ for which the following condition holds:*

for all $n \in \omega \backslash \{0\}$ and all algebras $\mathbf{B} \leqslant \mathbf{A} \leqslant \underline{\mathbf{M}}^n$, there is a subretraction $\varphi : \underline{\mathbf{M}}^n \twoheadrightarrow \mathbf{C}$, with $\mathbf{B} \leqslant \mathbf{C} \leqslant \underline{\mathbf{M}}^n$, such that $\varphi(A) \subseteq A$ and $|C| \leqslant f(|B|)$.

Then $\underline{\mathbf{M}}$ has enough algebraic operations.

Proof Define $\mathcal{A} := \mathbb{ISP}(\underline{\mathbf{M}})$. Let $\mathbf{B} \leqslant \mathbf{A} \leqslant \underline{\mathbf{M}}^n$, for some $n \in \omega\backslash\{0\}$, and let $h : \mathbf{A} \to \underline{\mathbf{M}}$ be a homomorphism. By assumption, there is a subretraction $\varphi : \underline{\mathbf{M}}^n \twoheadrightarrow \mathbf{C}$, with $\mathbf{B} \leqslant \mathbf{C} \leqslant \underline{\mathbf{M}}^n$, such that $\varphi(A) \subseteq A$ and $|C| \leqslant f(|B|)$. Using Lemma 4.1.1, there is a non-empty subset Z of $\mathcal{A}(\underline{\mathbf{M}}^n, \underline{\mathbf{M}})$ such that Z separates the elements of C and $|Z| \leqslant |C|$. Now define the subset Y of $\mathcal{A}(\underline{\mathbf{M}}^n, \underline{\mathbf{M}})$ by

$$Y := \{\, z \circ \varphi \mid z \in Z \,\}.$$

We have

$$|Y| \leqslant |Z| \leqslant |C| \leqslant f(|B|).$$

To prove that $\underline{\mathbf{M}}$ has enough algebraic operations, it remains to find a homomorphism $h' : \sqcap Y(\mathbf{A}) \to \underline{\mathbf{M}}$ such that $h' \circ \sqcap Y{\restriction}_B = h{\restriction}_B$.

First, we want to define a homomorphism

$$\eta : \sqcap Y(\mathbf{A}) \to \varphi(\mathbf{A}) \quad \text{by} \quad \eta(\sqcap Y(a)) := \varphi(a).$$

To see that η is well defined, let $a_1, a_2 \in A$ with $\sqcap Y(a_1) = \sqcap Y(a_2)$. Then, for all $z \in Z$, we have $z \circ \varphi(a_1) = z \circ \varphi(a_2)$. But Z separates the elements of C. So it follows that $\varphi(a_1) = \varphi(a_2)$, whence η is well defined.

Since $\varphi(\mathbf{A}) \leqslant \mathbf{A}$, we can define the homomorphism $h' : \sqcap Y(\mathbf{A}) \to \underline{\mathbf{M}}$ by $h' := h \circ \eta$. For all $b \in B$, we have

$$h' \circ \sqcap Y(b) = h \circ \eta \circ \sqcap Y(b) = h \circ \varphi(b) = h(b),$$

as $B \subseteq C = \varphi(M^n)$ and φ is a subretraction. Hence $\underline{\mathbf{M}}$ has enough algebraic operations. ∎

Each finite linear unary algebra is dualisable, by Theorem 7.2.12. So our final result follows from the previous two lemmas and the EAO Theorem, 1.5.4.

7.3.3 Theorem *Every finite linear unary algebra is strongly dualisable.*

Epilogue

Dualisability is a rich and complex concept. In particular, dualisability of unary algebras is complex. We have shown that the dualisable and non-dualisable unary algebras are tightly entangled:

- non-dualisable unary algebras can be built from dualisable unary algebras via natural algebraic constructions, such as taking products, coproducts or homomorphic images (Table 5.1);
- there are many infinite ascending chains of unary algebras (under the subalgebra order) such that in each chain the algebras are alternately dualisable and non-dualisable (Theorem 7.1.2);
- there are arbitrarily long finite chains of unary algebras (under the inclusion order on their sets of operations) such that in each chain the algebras are alternately dualisable and non-dualisable (Theorem 6.2.3);
- adding a constant operation can alter the dualisability of a unary algebra, making a dualisable algebra non-dualisable, or making a non-dualisable algebra dualisable (Example 6.3.3 and Table 3.1).

The results above suggest that the class of unary algebras reflects much of the complexity of dualisability. Indeed, dualisability is more badly behaved within the class of unary algebras than it is in some other classes of 'more complicated' algebras. For example, dualisability is preserved under taking finite products, subalgebras and homomorphic images in the variety of commutative rings with identity [14] and the variety of p-semilattices (Theorem 5.2.4). Within the class of graph algebras, every finite entropic algebra is dualisable [23]. But there are entropic unary algebras that are non-dualisable (Lemma 5.3.7 and Example 6.4.1).

Despite the complexity of dualisability for unary algebras, we found some large classes of dualisable unary algebras. We used two different approaches for proving that particular unary algebras are dualisable. As an illustration of these approaches, contrast the two different ways we proved that the four-element unary algebra $\underline{\mathbf{M}} = \langle\{0,1,2,3\}; 0010, 0321\rangle$, drawn in Figure 2.2, is dualisable. Firstly, in Example 2.2.13, we showed that $\underline{\mathbf{M}}$ is dualisable because it has two very special binary homomorphisms. Secondly, in Theorem 3.2.10, we showed that the algebra $\underline{\mathbf{M}}$ is dualisable because it is a one-kernel algebra and therefore the quasi-variety $\mathbb{ISP}(\underline{\mathbf{M}})$ is very simple.

The first general approach for proving dualisability, developed in Chapter 2, is to focus exclusively on a few nice algebraic operations that we can use in an alter ego for the algebra. This alter-ego approach is associated with general conditions on the algebraic operations of the algebra:

- a finite algebra is dualisable provided it has a pair of algebraic lattice operations (Theorem 2.1.1);
- a finite algebra is dualisable provided each of its elements is a strong idempotent of a binary algebraic operation (Theorem 2.2.3).

In some sense, the alter-ego approach ignores most of the algebras in the quasi-variety.

The second approach for proving dualisability, used in Chapters 3 and 7, is to study *all* of the finite algebras in the quasi-variety and their homomorphisms into the generator, and to just let the alter ego take care of itself. This algebras-in-the-quasi-variety approach is associated with conditions on the term functions of the algebra:

- every finite one-kernel unary algebra is dualisable (Theorem 3.2.10);
- every finite linear unary algebra is dualisable (Theorem 7.2.12).

There is a natural generalisation of the two conditions above. Let us say that a unary algebra $\underline{\mathbf{M}}$ has a *chain of kernels* if the kernels of $\underline{\mathbf{M}}$ form a chain under set inclusion. It is easy to check that every one-kernel unary algebra has a chain of kernels, and that every linear unary algebra has a chain of kernels. It seems reasonable to conjecture that the finite unary algebras that have a chain of kernels may all be dualisable.

The secondary theme of this text was strong dualisability. We proved some results that we hope will lead to a better understanding of strong dualisability and the subtle difference between it and full dualisability:

- for three-element unary algebras, strong dualisability is equivalent to full dualisability and to a weak form of injectivity (Theorem 4.0.1);

- every finite zero-kernel or one-kernel unary algebra is strongly dualisable (Theorem 4.1.3);
- every finite linear unary algebra is strongly dualisable (Theorem 7.3.3).

In the appendix that follows, we prove in full the two general theorems (1.5.3 and 1.5.4) used to obtain these results. Moreover, we introduce the concept of the 'height' of a finite algebra, and use it to give a new characterisation of strong dualisability. This provides a natural and transparent path to R. Willard's concept of the 'rank' of a finite algebra, a somewhat technical sufficient condition for strong dualisability.

As we write, the theory of natural dualities continues to evolve. For example, recent work on the Full versus Strong Problem is improving our understanding of full dualities [18–21]. Meanwhile, new topics are arising from within the theory. Workable descriptions of dual categories are very important to the utility of duality theory. Accordingly, there has been a recent push to develop a general theory of the axiomatic description of finitely generated topological quasi-varieties [10, 9, 28, 11]. This topic has taken on a life of its own, quite independent of its roots in the theory of natural dualities.

Duality theory is also evolving in more fundamental directions. The definition of an alter ego can be extended by allowing into the type non-finitary algebraic operations and relations. This leads to the question: 'Is there a finite algebra that cannot be dualised using only finitary relations, but that can be dualised using relations of arity less than κ, for some infinite ordinal κ?' Similar questions apply to full and to strong dualisability. Several papers have appeared that consider dualisability in these infinitary versions [41, 45, 26, 14].

Up to term equivalence and isomorphism, there are only 2^ω finite algebras. Thus, there is a smallest ordinal μ such that every finite algebra that is dualisable at all (via a set of possibly infinitary relations) is dualisable via relations of arity less than μ. The ordinal μ is called the *Hanf number for dualisability*. There are known bounds on the Hanf number for dualisability within particular classes of algebras, but little is known in general [45].

There is still a great deal of scope for investigating dualisability and strong dualisability through unary algebras. There are many unsolved problems in natural duality theory for which unary algebras might be a valuable source of examples and counterexamples: for instance, the Finite Type Problem and the Full versus Strong Problem. It may even be possible to prove that, within the class of unary algebras, dualisability is undecidable.

A

Appendix: Strong dualisability

Two important general results on strong dualisability were stated in Chapter 1 and used extensively in Chapters 4 and 7; here we prove them. Along the way, we find that it is natural to introduce a new concept, the 'height' of a finite algebra. This provides us with a characterisation of strong duality and leads readily to the significant but technical concept of 'rank'.

When is a duality strong? Assume we have established a natural duality based on a finite algebra $\underline{\mathbf{M}}$ and an alter ego $\underset{\sim}{\mathbf{M}}$. To find out whether this duality is strong, via the definition, we need to check whether every closed substructure of an arbitrary non-zero power of $\underset{\sim}{\mathbf{M}}$ is term closed relative to $\underline{\mathbf{M}}$. We would prefer to have alternative, simpler conditions under which a duality is strong. More particularly, we want finitary conditions.

Clark, Idziak, Sabourin, Szabó and Willard [14] have a characterisation of strong duality that, although not finitary, does avoid the concept of term closure; see Theorem 1.5.3. We relied on this characterisation in Chapter 4, to show that some three-element unary algebras are not strongly dualisable. In Section A.2, we present a direct proof of their characterisation.

Starting from this characterisation of strong duality, we begin a journey that takes us first to the concept of 'enough algebraic operations', a concept that provides a finitary sufficient condition under which a dualisable algebra is strongly dualisable; see the EAO Theorem, 1.5.4. This condition, introduced by Lampe, McNulty and Willard [45], was used extensively in Chapters 4 and 7. In Section A.4, we present a direct proof that every dualisable algebra with enough algebraic operations is strongly dualisable. Indeed, we put the concept in a more general setting to obtain a stronger result.

The concept of 'enough algebraic operations' was born of R. Willard's more technical notion of 'rank' [65]. We finish this text with a new way to motivate and develop rank. En route, we introduce the new concept of the 'height' of a finite algebra, and use this concept to give a new necessary and sufficient condition under which a duality is strong. A small step from height takes us to our final destination, the notion of rank, which provides a finitary sufficient (and nearly necessary) condition under which a duality is strong.

A.1 Term-closed sets

The concept of term-closed sets, together with some of their important properties, was first presented by Clark and Krauss [15]. We briefly introduced term-closed sets in Chapter 1. Since they play a pivotal role in the theory of full and strong dualities, we give a slightly expanded introduction here.

Fix a finite algebra $\underline{\mathbf{M}}$ and a non-empty set S. Let $F_{\underline{\mathbf{M}}}(S)$ denote the set of all S-ary term functions of $\underline{\mathbf{M}}$. For each $s \in S$, we use $\pi_s : M^S \to M$ to denote the sth projection function. Then $F_{\underline{\mathbf{M}}}(S)$ is the underlying set of the subalgebra $\mathbf{F}_{\underline{\mathbf{M}}}(S)$ of $\underline{\mathbf{M}}^{M^S}$ generated by $\{\, \pi_s : M^S \to M \mid s \in S \,\}$. In fact, the algebra $\mathbf{F}_{\underline{\mathbf{M}}}(S)$ is the S-generated free algebra in the quasi-variety $\mathbb{ISP}(\underline{\mathbf{M}})$, with the projection functions as the free generators.

A subset X of M^S is said to be **term closed (relative to $\underline{\mathbf{M}}$)** if, for each $y \in M^S \backslash X$, there are term functions $t_1, t_2 \in F_{\underline{\mathbf{M}}}(S)$ that agree on X but differ at y.

A.1.1 Example Let $\underline{\mathbf{M}}$ be a finite algebra and let $\mathbf{A}$ belong to the quasi-variety $\mathcal{A} := \mathbb{ISP}(\underline{\mathbf{M}})$. Then the set $\mathcal{A}(\mathbf{A}, \underline{\mathbf{M}})$ of all homomorphisms from $\mathbf{A}$ into $\underline{\mathbf{M}}$ is a term-closed subset of M^A. To see this, choose some $y \in M^A \backslash \mathcal{A}(\mathbf{A}, \underline{\mathbf{M}})$. Since $y : A \to M$ is not a homomorphism, there must exist an n-ary term τ of the type of $\underline{\mathbf{M}}$, for some $n \in \omega$, and $a_0, \dots, a_{n-1} \in A$ such that

$$y\big(\tau(a_0, \dots, a_{n-1})\big) \neq \tau\big(y(a_0), \dots, y(a_{n-1})\big).$$

The A-ary term functions

$$t_1 := \pi_{\tau(a_0,\dots,a_{n-1})} \quad \text{and} \quad t_2 := \tau(\pi_{a_0}, \dots, \pi_{a_{n-1}})$$

of $\underline{\mathbf{M}}$ agree on $\mathcal{A}(\mathbf{A}, \underline{\mathbf{M}})$ but differ at y.

Our first easy lemma gives an alternative definition of term-closed sets.

A.1.2 Lemma *Let $\underline{\mathbf{M}}$ be a finite algebra and let S be a non-empty set. A subset X of M^S is term closed if and only if it is an intersection of equalisers of S-ary term functions of $\underline{\mathbf{M}}$.*

Using the preceding lemma, the set of all term-closed subsets of M^S is closed under arbitrary intersections. Thus, for each subset X of M^S, there is a smallest term-closed subset of M^S containing X, which we call the **term closure of** X **(relative to $\underline{\mathbf{M}}$)** and denote by $\mathrm{tc}_{\underline{\mathbf{M}}}(X)$. We shall be using the description of term closure provided by the next easy lemma.

A.1.3 Lemma *Let $\underline{\mathbf{M}}$ be a finite algebra and let S be a non-empty set. For each subset X of M^S, the term closure $\mathrm{tc}_{\underline{\mathbf{M}}}(X)$ consists of all $y \in M^S$ that satisfy*

$$t_1\restriction_X = t_2\restriction_X \implies t_1(y) = t_2(y),$$

for all $t_1, t_2 \in F_{\underline{\mathbf{M}}}(S)$.

The following is a special case of a result in the Clark–Davey text [8, 3.1.3].

A.1.4 Lemma *Let $\underline{\mathbf{M}}$ be a finite algebra and let $\underset{\sim}{\mathbf{M}}$ be an alter ego of $\underline{\mathbf{M}}$. Let X be a subset of M^S, for some non-empty set S. Then the term closure $\mathrm{tc}_{\underline{\mathbf{M}}}(X)$ is the underlying set of a closed substructure of $\underset{\sim}{\mathbf{M}}^S$.*

Proof We will sketch the proof. The term closure $\mathrm{tc}_{\underline{\mathbf{M}}}(X)$ is an intersection of equalisers of S-ary term functions of $\underline{\mathbf{M}}$. Every S-ary term function of $\underline{\mathbf{M}}$ has a finite support, and so is continuous. Therefore an equaliser of term functions is topologically closed, whence $\mathrm{tc}_{\underline{\mathbf{M}}}(X)$ is topologically closed. It is easy to see that $\mathrm{tc}_{\underline{\mathbf{M}}}(X)$ is closed under all the algebraic partial operations on $\underline{\mathbf{M}}$, since term functions and homomorphisms commute. ∎

When working with strong dualities, we need to take some care with regard to nullary operations. Our next lemma will clarify the situation, but first we introduce some notation.

Let $\underset{\sim}{\mathbf{M}} = \langle M; G, H, R, \mathcal{T}\rangle$ be an alter ego for our finite algebra $\underline{\mathbf{M}}$. We use $[G \cup H]$ to denote the enriched partial clone generated by $G \cup H$, and refer to it as the **enriched partial clone of $\underset{\sim}{\mathbf{M}}$**. The set $[G \cup H]$ is generated by composition from the operations in G, the partial operations in H, and all finitary projections on M. (When composing partial operations, the maximum possible domain is taken for the composition.) The adjective *enriched* is used to emphasise the fact that the usual notion of partial clone has been enriched by allowing nullary operations. For all $k \in \omega$, the set of k-ary partial operations in $[G \cup H]$ is denoted by $[G \cup H]_k$. We refer to the text by Clark and Davey [8] for the formal definitions.

According to our naming convention, an algebraic operation on $\underline{\mathbf{M}}$ must be a total operation, but an algebraic partial operation on $\underline{\mathbf{M}}$ does not have to be a proper partial operation. All operations, including nullaries, are included amongst the partial operations. For emphasis, we will occasionally refer to algebraic operations on $\underline{\mathbf{M}}$ as *algebraic total operations on* $\underline{\mathbf{M}}$.

The following easy lemma is a slightly expanded version of a result from Clark and Davey's text [8, 3.1.2]. Note that the empty structure belongs to the topological quasi-variety $\mathbb{IS}_c\mathbb{P}^+(\underset{\sim}{\mathrm{M}})$ if and only if there are no nullary operations in the type of $\underset{\sim}{\mathrm{M}}$.

A.1.5 Lemma *Let* $\underset{\sim}{\mathrm{M}} = \langle M; G, H, R, \mathcal{T} \rangle$ *be an alter ego of a finite algebra* $\underline{\mathbf{M}}$. *Then the following are equivalent*:

(i) *for each non-empty set* S, *the zero-generated substructure of* $\underset{\sim}{\mathrm{M}}^S$ *is term closed*;

(ii) *the dual* $\mathrm{D}(\mathbf{1})$ *is the zero-generated substructure of* $\underset{\sim}{\mathrm{M}}^1$, *where* $\mathbf{1}$ *is the one-element algebra* $\underline{\mathbf{M}}^\varnothing$;

(iii) *every nullary algebraic operation on* $\underline{\mathbf{M}}$ *belongs to* $[G \cup H]_0$;

(iv) *every element of* M *that determines a one-element subalgebra of* $\underline{\mathbf{M}}$ *is the value of a nullary term function of* $\underset{\sim}{\mathrm{M}}$.

For brevity, we say that $\underset{\sim}{\mathrm{M}}$ **strongly dualises** $\underline{\mathbf{M}}$ if the structure $\underset{\sim}{\mathrm{M}}$ yields a strong duality on the quasi-variety $\mathbb{ISP}(\underline{\mathbf{M}})$.

A.1.6 Corollary *Let* $\underset{\sim}{\mathrm{M}}$ *be an alter ego of a finite algebra* $\underline{\mathbf{M}}$. *If* $\underset{\sim}{\mathrm{M}}$ *strongly dualises* $\underline{\mathbf{M}}$, *then each element of* M *that determines a one-element subalgebra of* $\underline{\mathbf{M}}$ *must be the value of a nullary term function of* $\underset{\sim}{\mathrm{M}}$.

A.2 Term closure via duals

In this section, we present a proof of the characterisation of strong duality, obtained by Clark, Idziak, Sabourin, Szabó and Willard [14], that is given in Theorem 1.5.3. We shall take a direct route to this result that requires the notion of term-closed sets only. An alternative route, via hom-closed sets, is given by Davey, Haviar and Willard [21].

Recall that, whenever we have $\mathbf{A} \leqslant \underline{\mathbf{M}}^X$ and $x \in X$, we use $\rho_x : A \to M$ to denote the restriction of the projection $\pi_x : M^X \to M$.

A.2.1 Lemma [14, 3.9] *Let* $\underline{\mathbf{M}}$ *be a finite algebra and let* $\underset{\sim}{\mathrm{M}}$ *be an alter ego of* $\underline{\mathbf{M}}$. *Let* X *be a subset of* M^S, *for some non-empty set* S. *Then there exists a subalgebra* $\mathbf{A}$ *of* $\underline{\mathbf{M}}^X$ *and an embedding* $\nu : \mathrm{D}(\mathbf{A}) \hookrightarrow \underset{\sim}{\mathrm{M}}^S$ *such that*

(i) $\nu(\rho_x) = x$, *for all* $x \in X$, *and*

(ii) *the image of* ν *is the term closure of* X *in* M^S.

Proof Define the quasi-variety $\mathcal{A} := \mathbb{ISP}(\underline{\mathbf{M}})$. The set

$$A := \{ t{\restriction}_X \mid t \in F_{\underline{\mathbf{M}}}(S) \}$$

forms a subalgebra $\mathbf{A}$ of $\underline{\mathbf{M}}^X$. (This is true even if X is empty.) We want to show that we can define an embedding $\nu : \mathrm{D}(\mathbf{A}) \hookrightarrow \underset{\sim}{\mathbf{M}}^S$ by

$$\nu(z)(s) := z(\pi_s{\restriction}_X), \quad \text{for all } z \in \mathcal{A}(\mathbf{A}, \underline{\mathbf{M}}) \text{ and } s \in S.$$

It is straightforward, but tedious, to prove this directly. Alternatively, we can use a few basic results from the text by Clark and Davey [8].

It is easy to check that $\varphi : \mathbf{F}_{\underline{\mathbf{M}}}(S) \twoheadrightarrow \mathbf{A}$, defined by $\varphi(t) := t{\restriction}_X$, is a surjective homomorphism. So the morphism $\mathrm{D}(\varphi) : \mathrm{D}(\mathbf{A}) \hookrightarrow \mathrm{D}(\mathbf{F}_{\underline{\mathbf{M}}}(S))$ is an embedding [8, 1.5.3]. Let $\psi : \mathrm{D}(\mathbf{F}_{\underline{\mathbf{M}}}(S)) \hookrightarrow\!\!\!\!\rightarrow \underset{\sim}{\mathbf{M}}^S$ be the isomorphism given by $\psi(w)(s) := w(\pi_s)$, for all $w \in \mathcal{A}(\mathbf{F}_{\underline{\mathbf{M}}}(S), \underline{\mathbf{M}})$ and $s \in S$ [8, 2.2.1]. We have now constructed the embedding $\psi \circ \mathrm{D}(\varphi) : \mathrm{D}(\mathbf{A}) \hookrightarrow \underset{\sim}{\mathbf{M}}^S$. For all $z \in \mathcal{A}(\mathbf{A}, \underline{\mathbf{M}})$ and $s \in S$, we have

$$\nu(z)(s) = z(\pi_s{\restriction}_X) = z \circ \varphi(\pi_s) = \psi(z \circ \varphi)(s) = \psi \circ \mathrm{D}(\varphi)(z)(s).$$

Thus $\nu = \psi \circ \mathrm{D}(\varphi)$, whence ν is an embedding.

To prove claim (i), let $x \in X$. (For (i), we only need to consider the case where X is non-empty.) For all $s \in S$, we have

$$\nu(\rho_x)(s) = \rho_x(\pi_s{\restriction}_X) = (\pi_s{\restriction}_X)(x) = \pi_s(x) = x(s).$$

So $\nu(\rho_x) = x$, and therefore (i) holds.

To prove claim (ii), we begin by showing that $\mathrm{tc}_{\underline{\mathbf{M}}}(X) \subseteq \nu(\mathcal{A}(\mathbf{A}, \underline{\mathbf{M}}))$. Let $y \in \mathrm{tc}_{\underline{\mathbf{M}}}(X) \subseteq M^S$. Then, for all $t_1, t_2 \in F_{\underline{\mathbf{M}}}(S)$, we have

$$t_1{\restriction}_X = t_2{\restriction}_X \implies t_1(y) = t_2(y),$$

by Lemma A.1.3. It follows that $z_y : \mathbf{A} \to \underline{\mathbf{M}}$, given by $z_y(t{\restriction}_X) := t(y)$, for all $t \in F_{\underline{\mathbf{M}}}(S)$, is a well-defined homomorphism. So

$$\nu(z_y)(s) = z_y(\pi_s{\restriction}_X) = \pi_s(y) = y(s),$$

for all $s \in S$, whence $\nu(z_y) = y$. Thus $y \in \nu(\mathcal{A}(\mathbf{A}, \underline{\mathbf{M}}))$.

Finally, we show that $\nu(\mathcal{A}(\mathbf{A}, \underline{\mathbf{M}})) \subseteq \mathrm{tc}_{\underline{\mathbf{M}}}(X)$. Let $y \in \nu(\mathcal{A}(\mathbf{A}, \underline{\mathbf{M}}))$; say $y = \nu(z)$, where $z \in \mathcal{A}(\mathbf{A}, \underline{\mathbf{M}})$. In order to establish that $y \in \mathrm{tc}_{\underline{\mathbf{M}}}(X)$, we must show that, for all $t_1, t_2 \in F_{\underline{\mathbf{M}}}(S)$, we have

$$t_1{\restriction}_X = t_2{\restriction}_X \implies t_1(y) = t_2(y).$$

Assume that t_1 and t_2 are S-ary term functions of $\underline{\mathbf{M}}$. Then there is some $n \in \omega$ for which there are n-ary terms τ_1 and τ_2 of the type of $\underline{\mathbf{M}}$ and $s_0, \dots, s_{n-1} \in S$ such that

$$t_1 = \tau_1(\pi_{s_0}, \dots, \pi_{s_{n-1}}) \quad \text{and} \quad t_2 = \tau_2(\pi_{s_0}, \dots, \pi_{s_{n-1}}).$$

Therefore, since $y = \nu(z) \in M^S$, we have

$$\begin{aligned} t_1(y) &= \tau_1(\pi_{s_0}, \dots, \pi_{s_{n-1}})(y) = \tau_1\big(y(s_0), \dots, y(s_{n-1})\big) \\ &= \tau_1\big(\nu(z)(s_0), \dots, \nu(z)(s_{n-1})\big) = \tau_1\big(z(\pi_{s_0}{\restriction}_X), \dots, z(\pi_{s_{n-1}}{\restriction}_X)\big) \\ &= z\big(\tau_1(\pi_{s_0}{\restriction}_X, \dots, \pi_{s_{n-1}}{\restriction}_X)\big) = z\big(\tau_1(\pi_{s_0}, \dots, \pi_{s_{n-1}}){\restriction}_X\big) \\ &= z(t_1{\restriction}_X). \end{aligned}$$

Similarly, $t_2(y) = z(t_2{\restriction}_X)$. So, if $t_1{\restriction}_X = t_2{\restriction}_X$, then $t_1(y) = t_2(y)$. Thus $y \in \mathrm{tc}_{\underline{\mathbf{M}}}(X)$. We have shown that $\mathrm{tc}_{\underline{\mathbf{M}}}(X) = \nu(\mathcal{A}(\mathbf{A}, \underline{\mathbf{M}}))$. Hence claim (ii) holds. ∎

The following lemma can be extracted from the Clark–Idziak–Sabourin–Szabó–Willard paper [14], but was first stated explicitly by Davey, Haviar and Willard [21].

A.2.2 Lemma *Let $\underline{\mathbf{M}}$ be a finite algebra and define $\mathcal{A} := \mathbb{ISP}(\underline{\mathbf{M}})$. Let $\mathbf{A}$ be a subalgebra of $\underline{\mathbf{M}}^I$, for some set I. For each subset X of $\mathcal{A}(\mathbf{A}, \underline{\mathbf{M}})$, the following are equivalent:*

(i) *X is term closed in M^A and contains the set $\{\, \rho_i : A \to M \mid i \in I \,\}$ of projections;*

(ii) *X is term closed in M^A and separates the elements of A;*

(iii) *X is equal to $\mathcal{A}(\mathbf{A}, \underline{\mathbf{M}})$.*

Proof Let X be a subset of $\mathcal{A}(\mathbf{A}, \underline{\mathbf{M}})$. The implication (i) $\Rightarrow$ (ii) is trivial. Since $\mathcal{A}(\mathbf{A}, \underline{\mathbf{M}})$ is term closed in M^A, by Example A.1.1, we have (iii) $\Rightarrow$ (i). To prove that $\neg$(iii) $\Rightarrow$ $\neg$(ii), assume that $X \neq \mathcal{A}(\mathbf{A}, \underline{\mathbf{M}})$ and that X is term closed in M^A. We need to show that the homomorphisms in X do not separate the elements of A.

There is some $y \in \mathcal{A}(\mathbf{A}, \underline{\mathbf{M}}) \backslash X$. Since X is term closed in M^A, there are A-ary term functions t_1 and t_2 of $\underline{\mathbf{M}}$ that agree on X but differ at y. This implies that there exists some $n \in \omega$ for which there are n-ary terms τ_1 and τ_2 of the type of $\underline{\mathbf{M}}$ and $a_0, \dots, a_{n-1} \in A$ such that

$$t_1 = \tau_1(\pi_{a_0}, \dots, \pi_{a_{n-1}}) \quad \text{and} \quad t_2 = \tau_2(\pi_{a_0}, \dots, \pi_{a_{n-1}}).$$

Define the elements $b_1 := \tau_1(a_0, \dots, a_{n-1})$ and $b_2 := \tau_2(a_0, \dots, a_{n-1})$ of A. Then, for all $z \in \mathcal{A}(\mathbf{A}, \underline{\mathbf{M}})$, we have

$$\begin{aligned} z(b_1) &= z\big(\tau_1(a_0, \dots, a_{n-1})\big) = \tau_1\big(z(a_0), \dots, z(a_{n-1})\big) \\ &= \tau_1\big(\pi_{a_0}(z), \dots, \pi_{a_{n-1}}(z)\big) = \tau_1(\pi_{a_0}, \dots, \pi_{a_{n-1}})(z) = t_1(z) \end{aligned}$$

and, by symmetry, $z(b_2) = t_2(z)$.

For all $x \in X \subseteq \mathcal{A}(\mathbf{A}, \underline{\mathbf{M}})$, we have

$$x(b_1) = t_1(x) = t_2(x) = x(b_2),$$

as t_1 and t_2 agree on X. So b_1 and b_2 are not separated by the maps in X. But

$$y(b_1) = t_1(y) \neq t_2(y) = y(b_2),$$

which implies that $b_1 \neq b_2$. Therefore X does not separate the elements of A, giving $\neg$ (iii) $\Rightarrow$ $\neg$ (ii). ∎

By combining the previous two lemmas, we obtain an important result due to Clark, Idziak, Sabourin, Szabó and Willard.

A.2.3 Dual Generation Theorem [14, 4.8] *Let $\underline{\mathbf{M}}$ be a finite algebra and let $\underset{\sim}{\mathbf{M}}$ be an alter ego of $\underline{\mathbf{M}}$. The following are equivalent:*

(i) *each closed substructure of each non-zero power of $\underset{\sim}{\mathbf{M}}$ is term closed;*

(ii) *for each set I and each subalgebra $\mathbf{A}$ of $\underline{\mathbf{M}}^I$, the closed substructure of $\underset{\sim}{\mathbf{M}}^A$ generated by the set $\{\, \rho_i : A \to M \mid i \in I \,\}$ of projections is term closed;*

(iii) *for each set I and each subalgebra $\mathbf{A}$ of $\underline{\mathbf{M}}^I$, the closed substructure of $\mathrm{D}(\mathbf{A})$ generated by the set $\{\, \rho_i : A \to M \mid i \in I \,\}$ of projections is $\mathrm{D}(\mathbf{A})$ itself;*

(iv) *for each algebra $\mathbf{A}$ in $\mathbb{ISP}(\underline{\mathbf{M}})$, the only closed substructure of $\mathrm{D}(\mathbf{A})$ that separates the elements of A is $\mathrm{D}(\mathbf{A})$ itself.*

Proof The two implications (i) $\Rightarrow$ (ii) and (iv) $\Rightarrow$ (iii) are easy. The implication (ii) $\Rightarrow$ (iii) follows from (i) $\Rightarrow$ (iii) of the previous lemma. The implication (i) $\Rightarrow$ (iv) follows from (ii) $\Rightarrow$ (iii) of the previous lemma. It remains to prove that (iii) $\Rightarrow$ (i).

Assume (iii), and let $\mathbf{X}$ be a closed substructure of $\underset{\sim}{\mathbf{M}}^S$, for some non-empty set S. Let $\mathbf{A} \leqslant \underline{\mathbf{M}}^X$ and $\nu : \mathrm{D}(\mathbf{A}) \hookrightarrow \underset{\sim}{\mathbf{M}}^S$ be as given by Lemma A.2.1. Then $\nu^{-1}(\mathbf{X})$ is a closed substructure of $\mathrm{D}(\mathbf{A})$. Since $\nu^{-1}(\mathbf{X})$ contains the projections, by A.2.1(i), our assumption tells us that $\nu^{-1}(\mathbf{X}) = \mathrm{D}(\mathbf{A})$. Since $\mathbf{X} \leqslant \nu(\mathrm{D}(\mathbf{A}))$, again by A.2.1(i), it follows that $\mathbf{X} = \nu(\mathrm{D}(\mathbf{A}))$. Thus X is term closed, by A.2.1(ii), whence condition (i) holds. ∎

We can obtain Theorem 1.5.3 as a corollary of the preceding theorem, just by using the definition of a strong duality.

A.3 The Inverse Limit Lemma at work

Our ultimate aim is to produce a sufficient condition for a duality to be strong that depends only upon the finite algebras in the quasi-variety. The Inverse Limit Lemma will play a vital role in this process.

Let $\boldsymbol{\mathcal{D}} = \langle \mathcal{D}; \leqslant \rangle$ be a directed ordered set. (This means that the set $\mathcal{D}$ is non-empty and that each pair of elements in $\mathcal{D}$ has an upper bound in $\boldsymbol{\mathcal{D}}$.) Assume that we have a collection of sets $\{ F_i \mid i \in \mathcal{D} \}$ indexed by $\mathcal{D}$. Assume that we also have a collection of connecting maps $\{ \gamma_{ji} : F_j \to F_i \mid j \geqslant i \text{ in } \boldsymbol{\mathcal{D}} \}$ such that

$$\gamma_{ii} = \mathrm{id}_{F_i} \quad \text{and} \quad k \geqslant j \geqslant i \implies \gamma_{ji} \circ \gamma_{kj} = \gamma_{ki},$$

for all $i, j, k \in \mathcal{D}$. Such a system of sets and maps is called an **inverse system**. The **inverse limit** of this system $\boldsymbol{\mathcal{S}}$ is defined to be the subset of the product $\prod\{F_i \mid i \in \mathcal{D}\}$ given by

$$\varprojlim \boldsymbol{\mathcal{S}} := \big\{ z \in \textstyle\prod\{F_i \mid i \in \mathcal{D}\} \bigm| (\forall i, j \in \mathcal{D})\big(j \geqslant i \Rightarrow \gamma_{ji}(z(j)) = z(i)\big) \big\}.$$

(Inverse limits also have a universal-mapping characterisation [46].)

The following result is an easy consequence of Tychonoff's Theorem on products of compact topological spaces [8, 1.3.3].

A.3.1 Inverse Limit Lemma *The inverse limit of an inverse system of finite non-empty sets is non-empty.*

Next we will give a hand-wavy description of a typical strategy for using the Inverse Limit Lemma to lift a property up from finite algebras to arbitrary algebras. Throughout the rest of this appendix, we shall write $\mathbf{B} \ll \mathbf{A}$ to denote the fact that $\mathbf{B}$ is a finite subalgebra of $\mathbf{A}$.

A.3.2 Inverse Limit Lemma Strategy Let $\underline{\mathbf{M}}$ be a finite algebra and let $\mathbf{A}$ be an infinite algebra in $\mathbb{ISP}(\underline{\mathbf{M}})$.

- Assume that, for all $\mathbf{B} \ll \mathbf{A}$, we have a finite non-empty set $F_{\mathbf{B}}$ of objects defined on $\mathbf{B}$. For example: for each $\mathbf{B} \ll \mathbf{A}$, the set $F_{\mathbf{B}}$ might consist of a finite number of tuples of homomorphisms from $\mathbf{B}$ to $\underline{\mathbf{M}}$.
- Assume further that, given $\mathbf{B} \leqslant \mathbf{C} \ll \mathbf{A}$, every element x of $F_{\mathbf{C}}$ can be 'restricted' to $\mathbf{B}$, and thereby produce an element $\gamma_{\mathbf{CB}}(x)$ of $F_{\mathbf{B}}$. In our example: for $\mathbf{B} \leqslant \mathbf{C} \ll \mathbf{A}$, we define the connecting map $\gamma_{\mathbf{CB}} : F_{\mathbf{C}} \to F_{\mathbf{B}}$ by $\gamma_{\mathbf{CB}}\big((x_0, \dots, x_{k-1})\big) := (x_0{\restriction_B}, \dots, x_{k-1}{\restriction_B})$.

- The quasi-variety $\mathbb{ISP}(\underline{\mathbf{M}})$ is locally finite, and so every finitely generated subalgebra of $\mathbf{A}$ is finite. Thus $\mathcal{D} := \{ \mathbf{B} \mid \mathbf{B} \ll \mathbf{A} \}$ forms a directed set $\mathcal{D}$ under the subalgebra order. The collection $\{ F_{\mathbf{B}} \mid \mathbf{B} \ll \mathbf{A} \}$, together with the connecting maps $\{ \gamma_{\mathbf{CB}} \mid \mathbf{B} \leqslant \mathbf{C} \ll \mathbf{A} \}$, forms an inverse system $\boldsymbol{\mathcal{S}}$ of finite non-empty sets. So the inverse limit $\varprojlim \boldsymbol{\mathcal{S}}$ is non-empty.
- The algebra $\mathbf{A}$ in $\mathbb{ISP}(\underline{\mathbf{M}})$ is the union of its finite subalgebras. Since $\mathcal{D}$ is directed, an element of the inverse limit $\varprojlim \boldsymbol{\mathcal{S}} \subseteq \prod\{ F_{\mathbf{B}} \mid \mathbf{B} \ll \mathbf{A} \}$ will determine an object defined on $\mathbf{A}$. In our example: if $(z_{\mathbf{B},0}, \dots, z_{\mathbf{B},k-1})_{\mathbf{B}\in\mathcal{D}}$ is an element of the inverse limit, then we can define the k-tuple $(z_0, \dots, z_{k-1})$ of homomorphisms from $\mathbf{A}$ to $\underline{\mathbf{M}}$ by

$$z_j := \bigcup\{ z_{\mathbf{B},j} \mid \mathbf{B} \ll \mathbf{A} \},$$

for all $j \in \{0, \dots, k-1\}$.

This strategy has already been applied successfully in the theory of natural dualities. The text by Clark and Davey gives two proofs of the Duality Compactness Theorem, 1.4.2, one due to R. Willard [8, 2.2.11] and the other due to L. Zádori [8, 10.6.4]. The two proofs are quite different, but both use variants of this Inverse Limit Lemma Strategy.

Now let $\underset{\sim}{\mathbf{M}} = \langle M; G, H, R, \mathcal{T} \rangle$ be a structure that we hope will strongly dualise a finite algebra $\underline{\mathbf{M}}$. We want to show that each closed substructure of each non-zero power of $\underset{\sim}{\mathbf{M}}$ is term closed. Using the Dual Generation Theorem, A.2.3, we can focus our attention on the closed substructure generated by the projections within each dual. In the next section, we shall develop an internal description of the generation of closed substructures within powers of $\underset{\sim}{\mathbf{M}}$. We will need to 'construct' maps that belong to the algebraic closure of the topological closure of a subset of the power. We end this section with an application of the Inverse Limit Lemma Strategy that will do most of the work for us.

First, consider a subset Z of some structure $\mathbf{X} \leqslant \underset{\sim}{\mathbf{M}}^S$, where S is a non-empty set. We denote the closure of Z under the operations and partial operations in $G^{\mathbf{X}} \cup H^{\mathbf{X}}$ by $[G \cup H](Z)$, and refer to it as the **algebraic closure of** Z **(in X)**. This notation is consistent with our definition of $[G \cup H]$ as the enriched partial clone of $\underset{\sim}{\mathbf{M}}$, since the closure of Z under $\{ p^{\mathbf{X}} \mid p \in [G \cup H] \}$ is equal to the closure of Z under $G^{\mathbf{X}} \cup H^{\mathbf{X}}$.

As usual, the topological closure of Z in $\mathbf{X}$ will be denoted by $\overline{Z}$. Indeed, whenever $Z \subseteq M^S$, for some finite set M, we will use $\overline{Z}$ to denote the topological closure of Z under the product topology induced on M^S by giving the set M the discrete topology. The following lemma encompasses all we shall need to know concerning topological closure in duals.

A.3.3 Topological Closure Lemma [8, B.6] *Let $\underline{\mathbf{M}}$ be a finite algebra and let $\mathbf{A}$ belong to $\mathcal{A} := \mathbb{ISP}(\underline{\mathbf{M}})$. Assume $Z \subseteq \mathcal{A}(\mathbf{A}, \underline{\mathbf{M}})$ and $x \in \mathcal{A}(\mathbf{A}, \underline{\mathbf{M}})$. The following are equivalent:*

(i) $x \in \overline{Z}$;

(ii) *for each finite subset B of A, there exists $z \in Z$ such that $x{\restriction}_B = z{\restriction}_B$;*

(iii) *for all $\mathbf{B} \ll \mathbf{A}$, there exists $z \in Z$ such that $x{\restriction}_B = z{\restriction}_B$.*

Proof The equivalence of (i) and (ii) is a simple consequence of the definition of the product topology on M^A, when we take the discrete topology on M. Since $\underline{\mathbf{M}}$ is finite and $\mathbf{A} \in \mathbb{ISP}(\underline{\mathbf{M}})$, every finitely generated subalgebra of $\mathbf{A}$ is finite. Hence (ii) is equivalent to (iii). ∎

The following simple lemma allows us to give a diagrammatic interpretation of the conditions that arise. For every $\mathbf{A} \in \mathbb{ISP}(\underline{\mathbf{M}})$ and all homomorphisms $z_0, \ldots, z_{k-1} : \mathbf{A} \to \underline{\mathbf{M}}$, where $k \in \omega$, we can define the homomorphism

$$\prod_{j\in k} z_j : \mathbf{A} \to \underline{\mathbf{M}}^k \quad \text{by} \quad \Big(\prod_{j\in k} z_j\Big)(a) := \big(z_0(a), \ldots, z_{k-1}(a)\big).$$

Note that here, as in Chapter 6, we view a natural number $k \in \omega$ as the set $\{0, \ldots, k-1\}$. For each $k \in \omega$ and $p \in [G \cup H]_k$, we use $\mathbf{dom}(p)$ to denote the subalgebra of $\underline{\mathbf{M}}^k$ with underlying set $\mathrm{dom}(p)$, the domain of p.

A.3.4 Diagram Lemma *Let $\underline{\mathbf{M}}$ be a finite algebra, let $\underset{\sim}{\mathrm{M}} = \langle M; G, H, R, \mathcal{T}\rangle$ be an alter ego of $\underline{\mathbf{M}}$ and define $\mathcal{A} := \mathbb{ISP}(\underline{\mathbf{M}})$. Let $\mathbf{B} \leqslant \mathbf{C} \leqslant \mathbf{A}$ in $\mathcal{A}$ and let $y : \mathbf{B} \to \underline{\mathbf{M}}$ be a homomorphism. Assume that $p \in [G \cup H]_k$, for some $k \in \omega$, and that $z_0, \ldots, z_{k-1} \in \mathcal{A}(\mathbf{A}, \underline{\mathbf{M}})$. Then the following are equivalent:*

(i) $p^{\mathrm{D}(\mathbf{C})}(z_0{\restriction}_C, \ldots, z_{k-1}{\restriction}_C) : \mathbf{C} \to \underline{\mathbf{M}}$ *is a well-defined extension of y;*

(ii) $\big(\prod_{j\in k} z_j\big)(C) \subseteq \mathrm{dom}(p)$ *and* $y = p \circ \big(\prod_{j\in k} z_j\big){\restriction}_B$;

(iii) *the diagram*

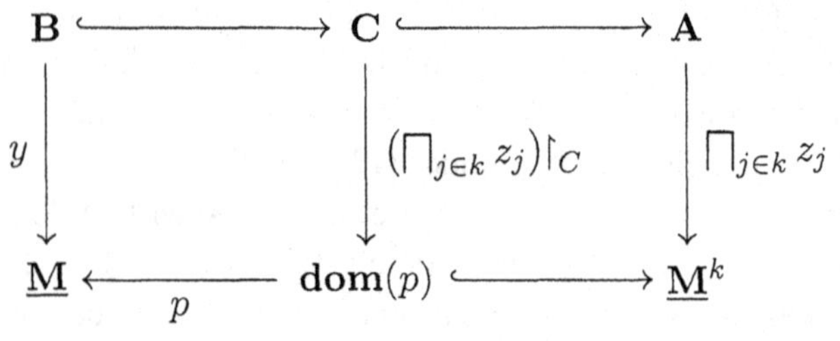

is well defined and commuting.

A.3.5 Much Ado About Nullaries In this appendix, we will often be including the nullary case within the general case. To see how the previous lemma is interpreted when $k = 0$, assume $p : \underline{\mathbf{M}}^{\varnothing} \to \underline{\mathbf{M}}$ is a nullary algebraic operation on $\underline{\mathbf{M}}$. The value of the nullary operation

$$p^{\mathrm{D}(\mathbf{C})} : \mathcal{A}(\mathbf{C}, \underline{\mathbf{M}})^{\varnothing} \to \mathcal{A}(\mathbf{C}, \underline{\mathbf{M}})$$

is the constant homomorphism $p^{\mathrm{D}(\mathbf{C})}(\varnothing) : \mathbf{C} \to \underline{\mathbf{M}}$ with the same value as p. So condition (i) of the previous lemma reduces to the statement that y is a constant homomorphism with the same value as p.

The homomorphism $\sqcap\varnothing : \mathbf{A} \to \underline{\mathbf{M}}^{\varnothing}$ is the unique map from A into the set $M^{\varnothing} = \{\varnothing\}$. So the first part of (ii) reduces to the true statement

$$(\sqcap\varnothing)(C) \subseteq \operatorname{dom}(p) = M^{\varnothing}.$$

The second part of (ii) is equivalent to (i), since $p \circ (\sqcap\varnothing){\restriction}_C : \mathbf{C} \to \underline{\mathbf{M}}$ is a constant homomorphism with the same value as p.

A.3.6 Lemma *Let $\underline{\mathbf{M}}$ be a finite algebra and let $\underset{\sim}{\mathbf{M}} = \langle M; G, H, R, \mathcal{T}\rangle$ be an alter ego of $\underline{\mathbf{M}}$. Let $\mathbf{A}$ belong to $\mathcal{A} := \mathbb{ISP}(\underline{\mathbf{M}})$ and let Z be a subset of $\mathcal{A}(\mathbf{A}, \underline{\mathbf{M}})$. Let $y : \mathbf{B} \to \underline{\mathbf{M}}$ be a homomorphism, for some $\mathbf{B} \ll \mathbf{A}$. Assume there exists $\ell \in \omega$ for which the following condition holds:*

> *for all $\mathbf{C}$ with $\mathbf{B} \leqslant \mathbf{C} \ll \mathbf{A}$, there exists $k \in \omega$ with $k \leqslant \ell$, homomorphisms $z_0, \dots, z_{k-1} : \mathbf{A} \to \underline{\mathbf{M}}$ in Z, and a partial operation $p \in [G \cup H]_k$ such that $p^{\mathrm{D}(\mathbf{C})}(z_0{\restriction}_C, \dots, z_{k-1}{\restriction}_C) : \mathbf{C} \to \underline{\mathbf{M}}$ is a well-defined extension of y.*

$$\begin{array}{ccccc}
\mathbf{B} & \hookrightarrow & \mathbf{C} & \hookrightarrow & \mathbf{A} \\
{\scriptstyle y}\downarrow & & \downarrow{\scriptstyle (\sqcap_{j\in k} z_j){\restriction}_C} & & \downarrow{\scriptstyle \sqcap_{j\in k} z_j} \\
\underline{\mathbf{M}} & \underset{p}{\longleftarrow} & \operatorname{dom}(p) & \hookrightarrow & \underline{\mathbf{M}}^k
\end{array}$$

Then there exists an extension $w : \mathbf{A} \to \underline{\mathbf{M}}$ of y such that $w \in [G \cup H](\overline{Z})$.

Proof We shall apply the Inverse Limit Lemma Strategy. But instead of using the set of all finite subalgebras of $\mathbf{A}$ as the universe of our directed set $\mathcal{D}$, we use the set

$$\mathcal{D} := \{\, \mathbf{C} \mid \mathbf{B} \leqslant \mathbf{C} \ll \mathbf{A} \,\}$$

of finite subalgebras of $\mathbf{A}$ that contain $\mathbf{B}$.

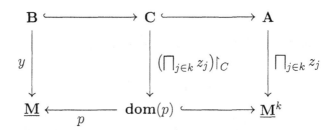

Consider any $\mathbf{C} \in \mathcal{D}$. Define $F_{\mathbf{C}}$ to be the set of all pairs

$$\big((z_0{\restriction}_C, \dots, z_{k-1}{\restriction}_C), p\big), \text{ with } k \leqslant \ell,\ z_0, \dots, z_{k-1} \in Z \text{ and } p \in [G \cup H]_k,$$

such that the conditions

$$\Big(\prod_{j\in k} z_j\Big)(C) \subseteq \operatorname{dom}(p) \quad\text{and}\quad y = p \circ \Big(\prod_{j \in k} z_j\Big){\restriction}_B$$

hold. Using the Diagram Lemma, A.3.4, we know that the set $F_{\mathbf{C}}$ is non-empty. As $\mathbf{C}$ is finite and $[G \cup H]_k$ is finite, for all $k \leqslant \ell$, the set $F_{\mathbf{C}}$ must be finite.

For all $\mathbf{C}, \mathbf{D} \in \mathcal{D}$ with $\mathbf{D} \geqslant \mathbf{C}$, we may define a map $\gamma_{\mathbf{DC}} : F_{\mathbf{D}} \to F_{\mathbf{C}}$ by

$$\gamma_{\mathbf{DC}}\big(((z_0{\restriction}_D, \dots, z_{k-1}{\restriction}_D), p)\big) := \big((z_0{\restriction}_C, \dots, z_{k-1}{\restriction}_C), p\big).$$

We have created an inverse system of finite non-empty sets $\{\, F_{\mathbf{C}} \mid \mathbf{C} \in \mathcal{D} \,\}$ with connecting maps $\{\, \gamma_{\mathbf{DC}} \mid \mathbf{D} \geqslant \mathbf{C} \text{ in } \boldsymbol{\mathcal{D}} \,\}$. By the Inverse Limit Lemma, A.3.1, this system has a non-empty inverse limit.

Now let

$$\big((z_{\mathbf{C},0}{\restriction}_C, \dots, z_{\mathbf{C},k_{\mathbf{C}}-1}{\restriction}_C), p_{\mathbf{C}}\big)_{\mathbf{C}\in\mathcal{D}}$$

be an element of the inverse limit. Since $\boldsymbol{\mathcal{D}}$ is a directed set, the definition of the connecting maps guarantees that $k_{\mathbf{C}} = k_{\mathbf{D}}$ and $p_{\mathbf{C}} = p_{\mathbf{D}}$, for all $\mathbf{C}, \mathbf{D} \in \mathcal{D}$. Denote these common values by k and p, respectively.

It is easy to check that, for each $j \in k$, we can define the homomorphism

$$w_j : \mathbf{A} \to \underline{\mathbf{M}} \quad\text{by}\quad w_j := \bigcup_{\mathbf{C}\in\mathcal{D}} (z_{\mathbf{C},j}{\restriction}_C),$$

and that

$$\Big(\prod_{j\in k} w_j\Big)(A) \subseteq \operatorname{dom}(p) \quad\text{and}\quad y = p \circ \Big(\prod_{j\in k} w_j\Big){\restriction}_B.$$

So, using the Diagram Lemma, it follows that the diagram

$$\begin{array}{ccccc}
\mathbf{B} & \hookrightarrow & \mathbf{A} & & \\
{\scriptstyle y}\downarrow & & \downarrow{\scriptstyle \prod_{j\in k} w_j} & & \\
\underline{\mathbf{M}} & \xleftarrow[p]{} & \mathbf{dom}(p) & \hookrightarrow & \underline{\mathbf{M}}^k
\end{array}$$

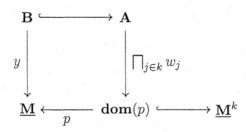

is well defined and commuting, and that $w := p^{\mathrm{D}(\mathbf{A})}(w_0, \dots, w_{k-1}) : \mathbf{A} \to \underline{\mathbf{M}}$ is a well-defined extension of $y : \mathbf{B} \to \underline{\mathbf{M}}$. It remains to show that w belongs to $[G \cup H](\overline{Z})$. As $p \in [G \cup H]_k$, it is enough to show that $w_0, \dots, w_{k-1} \in \overline{Z}$. We will use the Topological Closure Lemma, A.3.3.

Let $j \in \{0, \dots, k-1\}$ and let F be a finite subset of A. Define $\mathbf{C}$ to be the subalgebra of $\mathbf{A}$ generated by $B \cup F$. Since $\mathbf{C} \in \mathcal{D}$, we have $w_j{\restriction}_C = z_{\mathbf{C},j}{\restriction}_C$ and so $w_j{\restriction}_F = z_{\mathbf{C},j}{\restriction}_F$. As $z_{\mathbf{C},j} \in Z$, it follows that $w_j \in \overline{Z}$, by the Topological Closure Lemma. ∎

A.4 The height of a finite algebra

Fix a finite algebra $\underline{\mathbf{M}}$ and an alter ego $\underset{\sim}{\mathbf{M}} = \langle M; G, H, R, \mathcal{T} \rangle$. In order to apply the Dual Generation Theorem, A.2.3, we need conditions that will ensure that, for every $\mathbf{A} \in \mathbb{SP}(\underline{\mathbf{M}})$, the dual $\mathrm{D}(\mathbf{A})$ is generated by the projections. We take our first step towards this goal in this section. As an application of the ideas developed here, we will give a proof that a dualisable algebra with enough algebraic operations must be strongly dualisable; see the EAO Theorem, 1.5.4.

The basis of our approach will be a transfinite internal description of the closed substructure of a structure generated by a given subset.

A.4.1 Definition Let $\underline{\mathbf{M}}$ be a finite algebra and let $\underset{\sim}{\mathbf{M}} = \langle M; G, H, R, \mathcal{T} \rangle$ be an alter ego of $\underline{\mathbf{M}}$. Let $\mathbf{X} \leqslant \underset{\sim}{\mathbf{M}}^S$, for some non-empty set S, and let Y be a subset of X. We define an ordinal sequence $[Y]_0, [Y]_1, \dots, [Y]_\omega, \dots$ of subsets of X inductively, as follows. Define $[Y]_0 := \overline{Y}$ and, for each ordinal $\alpha > 0$, define

$$[Y]_\alpha := \overline{[G \cup H](\overline{Z})}, \quad \text{where } Z := \bigcup \{\, [Y]_\beta \mid \beta < \alpha \,\}.$$

This sequence is non-decreasing. So the construction simplifies in the case of successor ordinals: for each ordinal α, we have $[Y]_{\alpha+1} = \overline{[G \cup H]([Y]_\alpha)}$.

The proof of the following lemma is straightforward; indeed, the lemma is a slightly modified version of Exercise 3.4 in the text by Clark and Davey [8].

A.4.2 Lemma *Let $\underline{\mathbf{M}}$ be a finite algebra and let $\underset{\sim}{\mathbf{M}} = \langle M; G, H, R, \mathcal{T} \rangle$ be an alter ego of $\underline{\mathbf{M}}$. Let $\mathbf{X} \leqslant \underset{\sim}{\mathbf{M}}^S$, for some non-empty set S, and let Y be a subset of X. Define the non-decreasing ordinal sequence $[Y]_0, [Y]_1, \dots$ as in A.4.1.*

(i) *There is an ordinal α such that $[Y]_\alpha = [Y]_\beta$, for all $\beta > \alpha$.*

(ii) *If $[Y]_\alpha = [Y]_\beta$, with $\beta > \alpha$, then $[Y]_\alpha$ is the underlying set of the closed substructure of $\mathbf{X}$ generated by Y.*

(iii) *If $H = \varnothing$, then $[Y]_1 = [Y]_\beta$, for all $\beta > 1$.*

Given a subalgebra $\mathbf{A}$ of a power of $\underline{\mathbf{M}}$, this lemma leads us naturally to a measure of the distance of a homomorphism in $\mathrm{D}(\mathbf{A})$ from the projections, which we will refer to as its height.

A.4.3 Height in Duals Let $\underset{\sim}{\mathbf{M}}$ be an alter ego of a finite algebra $\underline{\mathbf{M}}$. Let $\mathbf{A}$ be a subalgebra of $\underline{\mathbf{M}}^I$, for some set I, and define $\varrho(\mathbf{A})$ to be the set of projections:

$$\varrho(\mathbf{A}) := \{\, \rho_i : \mathbf{A} \to \underline{\mathbf{M}} \mid i \in I \,\}.$$

Consider the non-decreasing ordinal sequence $[\varrho(\mathbf{A})]_0, [\varrho(\mathbf{A})]_1, \dots$ of subsets of $\mathrm{D}(\mathbf{A})$, as given by Definition A.4.1. Let $h : \mathbf{A} \to \underline{\mathbf{M}}$ be a homomorphism in $\mathrm{D}(\mathbf{A})$.

- For each ordinal α, the homomorphism h has **height at most** α **in** $\mathrm{D}(\mathbf{A})$ if h belongs to $[\varrho(\mathbf{A})]_\alpha$.
- If h has height at most α in $\mathrm{D}(\mathbf{A})$, for some ordinal α, then we say that h **has a height in** $\mathrm{D}(\mathbf{A})$. In this case, the **height of** h **in** $\mathrm{D}(\mathbf{A})$ is the least ordinal α such that h has height at most α in $\mathrm{D}(\mathbf{A})$.
- The homomorphism h has a height in $\mathrm{D}(\mathbf{A})$ if and only if it belongs to the closed substructure of $\mathrm{D}(\mathbf{A})$ generated by the set $\varrho(\mathbf{A})$ of projections.

There are three possibilities for the algebra $\underline{\mathbf{M}}$.

- If there is some ordinal α such that, for all $\mathbf{A} \in \mathbb{SP}(\underline{\mathbf{M}})$ and all $h : \mathbf{A} \to \underline{\mathbf{M}}$, the height of h in $\mathrm{D}(\mathbf{A})$ is at most α, then we say that $\underline{\mathbf{M}}$ **has height at most** α **relative to** $\underset{\sim}{\mathbf{M}}$. In this case, the **height of** $\underline{\mathbf{M}}$ **relative to** $\underset{\sim}{\mathbf{M}}$ is the least ordinal α such that $\underline{\mathbf{M}}$ has height at most α relative to $\underset{\sim}{\mathbf{M}}$.
- If there is some $\mathbf{A} \in \mathbb{SP}(\underline{\mathbf{M}})$ and $h : \mathbf{A} \to \underline{\mathbf{M}}$ such that h does not have a height in $\mathrm{D}(\mathbf{A})$, then we say that $\underline{\mathbf{M}}$ **does not have a height relative to** $\underset{\sim}{\mathbf{M}}$.
- Otherwise, for every $\mathbf{A} \in \mathbb{SP}(\underline{\mathbf{M}})$ and $h : \mathbf{A} \to \underline{\mathbf{M}}$, there is an ordinal α such that h has height α in $\mathrm{D}(\mathbf{A})$, but there is no bound on the class of such ordinals. In this case, we say that $\underline{\mathbf{M}}$ **has unbounded height relative to** $\underset{\sim}{\mathbf{M}}$.

There are two cases in which we say that $\underline{\mathbf{M}}$ **has a height relative to** $\underset{\sim}{\mathbf{M}}$: when $\underline{\mathbf{M}}$ has height α, for some ordinal α, and when $\underline{\mathbf{M}}$ has unbounded height.

We shall mostly find it easier to work with 'height at most' than with 'height', as this avoids the need to distinguish between successor and limit ordinals in proofs by transfinite induction.

Let H_ω be the set of all finitary algebraic partial operations on $\underline{\mathbf{M}}$. Then

$$\underset{\sim}{\mathbf{M}}_\Omega := \langle M; H_\omega, \mathcal{T} \rangle$$

is called the **strong brute-force alter ego of** $\underline{\mathbf{M}}$. If some alter ego strongly dualises $\underline{\mathbf{M}}$, then the strong brute-force alter ego $\underset{\sim}{\mathbf{M}}_\Omega$ strongly dualises $\underline{\mathbf{M}}$.

Similarly, for each ordinal α, if $\underline{\mathbf{M}}$ has height at most α relative to some alter ego, then $\underline{\mathbf{M}}$ must have height at most α relative to $\underset{\sim}{\mathbf{M}}_\Omega$. In future, we will simply say '**height**' instead of 'height relative to $\underset{\sim}{\mathbf{M}}_\Omega$'.

Height is completely straightforward for a homomorphism h in the dual of a finite algebra $\mathbf{A} \in \mathbb{SP}(\underline{\mathbf{M}})$. Since the topology on $\mathrm{D}(\mathbf{A})$ is discrete, the closed substructure of $\mathrm{D}(\mathbf{A})$ generated by $\varrho(\mathbf{A})$ has underlying set $[G \cup H](\varrho(\mathbf{A})) = [\varrho(\mathbf{A})]_1$. It follows that if h has a height in $\mathrm{D}(\mathbf{A})$, then it can be at most one. In fact, we can say a little more.

A.4.4 Lemma *Let $\underline{\mathbf{M}}$ be a finite algebra, let $\underset{\sim}{\mathbf{M}} = \langle M; G, H, R, \mathcal{T}\rangle$ be an alter ego of $\underline{\mathbf{M}}$ and let $\mathbf{A} \leqslant \underline{\mathbf{M}}^n$, for some $n \in \omega$.*

(i) *The closed substructure of $\mathrm{D}(\mathbf{A})$ generated by the set $\varrho(\mathbf{A})$ of projections consists of all the homomorphisms from $\mathbf{A}$ to $\underline{\mathbf{M}}$ that have an extension in $[G \cup H]_n$.*

(ii) *Let $h : \mathbf{A} \to \underline{\mathbf{M}}$ be a homomorphism. If h has an extension in $[G \cup H]_n$, then h has height at most 1 in $\mathrm{D}(\mathbf{A})$, else h does not have a height in $\mathrm{D}(\mathbf{A})$.*

Proof Let $h : \mathbf{A} \to \underline{\mathbf{M}}$ be a homomorphism. Since $\mathbf{A} \leqslant \underline{\mathbf{M}}^n$, we know that $\varrho(\mathbf{A}) = \{\rho_0, \dots, \rho_{n-1}\}$ and $a = (\rho_0(a), \dots, \rho_{n-1}(a))$, for each $a \in A$. It follows that

h belongs to $[G \cup H](\varrho(\mathbf{A}))$

$\iff$ there is some $p \in [G \cup H]_n$ for which $(\rho_0, \dots, \rho_{n-1}) \in \mathrm{dom}(p^{\mathrm{D}(\mathbf{A})})$ and $h = p^{\mathrm{D}(\mathbf{A})}(\rho_0, \dots, \rho_{n-1})$

$\iff$ there is some $p \in [G \cup H]_n$ for which $A \subseteq \mathrm{dom}(p)$ and $p{\restriction}_A = h$

$\iff$ h has an extension in $[G \cup H]_n$.

Since $\mathrm{D}(\mathbf{A})$ is finite, the underlying set of the closed substructure of $\mathrm{D}(\mathbf{A})$ generated by $\varrho(\mathbf{A})$ is equal to $[G \cup H](\varrho(\mathbf{A})) = [\varrho(\mathbf{A})]_1$. Both (i) and (ii) follow at once. ∎

Using Lemma A.4.2, we can reinterpret the Dual Generation Theorem, A.2.3.

A.4.5 Theorem *Let $\underline{\mathbf{M}}$ be a finite algebra and let $\underset{\sim}{\mathbf{M}}$ be an alter ego of $\underline{\mathbf{M}}$. The following are equivalent:*

(i) *each closed substructure of each non-zero power of $\underset{\sim}{\mathbf{M}}$ is term closed;*

(ii) *for each homomorphism $h : \mathbf{A} \to \underline{\mathbf{M}}$, where $\mathbf{A} \in \mathbb{SP}(\underline{\mathbf{M}})$, there exists an ordinal α such that h has height α in $\mathrm{D}(\mathbf{A})$;*

(iii) *the algebra $\underline{\mathbf{M}}$ has a height relative to $\underset{\sim}{\mathbf{M}}$.*

From the definition of strong duality, we have the following consequence.

A.4.6 Theorem *Let* $\underline{\mathbf{M}}$ *be a finite algebra and let* $\underset{\sim}{\mathbf{M}}$ *be an alter ego of* $\underline{\mathbf{M}}$.

(i) *The algebra* $\underline{\mathbf{M}}$ *is strongly dualised by* $\underset{\sim}{\mathbf{M}}$ *if and only if the algebra* $\underline{\mathbf{M}}$ *is dualised by* $\underset{\sim}{\mathbf{M}}$ *and has a height relative to* $\underset{\sim}{\mathbf{M}}$.

(ii) *The algebra* $\underline{\mathbf{M}}$ *is strongly dualisable if and only if it is dualisable and has a height.*

Height is defined in terms of the generation of closed substructures of duals from the projections. We shall next see that height is really about the generation of closed substructures in general.

A.4.7 Lemma *Let* $\underline{\mathbf{M}}$ *be a finite algebra and let* $\underset{\sim}{\mathbf{M}}$ *be an alter ego of* $\underline{\mathbf{M}}$. *Let* $\mathbf{X} \leqslant \underset{\sim}{\mathbf{M}}^S$, *for some non-empty set* S, *and let* Y *be a subset of* X.

(i) *Assume that* $\underline{\mathbf{M}}$ *has a height relative to* $\underset{\sim}{\mathbf{M}}$. *Then the closed substructure of* $\mathbf{X}$ *generated by* Y *has underlying set* $\mathrm{tc}_{\underline{\mathbf{M}}}(Y)$, *the term closure of* Y *relative to* $\underline{\mathbf{M}}$.

(ii) *Assume that* $\underline{\mathbf{M}}$ *has height* α *relative to* $\underset{\sim}{\mathbf{M}}$, *for some ordinal* α. *Then the closed substructure of* $\mathbf{X}$ *generated by* Y *has underlying set* $[Y]_\alpha$, *in the notation of A.4.1.*

Proof Define $\mathcal{A} := \mathbb{ISP}(\underline{\mathbf{M}})$. As $Y \subseteq X \subseteq M^S$, we can use Lemma A.2.1 to find a subalgebra $\mathbf{A}$ of $\underline{\mathbf{M}}^Y$ and an embedding $\nu : \mathrm{D}(\mathbf{A}) \hookrightarrow \underset{\sim}{\mathbf{M}}^S$ such that

$$\nu\big(\varrho(\mathbf{A})\big) = Y \quad \text{and} \quad \nu\big(\mathcal{A}(\mathbf{A}, \underline{\mathbf{M}})\big) = \mathrm{tc}_{\underline{\mathbf{M}}}(Y),$$

where $\varrho(\mathbf{A})$ is the set of projections $\{\, \rho_y : \mathbf{A} \to \underline{\mathbf{M}} \mid y \in Y \,\}$. Now define the ordinal sequences $[Y]_0, [Y]_1, \dots$ in $\mathbf{X}$ and $[\varrho(\mathbf{A})]_0, [\varrho(\mathbf{A})]_1, \dots$ in $\mathrm{D}(\mathbf{A})$, as in A.4.1.

Claim If there is an ordinal α such that $[\varrho(\mathbf{A})]_\alpha = \mathcal{A}(\mathbf{A}, \underline{\mathbf{M}})$, then the underlying set of the closed substructure of $\mathbf{X}$ generated by Y is $[Y]_\alpha = \mathrm{tc}_{\underline{\mathbf{M}}}(Y)$.

To prove the claim, assume that $[\varrho(\mathbf{A})]_\alpha = \mathcal{A}(\mathbf{A}, \underline{\mathbf{M}})$, for some ordinal α. Then, since ν is an embedding, we have

$$[Y]_\alpha = [\nu(\varrho(\mathbf{A}))]_\alpha = \nu\big([\varrho(\mathbf{A})]_\alpha\big) = \nu\big(\mathcal{A}(\mathbf{A}, \underline{\mathbf{M}})\big) = \mathrm{tc}_{\underline{\mathbf{M}}}(Y).$$

By Lemma A.1.4, the term closure $\mathrm{tc}_{\underline{\mathbf{M}}}(Y)$ forms a closed substructure of $\underset{\sim}{\mathbf{M}}^S$. So it follows, by the construction of $[Y]_\alpha$, that the underlying set of the closed substructure of $\mathbf{X}$ generated by Y is equal to $[Y]_\alpha = \mathrm{tc}_{\underline{\mathbf{M}}}(Y)$, as required.

We can now prove the lemma. For claim (i), assume that $\underline{\mathbf{M}}$ has a height relative to $\underset{\sim}{\mathbf{M}}$. Then there exists an ordinal α such that $[\varrho(\mathbf{A})]_\alpha = \mathcal{A}(\mathbf{A}, \underline{\mathbf{M}})$.

By the claim above, the underlying set of the closed substructure of $\mathbf{X}$ generated by Y is equal to $\mathrm{tc}_{\underline{\mathbf{M}}}(Y)$. For claim (ii), assume that $\underline{\mathbf{M}}$ has height α relative to $\underset{\sim}{\mathbf{M}}$, for some ordinal α. Then $[\varrho(\mathbf{A})]_\alpha = \mathcal{A}(\mathbf{A}, \underline{\mathbf{M}})$ again. Using the claim above, the underlying set of the closed substructure of $\mathbf{X}$ generated by Y is equal to $[Y]_\alpha$. ∎

A.4.8 Remark Let $\underset{\sim}{\mathbf{M}}$ be an alter ego of a finite algebra $\underline{\mathbf{M}}$. Claim (ii) of the previous lemma tells us that height is an accurate measure of the complexity of the generation of closed substructures in the dual category $\mathbb{IS}_c\mathbb{P}^+(\underset{\sim}{\mathbf{M}})$. Assume that $\underline{\mathbf{M}}$ has height α relative to $\underset{\sim}{\mathbf{M}}$, for some ordinal α. Then, for each $\mathbf{X} \in \mathbb{IS}_c\mathbb{P}^+(\underset{\sim}{\mathbf{M}})$, the closed substructure of $\mathbf{X}$ generated by a subset Y can be constructed from the generating set Y in at most α steps.

In order to understand the concept of height properly, we need to find an internal description of the set $\overline{[G \cup H](\overline{Z})}$, for each subset Z of $\mathrm{D}(\mathbf{A})$, as used in Definition A.4.1.

A.4.9 Lemma *Let $\underline{\mathbf{M}}$ be a finite algebra and let $\underset{\sim}{\mathbf{M}} = \langle M; G, H, R, \mathcal{T}\rangle$ be an alter ego of $\underline{\mathbf{M}}$. Given an algebra $\mathbf{A}$ in $\mathcal{A} := \mathbb{ISP}(\underline{\mathbf{M}})$, a subset Z of $\mathcal{A}(\mathbf{A}, \underline{\mathbf{M}})$ and a homomorphism $x \in \mathcal{A}(\mathbf{A}, \underline{\mathbf{M}})$, the following are equivalent:*

(i) $x \in \overline{[G \cup H](\overline{Z})}$;

(ii) *for all $\mathbf{B} \ll \mathbf{A}$, there is $\ell \in \omega$ for which the following condition holds:*

for all $\mathbf{C}$ with $\mathbf{B} \leqslant \mathbf{C} \ll \mathbf{A}$, there exists $k \in \omega$ with $k \leqslant \ell$, homomorphisms $z_0, \dots, z_{k-1} : \mathbf{A} \to \underline{\mathbf{M}}$ in Z, and a partial operation $p \in [G \cup H]_k$ such that $p^{\mathrm{D}(\mathbf{C})}(z_0{\restriction}_C, \dots, z_{k-1}{\restriction}_C) : \mathbf{C} \to \underline{\mathbf{M}}$ is a well-defined extension of $x{\restriction}_B$.

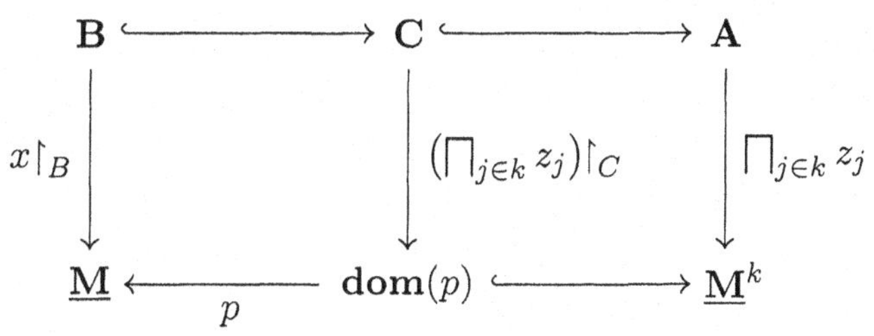

Proof We begin by proving that (i) implies (ii). Assume that $x \in \overline{[G \cup H](\overline{Z})}$ and let $\mathbf{B} \ll \mathbf{A}$. Then by the Topological Closure Lemma, A.3.3, there exists $w \in [G \cup H](\overline{Z})$ with $x{\restriction}_B = w{\restriction}_B$. Thus there exists $\ell \in \omega$, homomorphisms $w_0 \dots, w_{\ell-1} \in \overline{Z}$ and a partial operation $p \in [G \cup H]_\ell$ such that

$$(w_0, \dots, w_{\ell-1}) \in \mathrm{dom}(p^{\mathrm{D}(\mathbf{A})}) \quad \text{and} \quad x{\restriction}_B = p^{\mathrm{D}(\mathbf{A})}(w_0, \dots, w_{\ell-1}){\restriction}_B.$$

By the Diagram Lemma, A.3.4, this gives us

$$\Big(\prod_{j\in\ell} w_j\Big)(A) \subseteq \mathrm{dom}(p) \quad\text{and}\quad x{\restriction}_B = p\circ\Big(\prod_{j\in\ell} w_j\Big){\restriction}_B.$$

Now let $\mathbf{C}$ be an algebra with $\mathbf{B}\leqslant\mathbf{C}\ll\mathbf{A}$. For all $j\in\{0,\dots,\ell-1\}$, we have $w_j\in\overline{Z}$ and so, by the Topological Closure Lemma, there exists $z_j\in Z$ with $w_j{\restriction}_C = z_j{\restriction}_C$. Thus we have

$$\Big(\prod_{j\in\ell} z_j\Big)(C) = \Big(\prod_{j\in\ell} w_j\Big)(C) \subseteq \Big(\prod_{j\in\ell} w_j\Big)(A) \subseteq \mathrm{dom}(p).$$

Since $B\subseteq C$, we have $w_j{\restriction}_B = z_j{\restriction}_B$, for all $j\in\{0,\dots,\ell-1\}$. Thus

$$p\circ\Big(\prod_{j\in\ell} z_j\Big){\restriction}_B = p\circ\Big(\prod_{j\in\ell} w_j\Big){\restriction}_B = x{\restriction}_B.$$

Hence $p^{\mathrm{D}(\mathbf{C})}(z_0{\restriction}_C,\dots,z_{\ell-1}{\restriction}_C)\colon\mathbf{C}\to\underline{\mathbf{M}}$ is a well-defined extension of $x{\restriction}_B$, by the Diagram Lemma, whence condition (ii) holds.

Conversely, assume that condition (ii) holds. We will use the Topological Closure Lemma to prove that (i) holds. Let $\mathbf{B}\ll\mathbf{A}$. Then the assumptions of Lemma A.3.6 hold with $y := x{\restriction}_B$. Thus there exists a homomorphism $w\in[G\cup H](\overline{Z})$ such that $w{\restriction}_B = x{\restriction}_B$. It follows by the Topological Closure Lemma that $x\in\overline{[G\cup H](\overline{Z})}$. Hence (i) holds. ∎

In the next section, we will use the previous lemma to obtain a transfinite description of homomorphisms of height at most α in $\mathrm{D}(\mathbf{A})$, for each ordinal α and each $\mathbf{A}\in\mathbb{SP}(\underline{\mathbf{M}})$.

We finish this section by applying Lemma A.4.9 to show that every finite algebra with enough algebraic operations has height at most 2. This provides the promised proof of the EAO Theorem, 1.5.4. Indeed, we obtain a slightly stronger conclusion. Assume that $\underline{\mathbf{M}}$ has enough algebraic operations and define $\mathcal{A} := \mathbb{ISP}(\underline{\mathbf{M}})$. Then our next theorem says that, for every subalgebra $\mathbf{A}$ of a power of $\underline{\mathbf{M}}$, the set $\mathcal{A}(\mathbf{A},\underline{\mathbf{M}})$ can be obtained by:

- first closing the set $\varrho(\mathbf{A})$ of projections under all the algebraic *total* operations on $\underline{\mathbf{M}}$ and then closing topologically,
- next closing under all the algebraic *partial* operations on $\underline{\mathbf{M}}$ and then closing topologically again.

Since the proper partial operations are used only once, while the total operations are used twice, it might be said that having enough algebraic operations guarantees height $1\frac{1}{2}$. By working with height relative to a specific alter ego, we can prove a little more.

Consider an alter ego $\underset{\sim}{\mathbf{M}} = \langle M; G, H, R, \mathcal{T}\rangle$ of a finite algebra $\underline{\mathbf{M}}$. We say that $\underline{\mathbf{M}}$ has **enough algebraic operations relative to** $\underset{\sim}{\mathbf{M}}$ if $[G \cup H]_0$ contains all the nullary algebraic operations on $\underline{\mathbf{M}}$ and there is a map $f : \omega \to \omega$ for which the following condition holds:

> for all $n \in \omega\backslash\{0\}$, all $\mathbf{B} \leqslant \mathbf{C} \leqslant \underline{\mathbf{M}}^n$ and all non-constant homomorphisms $h : \mathbf{C} \to \underline{\mathbf{M}}$, there exists $k \in \omega$ with $k \leqslant f(|B|)$, total operations $g_0, \dots, g_{k-1} \in [G \cup H]_n$, and a partial operation $p \in [G \cup H]_k$ such that $p^{\mathrm{D}(\mathbf{C})}(g_0{\upharpoonright_C}, \dots, g_{k-1}{\upharpoonright_C}) : \mathbf{C} \to \underline{\mathbf{M}}$ is a well-defined extension of $h{\upharpoonright_B}$.

$$\begin{array}{ccccc} \mathbf{B} & \hookrightarrow & \mathbf{C} & \hookrightarrow & \underline{\mathbf{M}}^n \\ \Big\downarrow h{\upharpoonright_B} & & \Big\downarrow \big(\prod_{j\in k} g_j\big){\upharpoonright_C} & & \Big\downarrow \prod_{j\in k} g_j \\ \underline{\mathbf{M}} & \xleftarrow[p]{\qquad} & \mathbf{dom}(p) & \hookrightarrow & \underline{\mathbf{M}}^k \end{array}$$

Recall that $\underset{\sim}{\mathbf{M}}_\Omega$ denotes the strong brute-force alter ego of $\underline{\mathbf{M}}$. The enriched partial clone of $\underset{\sim}{\mathbf{M}}_\Omega$ certainly contains every nullary algebraic operation on $\underline{\mathbf{M}}$. It follows that the definition of 'enough algebraic operations' given on page 24 is equivalent to 'enough algebraic operations relative to $\underset{\sim}{\mathbf{M}}_\Omega$'.

A.4.10 Theorem *Let $\underline{\mathbf{M}}$ be a finite algebra and let $\underset{\sim}{\mathbf{M}} = \langle M; G, H, R, \mathcal{T}\rangle$ be an alter ego of $\underline{\mathbf{M}}$. Assume that $\underline{\mathbf{M}}$ has enough algebraic operations relative to $\underset{\sim}{\mathbf{M}}$. Then the height of $\underline{\mathbf{M}}$ relative to $\underset{\sim}{\mathbf{M}}$ is at most 2. In fact, for every $\mathbf{A} \in \mathbb{SP}(\underline{\mathbf{M}})$, the underlying set of $\mathrm{D}(\mathbf{A})$ is equal to $[\varrho(\mathbf{A})]_{1\frac{1}{2}}$, where*

$$[\varrho(\mathbf{A})]_{1\frac{1}{2}} := \overline{[G \cup H]\big([\varrho(\mathbf{A})]_{\frac{1}{2}}\big)}, \qquad [\varrho(\mathbf{A})]_{\frac{1}{2}} := \overline{[G \cup H]_{\text{total}}\big(\varrho(\mathbf{A})\big)}$$

and $[G \cup H]_{\text{total}}$ is the set of all total operations in $[G \cup H]$.

Proof Define the quasi-variety $\mathcal{A} := \mathbb{ISP}(\underline{\mathbf{M}})$. Assume that $\underline{\mathbf{M}}$ has enough algebraic operations relative to $\underset{\sim}{\mathbf{M}}$, via the bounding function $f : \omega \to \omega$. Now let $\mathbf{A} \leqslant \underline{\mathbf{M}}^I$, for some set I. It suffices to prove that

$$\mathcal{A}(\mathbf{A}, \underline{\mathbf{M}}) = \overline{[G \cup H](\overline{Z})}, \quad \text{where} \quad Z := [G \cup H]_{\text{total}}\big(\varrho(\mathbf{A})\big).$$

We will do this by showing that every homomorphism $x \in \mathcal{A}(\mathbf{A}, \underline{\mathbf{M}})$ satisfies condition (ii) of Lemma A.4.9. So choose $x \in \mathcal{A}(\mathbf{A}, \underline{\mathbf{M}})$. Let $\mathbf{B} \ll \mathbf{A}$ and define $\ell := f(|B|)$. Now let $\mathbf{C}$ be an algebra with $\mathbf{B} \leqslant \mathbf{C} \ll \mathbf{A}$.

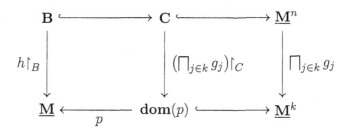

First assume that $x{\upharpoonright_C} : \mathbf{C} \to \underline{\mathbf{M}}$ is constant. Since $\underline{\mathbf{M}}$ has enough algebraic operations relative to $\underset{\sim}{\mathbf{M}}$, each nullary algebraic operation on $\underline{\mathbf{M}}$ belongs to the set $[G \cup H]_0$. So there is a nullary operation $p \in [G \cup H]_0$ with the same value as $x{\restriction}_C$. Therefore $p^{\mathrm{D}(\mathbf{C})}(\varnothing) : \mathbf{C} \to \underline{\mathbf{M}}$ is a well-defined extension of $x{\restriction}_B$.

Now we can assume that $x{\restriction}_C : \mathbf{C} \to \underline{\mathbf{M}}$ is not constant. The set I must be non-empty. Since $\mathbf{C}$ is finite, the set $\varrho(\mathbf{C}) = \{\, \rho_i{\restriction}_C \mid i \in I \,\}$ of projections is finite and non-empty. So there exists $n \in \omega\backslash\{0\}$ and $i_0, \dots, i_{n-1} \in I$ for which $\varrho(\mathbf{C}) = \{\rho_{i_0}{\restriction}_C, \dots, \rho_{i_{n-1}}{\restriction}_C\}$. We can define the embedding

$$\varphi : \mathbf{C} \hookrightarrow \underline{\mathbf{M}}^n \quad \text{by} \quad \varphi := \Big(\prod_{s \in n} \rho_{i_s}\Big){\restriction}_C.$$

Set $\mathbf{B}' := \varphi(\mathbf{B})$ and $\mathbf{C}' := \varphi(\mathbf{C})$. Then we have the following commutative diagram.

$$\begin{array}{ccccc}
\mathbf{B} & \hookrightarrow & \mathbf{C} & \hookrightarrow & \mathbf{A} \\
\varphi{\restriction}_B \downarrow & & \downarrow \varphi & & \downarrow \prod_{s \in n} \rho_{i_s} \\
\mathbf{B}' & \hookrightarrow & \mathbf{C}' & \hookrightarrow & \underline{\mathbf{M}}^n
\end{array}$$

Define the homomorphism $h : \mathbf{C}' \to \underline{\mathbf{M}}$ by $h := x{\restriction}_C \circ \varphi^{-1}$, where we have $\varphi^{-1} : \mathbf{C}' \hookrightarrow\!\!\!\!\to \mathbf{C}$. Since $x{\restriction}_C$ is non-constant, the map h is also non-constant.

We can now use the technical condition in the definition of enough algebraic operations relative to $\underset{\sim}{\mathrm{M}}$, with $\mathbf{B}$ and $\mathbf{C}$ replaced by $\mathbf{B}'$ and $\mathbf{C}'$. Therefore there exists $k \in \omega$ with $k \leqslant f(|B|) = \ell$, operations $g_0, \dots, g_{k-1} : \underline{\mathbf{M}}^n \to \underline{\mathbf{M}}$ in $[G \cup H]_{\text{total}}$, and a partial operation $p \in [G \cup H]_k$ such that

$$p^{\mathrm{D}(\mathbf{C}')}(g_0{\restriction}_{C'}, \dots, g_{k-1}{\restriction}_{C'}) : \mathbf{C}' \to \underline{\mathbf{M}}$$

is a well-defined extension of $h{\restriction}_{B'}$. This gives us

$$\Big(\prod_{j \in k} g_j\Big)(C') \subseteq \mathrm{dom}(p) \quad \text{and} \quad h{\restriction}_{B'} = p \circ \Big(\prod_{j \in k} g_j\Big){\restriction}_{B'},$$

by the Diagram Lemma, A.3.4.

Now, for each $j \in \{0, \dots, k-1\}$, define the homomorphism

$$z_j := g_j(\rho_{i_0}, \dots, \rho_{i_{n-1}}) : \mathbf{A} \to \underline{\mathbf{M}}.$$

Then $z_j \in [G \cup H]_{\text{total}}(\varrho(\mathbf{A})) = Z$, for all $j \in \{0, \dots, k-1\}$. It remains to show that $p^{\mathrm{D}(\mathbf{C})}(z_0{\restriction}_C, \dots, z_{k-1}{\restriction}_C) : \mathbf{C} \to \underline{\mathbf{M}}$ is a well-defined extension of $x{\restriction}_B$. To do this, we will use the Diagram Lemma again.

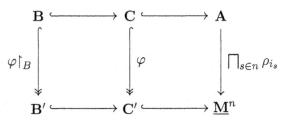

For all $c \in C$ and $j \in \{0, \dots, k-1\}$, we have

$$z_j(c) = g_j(\rho_{i_0}, \dots, \rho_{i_{n-1}})(c) = g_j \circ \Big(\prod_{s \in n} \rho_{i_s}\Big){\restriction_C}(c) = g_j \circ \varphi(c).$$

So

$$\Big(\prod_{j \in k} z_j\Big)(C) = \Big(\prod_{j \in k} g_j \circ \varphi\Big)(C) = \Big(\prod_{j \in k} g_j\Big)(C') \subseteq \mathrm{dom}(p).$$

Since $B \subseteq C$, we have $z_j{\restriction_B} = g_j{\restriction_{B'}} \circ \varphi{\restriction_B}$, for all $j \in \{0, \dots, k-1\}$, and therefore

$$p \circ \Big(\prod_{j \in k} z_j\Big){\restriction_B} = p \circ \prod_{j \in k} (g_j{\restriction_{B'}} \circ \varphi{\restriction_B}) = p \circ \Big(\prod_{j \in k} g_j\Big){\restriction_{B'}} \circ \varphi{\restriction_B}.$$

As $h{\restriction_{B'}} = p \circ (\prod_{j \in k} g_j){\restriction_{B'}}$, this gives

$$p \circ \Big(\prod_{j \in k} z_j\Big){\restriction_B} = h{\restriction_{B'}} \circ \varphi{\restriction_B} = (x{\restriction_C} \circ \varphi^{-1}){\restriction_{B'}} \circ \varphi{\restriction_B} = x{\restriction_B}.$$

So $p^{\mathbf{D}(\mathbf{C})}(z_0{\restriction_C}, \dots, z_{k-1}{\restriction_C}) : \mathbf{C} \to \underline{\mathbf{M}}$ is a well-defined extension of $x{\restriction_B}$, by the Diagram Lemma. We have shown that condition (ii) of Lemma A.4.9 holds for x, as required. ∎

The following theorem, which includes the EAO Theorem, 1.5.4, is an immediate corollary of Theorems A.4.6 and A.4.10.

A.4.11 Theorem

(i) *Each finite algebra with enough algebraic operations has height at most* 2.

(ii) *Each finite algebra with enough algebraic operations that is dualisable must be strongly dualisable.*

A.5 Reducing height to the finite level

Our aim is to arrive at R. Willard's concept of the rank of a finite algebra $\underline{\mathbf{M}}$. Rank provides a sufficient condition for the algebra $\underline{\mathbf{M}}$ to have height at most α, for some ordinal α, that involves only finite algebras in $\mathbb{ISP}(\underline{\mathbf{M}})$. As a first step towards rank, we concentrate on the finite restrictions of homomorphisms in the duals of algebras in $\mathbb{SP}(\underline{\mathbf{M}})$.

Assume that $\underset{\sim}{\mathbf{M}}$ is an alter ego of $\underline{\mathbf{M}}$, and let $\mathbf{A}$ be an algebra in $\mathbb{SP}(\underline{\mathbf{M}})$. For each $\mathbf{B} \ll \mathbf{A}$, we say that a homomorphism $y : \mathbf{B} \to \underline{\mathbf{M}}$ is a **fragment** of a homomorphism $x : \mathbf{A} \to \underline{\mathbf{M}}$ if we have $y = x{\restriction_B}$; in this case, we say that

y is a $\mathrm{D}(\mathbf{A})$-**fragment**. Informally, we will define $\mathrm{Frag}_{\mathrm{D}(\mathbf{A})}$ to be the set of all $\mathrm{D}(\mathbf{A})$-fragments.

More formally, we can consider $\mathrm{Frag}_{\mathrm{D}(\mathbf{A})}$ to be an inverse system in the topological quasi-variety $\mathbb{I}\mathbb{S}_{\mathrm{c}}\mathbb{P}^{+}(\underset{\sim}{\mathrm{M}})$. The set of objects is $\{\,\mathrm{D}(\mathbf{B}) \mid \mathbf{B} \ll \mathbf{A}\,\}$ and, for all $\mathbf{B} \leqslant \mathbf{C} \ll \mathbf{A}$, the connecting morphism $\gamma_{\mathbf{CB}} : \mathrm{D}(\mathbf{C}) \to \mathrm{D}(\mathbf{B})$ is given by $\gamma_{\mathbf{CB}}(x) := x{\restriction}_B$.

We shall now give a definition of the height of a fragment *in* $\mathrm{Frag}_{\mathrm{D}(\mathbf{A})}$, based on Lemma A.4.9. We then show that, for each ordinal α, a homomorphism has height at most α in $\mathrm{D}(\mathbf{A})$ if and only if all of its fragments have height at most α in $\mathrm{Frag}_{\mathrm{D}(\mathbf{A})}$.

A.5.1 Height of Fragments Let $\underset{\sim}{\mathrm{M}} = \langle M; G, H, R, \mathcal{T}\rangle$ be an alter ego of a finite algebra $\underline{\mathbf{M}}$. Let $\mathbf{A}$ be a subalgebra of $\underline{\mathbf{M}}^I$, for some set I, and let $y : \mathbf{B} \to \underline{\mathbf{M}}$ be a $\mathrm{D}(\mathbf{A})$-fragment. The height of y in $\mathrm{Frag}_{\mathrm{D}(\mathbf{A})}$ is defined by transfinite induction.

- The homomorphism y has **height at most** 0 **in** $\mathrm{Frag}_{\mathrm{D}(\mathbf{A})}$ if it is a projection, that is, if $y = \rho_i{\restriction}_B$, for some $i \in I$.
- For every ordinal $\alpha > 0$, the homomorphism y has **height at most** α **in** $\mathrm{Frag}_{\mathrm{D}(\mathbf{A})}$ if there is $\ell \in \omega$ for which the following condition holds:

  for all $\mathbf{C}$ with $\mathbf{B} \leqslant \mathbf{C} \ll \mathbf{A}$, there exists $k \in \omega$ with $k \leqslant \ell$, an ordinal $\beta < \alpha$, some $\mathrm{D}(\mathbf{A})$-fragments $v_0, \dots, v_{k-1} : \mathbf{C} \to \underline{\mathbf{M}}$ of height at most β in $\mathrm{Frag}_{\mathrm{D}(\mathbf{A})}$, and a partial operation $p \in [G \cup H]_k$ such that $p^{\mathrm{D}(\mathbf{C})}(v_0, \dots, v_{k-1}) : \mathbf{C} \to \underline{\mathbf{M}}$ is a well-defined extension of y.

$$
\begin{array}{ccccc}
\mathbf{B} & \hookrightarrow & \mathbf{C} & \hookrightarrow & \mathbf{A} \\
\big\downarrow{\scriptstyle y} & & \big\downarrow{\scriptstyle \sqcap_{j \in k} v_j} & & \\
\underline{\mathbf{M}} & \xleftarrow[p]{} & \mathbf{dom}(p) & \hookrightarrow & \underline{\mathbf{M}}^k
\end{array}
$$

In the special case that I is the empty set, it follows from this definition that a nullary operation $y : \underline{\mathbf{M}}^{\varnothing} \to \underline{\mathbf{M}}$ has height 1 in $\mathrm{Frag}_{\mathrm{D}(\mathbf{A})}$ if $y \in [G \cup H]_0$, and does not have a height in $\mathrm{Frag}_{\mathrm{D}(\mathbf{A})}$ otherwise.

As one would hope, for any ordinal α, the $\mathrm{D}(\mathbf{A})$-fragments of height at most α in $\mathrm{Frag}_{\mathrm{D}(\mathbf{A})}$ are precisely the fragments of homomorphisms of height at most α in $\mathrm{D}(\mathbf{A})$.

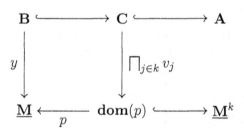

A.5.2 Lemma *Let $\underline{\mathbf{M}}$ be a finite algebra and let $\underset{\sim}{\mathbf{M}}$ be an alter ego of $\underline{\mathbf{M}}$. Assume that $\mathbf{A}$ is a subalgebra of $\underline{\mathbf{M}}^I$, for some set I, and let $y : \mathbf{B} \to \underline{\mathbf{M}}$ be a $\mathrm{D}(\mathbf{A})$-fragment. Then, for each ordinal α, the following are equivalent:*

(i) *the homomorphism y has height at most α in $\mathrm{Frag}_{\mathrm{D}(\mathbf{A})}$;*

(ii) *there is an extension $w : \mathbf{A} \to \underline{\mathbf{M}}$ of y that has height at most α in $\mathrm{D}(\mathbf{A})$.*

Proof The proof is a straightforward transfinite induction, once we rearrange the order of the universal quantifiers. Let $\underset{\sim}{\mathbf{M}} = \langle M; G, H, R, \mathcal{T} \rangle$ be an alter ego of $\underline{\mathbf{M}}$, and let $\mathbf{A} \leqslant \underline{\mathbf{M}}^I$, for some set I. For each ordinal α, consider the following claim.

$\mathcal{C}(\alpha)$ *For each $\mathbf{B} \ll \mathbf{A}$ and each $\mathrm{D}(\mathbf{A})$-fragment $y : \mathbf{B} \to \underline{\mathbf{M}}$, the homomorphism y has height at most α in $\mathrm{Frag}_{\mathrm{D}(\mathbf{A})}$ if and only if there is an extension $w : \mathbf{A} \to \underline{\mathbf{M}}$ of y that has height at most α in $\mathrm{D}(\mathbf{A})$.*

Then we can prove the lemma by showing that claim $\mathcal{C}(\alpha)$ holds, for each ordinal α.

Using the Topological Closure Lemma, A.3.3, it is easy to prove $\mathcal{C}(0)$. So let $\alpha > 0$, and assume that $\mathcal{C}(\beta)$ holds, for all ordinals $\beta < \alpha$. Define the ordinal sequence $[\varrho(\mathbf{A})]_0, [\varrho(\mathbf{A})]_1, \dots$ as in Definition A.4.1. Let $\mathbf{B} \ll \mathbf{A}$ and let $y : \mathbf{B} \to \underline{\mathbf{M}}$ be a $\mathrm{D}(\mathbf{A})$-fragment.

For the forward direction, assume that y has height at most α in $\mathrm{Frag}_{\mathrm{D}(\mathbf{A})}$, with the bound $\ell \in \omega$. To find an extension of y that has height at most α in $\mathrm{D}(\mathbf{A})$, we will show that the assumptions of Lemma A.3.6 hold, with

$$Z := \bigcup \{ [\varrho(\mathbf{A})]_\beta \mid \beta < \alpha \}.$$

Let $\mathbf{C}$ be an algebra such that $\mathbf{B} \leqslant \mathbf{C} \ll \mathbf{A}$. Since y has height at most α in $\mathrm{Frag}_{\mathrm{D}(\mathbf{A})}$, there is some $k \leqslant \ell$, an ordinal $\beta < \alpha$, some $\mathrm{D}(\mathbf{A})$-fragments $v_0, \dots, v_{k-1} : \mathbf{C} \to \underline{\mathbf{M}}$ of height at most β in $\mathrm{Frag}_{\mathrm{D}(\mathbf{A})}$, and a partial operation $p \in [G \cup H]_k$ such that $p^{\mathrm{D}(\mathbf{C})}(v_0, \dots, v_{k-1}) : \mathbf{C} \to \underline{\mathbf{M}}$ is a well-defined extension of y. We are assuming that claim $\mathcal{C}(\beta)$ holds. Therefore, for each $j \in \{0, \dots, k-1\}$, there is an extension $z_j : \mathbf{A} \to \underline{\mathbf{M}}$ of v_j that has height at most β in $\mathrm{D}(\mathbf{C})$. So we have $z_0, \dots, z_{k-1} \in [\varrho(\mathbf{A})]_\beta \subseteq Z$.

We have now established that the assumptions of Lemma A.3.6 hold. Hence there is an extension $w : \mathbf{A} \to \underline{\mathbf{M}}$ of y with $w \in [G \cup H](\overline{Z}) \subseteq [\varrho(\mathbf{A})]_\alpha$, as required.

For the reverse direction, assume that the $\mathrm{D}(\mathbf{A})$-fragment $y : \mathbf{B} \to \underline{\mathbf{M}}$ has an extension $w : \mathbf{A} \to \underline{\mathbf{M}}$ of height at most α in $\mathrm{D}(\mathbf{A})$. Then, by the definition of height at most α, we have

$$w \in [\varrho(\mathbf{A})]_\alpha = \overline{[G \cup H](\overline{Z})}, \quad \text{where} \quad Z := \bigcup \{ [\varrho(\mathbf{A})]_\beta \mid \beta < \alpha \}.$$

Now let $\ell \in \omega$ be the bound provided by Lemma A.4.9, where $x := w$. To prove that y has height at most α in $\mathrm{Frag}_{\mathrm{D}(\mathbf{A})}$, let $\mathbf{C}$ be an algebra with $\mathbf{B} \leqslant \mathbf{C} \ll \mathbf{A}$. Using Lemma A.4.9, there exists $k \leqslant \ell$, homomorphisms $z_0, \dots, z_{k-1} : \mathbf{A} \to \underline{\mathbf{M}}$ in Z, and a partial operation $p \in [G \cup H]_k$ such that $p^{\mathrm{D}(\mathbf{C})}(z_0{\restriction_C}, \dots, z_{k-1}{\restriction_C}) : \mathbf{C} \to \underline{\mathbf{M}}$ is a well-defined extension of $w{\restriction_B} = y$.

$$\begin{array}{ccccc} \mathbf{B} & \hookrightarrow & \mathbf{C} & \hookrightarrow & \mathbf{A} \\ \big\downarrow y & & \big\downarrow (\sqcap_{j\in k} z_j){\restriction_C} & & \big\downarrow \sqcap_{j\in k} z_j \\ \underline{\mathbf{M}} & \xleftarrow[p]{\quad} & \mathbf{dom}(p) & \hookrightarrow & \underline{\mathbf{M}}^k \end{array}$$

Since $Z = \bigcup\{\, [\varrho(\mathbf{A})]_\beta \mid \beta < \alpha \,\}$ and the sequence $[\varrho(\mathbf{A})]_0, [\varrho(\mathbf{A})]_1, \dots$ is non-decreasing, there must be some ordinal $\beta < \alpha$ such that $z_j \in [\varrho(\mathbf{A})]_\beta$, for all $j \in \{0, \dots, k-1\}$. So the homomorphisms $z_0, \dots, z_{k-1}$ all have height at most β in $\mathrm{D}(\mathbf{A})$. As $\mathcal{C}(\beta)$ holds, the $\mathrm{D}(\mathbf{A})$-fragment $v_j := z_j{\restriction_C} : \mathbf{C} \to \underline{\mathbf{M}}$ has height at most β in $\mathrm{Frag}_{\mathrm{D}(\mathbf{A})}$, for all $j \in \{0, \dots, k-1\}$. It now follows that y is a $\mathrm{D}(\mathbf{A})$-fragment of height at most α in $\mathrm{Frag}_{\mathrm{D}(\mathbf{A})}$. ∎

This lemma has, as an almost immediate corollary, a characterisation of the homomorphisms of height at most α in $\mathrm{D}(\mathbf{A})$ in terms of their fragments, for each ordinal α.

A.5.3 Theorem *Let $\underline{\mathbf{M}}$ be a finite algebra and let $\underset{\sim}{\mathbf{M}}$ be an alter ego of $\underline{\mathbf{M}}$. Assume that $\mathbf{A} \in \mathbb{SP}(\underline{\mathbf{M}})$ and let $x : \mathbf{A} \to \underline{\mathbf{M}}$ be a homomorphism. For every ordinal α, the following are equivalent:*

(i) *the homomorphism x has height at most α in $\mathrm{D}(\mathbf{A})$;*

(ii) *for all $\mathbf{B} \ll \mathbf{A}$, the homomorphism $x{\restriction_B}$ has height at most α in $\mathrm{Frag}_{\mathrm{D}(\mathbf{A})}$.*

Proof Define the non-decreasing ordinal sequence $[\varrho(\mathbf{A})]_0, [\varrho(\mathbf{A})]_1, \dots$ of subsets of $\mathrm{D}(\mathbf{A})$ as in Definition A.4.1, and let α be an ordinal. The implication (i) $\Rightarrow$ (ii) follows straight from the previous lemma.

To prove (ii) $\Rightarrow$ (i), assume that (ii) holds. The set $[\varrho(\mathbf{A})]_\alpha$ of all homomorphisms of height at most α in $\mathrm{D}(\mathbf{A})$ is topologically closed. So we can prove that $x \in [\varrho(\mathbf{A})]_\alpha$ using the Topological Closure Lemma, A.3.3. Let $\mathbf{B} \ll \mathbf{A}$. Then $x{\restriction_B}$ has height at most α in $\mathrm{Frag}_{\mathrm{D}(\mathbf{A})}$. Thus, by the previous lemma, there is some $w \in [\varrho(\mathbf{A})]_\alpha$ with $w{\restriction_B} = x{\restriction_B}$. Hence $x \in [\varrho(\mathbf{A})]_\alpha$, by the Topological Closure Lemma. ∎

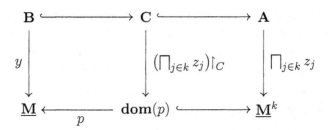

A.6 From height to rank

We have come to the final stage of our development. To introduce rank, we need to make a shift from finite subalgebras of arbitrary powers to subalgebras of finite powers.

To simplify our notation, we define the inverse of an embedding $\varphi : \mathbf{A} \hookrightarrow \mathbf{B}$ to be the unique homomorphism $\varphi^{-1} : \varphi(\mathbf{A}) \to \mathbf{A}$ satisfying $\varphi^{-1} \circ \varphi = \mathrm{id}_A$.

A.6.1 Associates of Fragments Let $\underline{\mathbf{M}}$ be a finite algebra and let $\mathbf{A}$ be a subalgebra of $\underline{\mathbf{M}}^I$, for some set I. Consider a $\mathrm{D}(\mathbf{A})$-fragment $y : \mathbf{B}_0 \to \underline{\mathbf{M}}$, where $\mathbf{B}_0 \ll \mathbf{A}$. We can find an associated partial operation $h : \mathbf{B} \to \underline{\mathbf{M}}$, for some $\mathbf{B} \leqslant \underline{\mathbf{M}}^n$ and $n \in \omega$, as follows.

Since the algebra $\mathbf{B}_0$ is finite, there is a finite subset $J = \{i_0, \dots, i_{n-1}\}$ of I, for some $n \in \omega$, such that $\varrho(\mathbf{B}_0) = \{\rho_{i_0}{\restriction}_{B_0}, \dots, \rho_{i_{n-1}}{\restriction}_{B_0}\}$. Thus, we can define the embedding

$$\varphi : \mathbf{B}_0 \hookrightarrow \underline{\mathbf{M}}^n \quad \text{by} \quad \varphi := \Big(\prod_{s \in n} \rho_{i_s}\Big){\restriction}_{B_0}.$$

Define $\mathbf{B} := \varphi(\mathbf{B}_0)$ and $h := y \circ \varphi^{-1} : \mathbf{B} \to \underline{\mathbf{M}}$. Then the following diagram commutes.

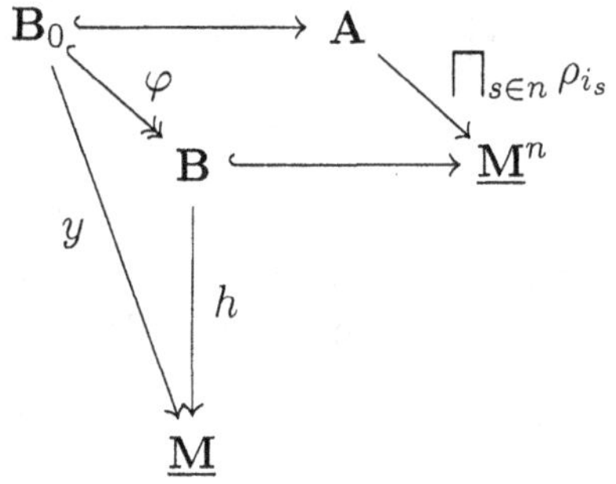

The n-ary partial operation $h : \mathbf{B} \to \underline{\mathbf{M}}$ is referred to as an **associate** of the $\mathrm{D}(\mathbf{A})$-fragment $y : \mathbf{B}_0 \to \underline{\mathbf{M}}$.

Now assume that $\mathbf{B}_0 \leqslant \mathbf{D}_0 \ll \mathbf{A}$ and that we have a $\mathrm{D}(\mathbf{A})$-fragment $y^+ : \mathbf{D}_0 \to \underline{\mathbf{M}}$ that extends $y : \mathbf{B}_0 \to \underline{\mathbf{M}}$. Having already constructed an associate of y via the embedding $\varphi : \mathbf{B}_0 \hookrightarrow \underline{\mathbf{M}}^n$, we wish to construct an associate $h^+ : \mathbf{D} \to \underline{\mathbf{M}}$ of y^+ that is compatible with φ, where $\mathbf{D} \leqslant \underline{\mathbf{M}}^{n+t}$ for some $t \in \omega$.

We can choose a finite extension $J^+ = \{i_0, \dots, i_{n-1}, i_n, \dots, i_{n+t-1}\}$ of J in I, for some $t \in \omega$, such that $\varrho(\mathbf{D}_0) = \{\rho_{i_0}{\restriction}_{D_0}, \dots, \rho_{i_{n+t-1}}{\restriction}_{D_0}\}$. Now we can define the embedding

$$\psi : \mathbf{D}_0 \hookrightarrow \underline{\mathbf{M}}^{n+t} \quad \text{by} \quad \psi := \Big(\prod_{s\in n+t} \rho_{i_s}\Big){\restriction}_{D_0}.$$

Define the subalgebra $\mathbf{D}$ of $\underline{\mathbf{M}}^{n+t}$ by $\mathbf{D} := \psi(\mathbf{D}_0)$, and define the two homomorphisms $h^+ := y^+ \circ \psi^{-1} : \mathbf{D} \to \underline{\mathbf{M}}$ and $\xi := \psi{\restriction}_{B_0} \circ \varphi^{-1} : \mathbf{B} \hookrightarrow \underline{\mathbf{M}}^{n+t}$. Then the following diagram commutes.

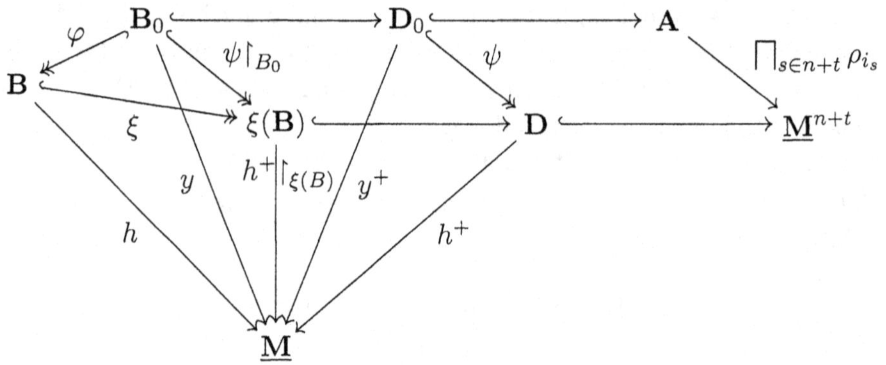

We say that h^+ is a φ-**compatible associate** for y^+.

Both of the isomorphisms $\varphi : \mathbf{B}_0 \hookrightarrow\!\!\!\!\rightarrow \mathbf{B}$ and $\psi : \mathbf{D}_0 \hookrightarrow\!\!\!\!\rightarrow \mathbf{D}$ are given by simply 'forgetting' some repeated coordinates. The fact that

$$\varrho(\mathbf{B}_0) = \{\rho_{i_0}{\restriction}_{B_0}, \dots, \rho_{i_{n-1}}{\restriction}_{B_0}\} \quad \text{and} \quad \varrho(\mathbf{D}_0) = \{\rho_{i_0}{\restriction}_{D_0}, \dots, \rho_{i_{n+t-1}}{\restriction}_{D_0}\}$$

therefore guarantees that the isomorphism $\xi := \psi{\restriction}_{B_0} \circ \varphi^{-1} : \mathbf{B} \hookrightarrow\!\!\!\!\rightarrow \xi(\mathbf{B})$ is given by coordinate repetition. In general, for all $n, t \in \omega$, a homomorphism $\xi : \underline{\mathbf{M}}^n \to \underline{\mathbf{M}}^{n+t}$ is called a **coordinate embedding** if there is a surjective map $\sigma : n + t \twoheadrightarrow n$ such that

$$\xi(a_0, \dots, a_{n-1}) = \big(a_{\sigma(0)}, \dots, a_{\sigma(n+t-1)}\big),$$

for all $(a_0, \dots, a_{n-1}) \in M^n$. The surjectivity of $\sigma : n + t \twoheadrightarrow n$ ensures that $\xi : \underline{\mathbf{M}}^n \to \underline{\mathbf{M}}^{n+t}$ is indeed an embedding.

By replacing fragments with their associates, we can transfer the definition of height for a fragment to a definition of rank for a finitary partial operation.

A.6.2 Rank Let $\underline{\mathbf{M}}$ be a finite algebra and let $\underset{\sim}{\mathbf{M}} = \langle M; G, H, R, \mathcal{T}\rangle$ be an alter ego of $\underline{\mathbf{M}}$. Let $h : \mathbf{B} \to \underline{\mathbf{M}}$ be an algebraic partial operation on $\underline{\mathbf{M}}$, where $\mathbf{B} \leqslant \underline{\mathbf{M}}^n$ for some $n \in \omega$. The rank of h is defined by transfinite induction.

- The partial operation h has **rank at most** 0 **relative to** $\underset{\sim}{\mathbf{M}}$ if it is a projection, that is, if $h = \pi_i{\restriction}_B$, for some $i \in n$.

- For each ordinal $\alpha > 0$, the partial operation h has **rank at most α relative to $\underset{\sim}{\mathrm{M}}$** if there exists $\ell \in \omega$ for which the following condition holds:

  whenever we have $t \in \omega$, a coordinate embedding $\xi : \underline{\mathbf{M}}^n \hookrightarrow \underline{\mathbf{M}}^{n+t}$ and algebras $\mathbf{C}$ and $\mathbf{D}$ such that $\xi(\mathbf{B}) \leqslant \mathbf{C} \leqslant \mathbf{D} \leqslant \underline{\mathbf{M}}^{n+t}$, and there is a partial operation $h^+ : \mathbf{D} \to \underline{\mathbf{M}}$ with $h^+{\restriction}_{\xi(B)} = h \circ \xi{\restriction}_B^{-1}$,

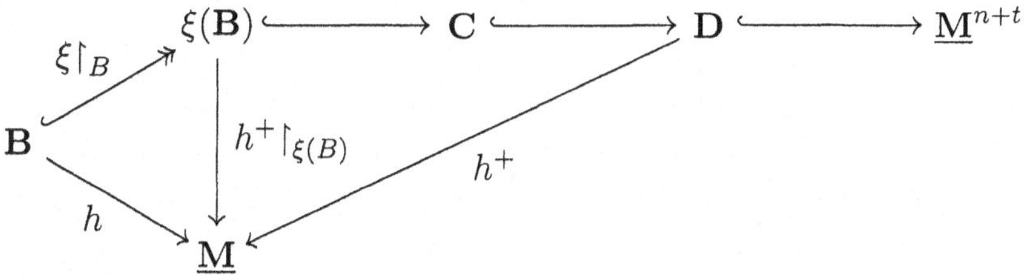

  then there exists $k \in \omega$ with $k \leqslant \ell$, an ordinal $\beta < \alpha$, partial operations $h_0, \dots, h_{k-1} : \mathbf{D} \to \underline{\mathbf{M}}$ such that the restrictions $h_0{\restriction}_C, \dots, h_{k-1}{\restriction}_C$ all have rank at most β relative to $\underset{\sim}{\mathrm{M}}$, and a partial operation $p \in [G \cup H]_k$ such that $p^{\mathbf{D}(\mathbf{C})}(h_0{\restriction}_C, \dots, h_{k-1}{\restriction}_C) : \mathbf{C} \to \underline{\mathbf{M}}$ is a well-defined extension of $h^+{\restriction}_{\xi(B)}$.

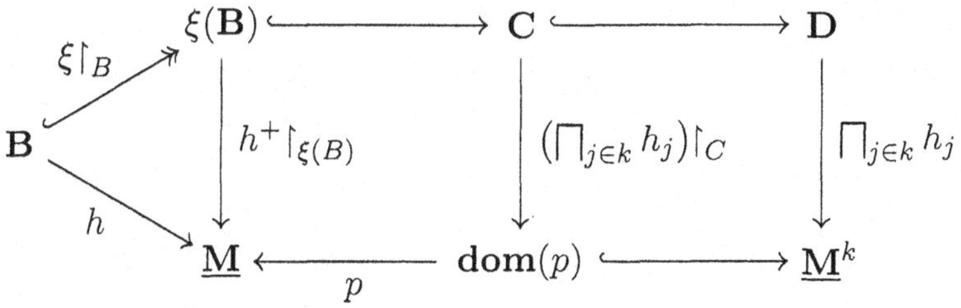

- If there is no ordinal α such that h has rank at most α relative to $\underset{\sim}{\mathrm{M}}$, then we say that h has **rank infinity relative to $\underset{\sim}{\mathrm{M}}$.**
- In the special case that $h : \underline{\mathbf{M}}^{\varnothing} \to \underline{\mathbf{M}}$ is a nullary operation, it follows from this definition that h has rank 1 relative to $\underset{\sim}{\mathrm{M}}$ if $h \in [G \cup H]_0$, and that h has rank infinity relative to $\underset{\sim}{\mathrm{M}}$ otherwise.
- If $\underline{\mathbf{M}}$ has a finitary algebraic partial operation of rank infinity relative to $\underset{\sim}{\mathrm{M}}$, then we say that **$\underline{\mathbf{M}}$ has rank infinity relative to $\underset{\sim}{\mathrm{M}}$**. Otherwise, the **rank of $\underline{\mathbf{M}}$ relative to $\underset{\sim}{\mathrm{M}}$** is defined to be the least ordinal α such that every finitary algebraic partial operation on $\underline{\mathbf{M}}$ has rank at most α relative to $\underset{\sim}{\mathrm{M}}$. (This least ordinal must exist, since there is only a countable number of finitary partial operations on a finite set.)

We simply say '**rank**' when we mean 'rank relative to $\underset{\sim}{\mathrm{M}}_\Omega$', where $\underset{\sim}{\mathrm{M}}_\Omega$ is the strong brute-force alter ego of the algebra $\underline{\mathbf{M}}$. This is the concept of rank introduced by R. Willard [65].

A.6.3 Remark In the definition of rank, the homomorphism $h : \mathbf{B} \to \underline{\mathbf{M}}$ represents a fragment of a homomorphism $x : \mathbf{A} \to \underline{\mathbf{M}}$, where the algebra $\mathbf{A}$ can be infinite. The algebra $\mathbf{D}$ should be thought of as a finite approximation to the algebra $\mathbf{A}$, and the homomorphism $h^+ : \mathbf{D} \to \underline{\mathbf{M}}$ as the corresponding approximation to the homomorphism $x : \mathbf{A} \to \underline{\mathbf{M}}$. The second diagram in A.6.1 explains the necessity of the coordinate embedding $\xi : \underline{\mathbf{M}}^n \hookrightarrow \underline{\mathbf{M}}^{n+t}$ in this transition.

A.6.4 Lemma *Let* $\underline{\mathbf{M}}$ *be a finite algebra and let* $\underset{\sim}{\mathbf{M}}$ *be an alter ego of* $\underline{\mathbf{M}}$. *Assume that* $\mathbf{A}$ *is a subalgebra of* $\underline{\mathbf{M}}^I$, *for some set* I, *and let* $y : \mathbf{B}_0 \to \underline{\mathbf{M}}$ *be a* $\mathrm{D}(\mathbf{A})$*-fragment, where* $\mathbf{B}_0 \ll \mathbf{A}$. *For each ordinal* α, *if* y *has an associate of rank at most* α *relative to* $\underset{\sim}{\mathbf{M}}$, *then* y *has height at most* α *in* $\mathrm{Frag}_{\mathrm{D}(\mathbf{A})}$.

Proof As per usual, this is proved by transfinite induction. Let $\underline{\mathbf{M}}$ be a finite algebra with alter ego $\underset{\sim}{\mathbf{M}} = \langle M; G, H, R, \mathcal{T}\rangle$. Let $\mathbf{A}$ be a subalgebra of $\underline{\mathbf{M}}^I$, for some set I. Then we want to prove the following claim, for all ordinals α.

$C(\alpha)$ *For each* $\mathbf{B}_0 \ll \mathbf{A}$ *and each* $\mathrm{D}(\mathbf{A})$*-fragment* $y : \mathbf{B}_0 \to \underline{\mathbf{M}}$, *if* y *has an associate of rank at most* α *relative to* $\underset{\sim}{\mathbf{M}}$, *then* y *has height at most* α *in* $\mathrm{Frag}_{\mathrm{D}(\mathbf{A})}$.

Let $\mathbf{B}_0 \ll \mathbf{A}$ and let $y : \mathbf{B}_0 \to \underline{\mathbf{M}}$ be a $\mathrm{D}(\mathbf{A})$-fragment. Then $y = x{\restriction}_{B_0}$, for some $x : \mathbf{A} \to \underline{\mathbf{M}}$. Now let $h : \mathbf{B} \to \underline{\mathbf{M}}$ be an associate of y, where $\mathbf{B} \leqslant \underline{\mathbf{M}}^n$ and $n \in \omega$. Then, using A.6.1, there is a subset $\{i_0, \dots, i_{n-1}\}$ of I for which $\varphi : \mathbf{B}_0 \hookrightarrow\!\!\!\!\to \mathbf{B}$, given by $\varphi := (\prod_{s\in n} \rho_{i_s}){\restriction}_{B_0}$, is an isomorphism and $h = y \circ \varphi^{-1}$.

To see that $C(0)$ holds, assume that h has rank 0 relative to $\underset{\sim}{\mathbf{M}}$. Then h is a projection; say $h = \pi_j{\restriction}_B$, where $j \in n$. So we have

$$y = h \circ \varphi = \pi_j{\restriction}_B \circ \Big(\prod_{s\in n} \rho_{i_s}\Big){\restriction}_{B_0} = \rho_{i_j}{\restriction}_{B_0}.$$

Hence y has height 0 in $\mathrm{Frag}_{\mathrm{D}(\mathbf{A})}$, whence $C(0)$ holds.

Now let α be an ordinal greater than 0 and assume that $C(\beta)$ holds, for all ordinals $\beta < \alpha$. Assume that the associate $h : \mathbf{B} \to \underline{\mathbf{M}}$ of y has rank at most α relative to $\underset{\sim}{\mathbf{M}}$. Let $\ell \in \omega$ be the bound given by the definition, A.6.2. We now want to show that $y : \mathbf{B}_0 \to \underline{\mathbf{M}}$ has height at most α in $\mathrm{Frag}_{\mathrm{D}(\mathbf{A})}$, using the same bound ℓ.

Choose $\mathbf{C}_0$ with $\mathbf{B}_0 \leqslant \mathbf{C}_0 \ll \mathbf{A}$. We need to find $k \in \omega$ with $k \leqslant \ell$, an ordinal $\beta < \alpha$, some $\mathrm{D}(\mathbf{A})$-fragments $v_0, \dots, v_{k-1} : \mathbf{C}_0 \to \underline{\mathbf{M}}$ of height at most β in $\mathrm{Frag}_{\mathrm{D}(\mathbf{A})}$, and a partial operation $p \in [G \cup H]_k$, as required by the definition of fragment height, A.5.1. To achieve this, we apply a variant of the Inverse Limit Lemma Strategy, A.3.2.

We aim to set up an inverse system indexed by the set

$$\mathcal{D} := \{\, \mathbf{D}_0 \mid \mathbf{C}_0 \leqslant \mathbf{D}_0 \ll \mathbf{A} \,\},$$

which forms a directed set $\mathcal{D}$ under the subalgebra order. To this end, let $\mathbf{D}_0$ be an arbitrary algebra with $\mathbf{B}_0 \leqslant \mathbf{C}_0 \leqslant \mathbf{D}_0 \ll \mathbf{A}$. Define $F_{\mathbf{D}_0}$ to be the set of all tuples $(w_0, \dots, w_{k-1})$ of homomorphisms from $\mathbf{D}_0$ to $\underline{\mathbf{M}}$, with $k \in \omega$ and $k \leqslant \ell$, for which the following three conditions hold.

(i) There exists a φ-compatible associate $h^+ : \mathbf{D} \to \underline{\mathbf{M}}$ of $x{\restriction}_{D_0}$, for some $\mathbf{D} \leqslant \underline{\mathbf{M}}^{n+t}$ and $t \in \omega$, with corresponding isomorphism $\psi : \mathbf{D}_0 \hookrightarrow\!\!\!\!\!\rightarrow \mathbf{D}$ and coordinate embedding $\xi : \underline{\mathbf{M}}^n \hookrightarrow \underline{\mathbf{M}}^{n+t}$.

(ii) Define $\mathbf{C} := \psi(\mathbf{C}_0)$ and, for all $j \in \{0, \dots, k-1\}$, define the associate $h_j : \mathbf{D} \to \underline{\mathbf{M}}$ of $w_j : \mathbf{D}_0 \to \underline{\mathbf{M}}$ by $h_j := w_j \circ \psi^{-1}$. Then there is an ordinal $\beta < \alpha$ such that, for all $j \in \{0, \dots, k-1\}$, the restriction $h_j{\restriction}_C$ has rank at most β relative to $\underset{\sim}{\mathbf{M}}$.

(iii) There is $p \in [G \cup H]_k$ such that $p^{\mathbf{D}(\mathbf{C})}(h_0{\restriction}_C, \dots, h_{k-1}{\restriction}_C) : \mathbf{C} \to \underline{\mathbf{M}}$ is a well-defined extension of $h^+{\restriction}_{\xi(B)}$.

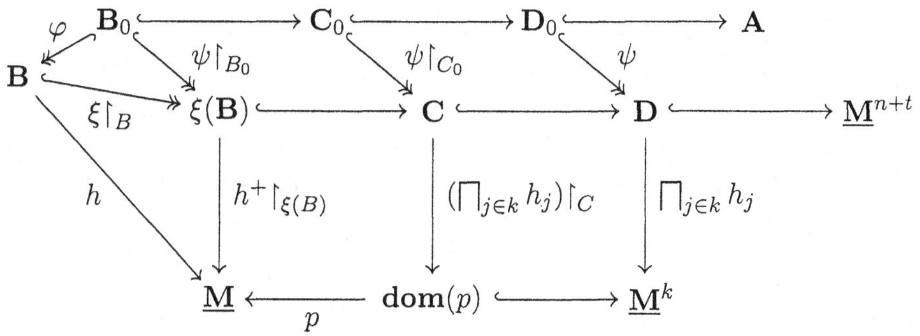

The set $F_{\mathbf{D}_0}$ defined above is non-empty, as $h : \mathbf{B} \to \underline{\mathbf{M}}$ has rank at most α relative to $\underset{\sim}{\mathbf{M}}$. The set $F_{\mathbf{D}_0}$ is finite, as $\mathbf{D}_0$ is finite.

Now that we have defined the collection of sets $\{\, F_{\mathbf{D}_0} \mid \mathbf{D}_0 \in \mathcal{D} \,\}$, we need to set up the connecting maps. Let $\mathbf{E}_0 \geqslant \mathbf{D}_0$ in $\mathcal{D}$. It is straightforward to check that we may define the map $\gamma_{\mathbf{E}_0\mathbf{D}_0} : F_{\mathbf{E}_0} \to F_{\mathbf{D}_0}$ by

$$\gamma_{\mathbf{E}_0\mathbf{D}_0}\big((w_0, \dots, w_{k-1})\big) := (w_0{\restriction}_{D_0}, \dots, w_{k-1}{\restriction}_{D_0}),$$

for all $(w_0, \dots, w_{k-1}) \in F_{\mathbf{E}_0}$.

The sets $\{\, F_{\mathbf{D}_0} \mid \mathbf{D}_0 \in \mathcal{D} \,\}$ and connecting maps $\{\, \gamma_{\mathbf{E}_0\mathbf{D}_0} \mid \mathbf{E}_0 \geqslant \mathbf{D}_0 \text{ in } \mathcal{D} \,\}$ form an inverse system of finite non-empty sets. By Lemma A.3.1, this system has a non-empty inverse limit. So choose an element of the inverse limit, say

$$\big(w_{\mathbf{D}_0,0}, \dots, w_{\mathbf{D}_0,k_{\mathbf{D}_0}-1}\big)_{\mathbf{D}_0 \in \mathcal{D}}.$$

As $\mathcal{D}$ is directed, the definition of the connecting maps ensures that $k_{\mathbf{D}_0} = k_{\mathbf{E}_0}$, for all $\mathbf{D}_0, \mathbf{E}_0 \in \mathcal{D}$. Let k denote this common value. Then we have $k \leqslant \ell$, by construction.

For each $j \in \{0, \dots, k-1\}$, we can define the homomorphism

$$z_j : \mathbf{A} \to \underline{\mathbf{M}} \quad \text{by} \quad z_j := \bigcup_{\mathbf{D}_0 \in \mathcal{D}} w_{\mathbf{D}_0, j}.$$

As $\mathbf{C}_0 \in \mathcal{D}$, we now have

$$\big(z_0{\restriction}_{C_0}, \dots, z_{k-1}{\restriction}_{C_0}\big) = \big(w_{\mathbf{C}_0,0}, \dots, w_{\mathbf{C}_0,k-1}\big) \in F_{\mathbf{C}_0}.$$

Thus there exists a φ-compatible associate $h^+ : \mathbf{C} \to \underline{\mathbf{M}}$ of $x{\restriction}_{C_0}$, for some $\mathbf{C} \leqslant \underline{\mathbf{M}}^{n+t}$ and $t \in \omega$, with corresponding isomorphism $\psi : \mathbf{C}_0 \hookrightarrow\!\!\!\to \mathbf{C}$ and coordinate embedding $\xi : \underline{\mathbf{M}}^n \hookrightarrow \underline{\mathbf{M}}^{n+t}$. For all $j \in \{0, \dots, k-1\}$, define the associate $h_j : \mathbf{C} \to \underline{\mathbf{M}}$ of $z_j{\restriction}_{C_0} : \mathbf{C}_0 \to \underline{\mathbf{M}}$ by $h_j := z_j{\restriction}_{C_0} \circ \psi^{-1}$. Then there is an ordinal $\beta < \alpha$ such that, for all $j \in \{0, \dots, k-1\}$, the partial operation $h_j : \mathbf{C} \to \underline{\mathbf{M}}$ has rank at most β relative to $\underset{\sim}{\mathbf{M}}$. There is also $p \in [G \cup H]_k$ such that $p^{\mathrm{D}(\mathbf{C})}(h_0, \dots, h_{k-1}) : \mathbf{C} \to \underline{\mathbf{M}}$ is a well-defined extension of $h^+{\restriction}_{\xi(B)}$.

Since we are assuming $\mathcal{C}(\beta)$ holds, it now follows that the homomorphisms $z_0{\restriction}_{C_0}, \dots, z_{k-1}{\restriction}_{C_0} : \mathbf{C}_0 \to \underline{\mathbf{M}}$ are $\mathrm{D}(\mathbf{A})$-fragments of height at most β in $\mathrm{Frag}_{\mathrm{D}(\mathbf{A})}$. To prove that y has height at most α in $\mathrm{Frag}_{\mathrm{D}(\mathbf{A})}$, it remains to show that $p^{\mathrm{D}(\mathbf{C}_0)}(z_0{\restriction}_{C_0}, \dots, z_{k-1}{\restriction}_{C_0}) : \mathbf{C}_0 \to \underline{\mathbf{M}}$ is a well-defined extension of y.

As $p^{\mathrm{D}(\mathbf{C})}(h_0, \dots, h_{k-1}) : \mathbf{C} \to \underline{\mathbf{M}}$ is a well-defined extension of $h^+{\restriction}_{\xi(B)}$, we can use the Diagram Lemma, A.3.4, and the second picture in A.6.1 to get

$$\Big(\prod_{j \in k} z_j\Big)(C_0) = \Big(\prod_{j \in k} h_j \circ \psi\Big)(C_0) = \Big(\prod_{j \in k} h_j\Big)(C) \subseteq \mathrm{dom}(p)$$

and

$$p \circ \Big(\prod_{j \in k} z_j\Big){\restriction}_{B_0} = p \circ \Big(\prod_{j \in k} h_j \circ \psi\Big){\restriction}_{B_0} = p \circ \Big(\prod_{j \in k} h_j \circ \xi \circ \varphi\Big)$$
$$= p \circ \Big(\prod_{j \in k} h_j\Big){\restriction}_{\xi(B)} \circ \xi \circ \varphi = h^+ \circ \xi \circ \varphi = h \circ \varphi = y.$$

So $p^{\mathrm{D}(\mathbf{C}_0)}(z_0{\restriction}_{C_0}, \dots, z_{k-1}{\restriction}_{C_0}) : \mathbf{C}_0 \to \underline{\mathbf{M}}$ is a well-defined extension of y, by the Diagram Lemma. We have proved that the $\mathrm{D}(\mathbf{A})$-fragment y has height at most α in $\mathrm{Frag}_{\mathrm{D}(\mathbf{A})}$. Thus $\mathcal{C}(\alpha)$ holds, as required. ■

We are now able to link rank to height.

A.6.5 Theorem *Let $\underline{\mathbf{M}}$ be a finite algebra and let $\underset{\sim}{\mathbf{M}}$ be an alter ego of $\underline{\mathbf{M}}$. For each ordinal α, if $\underline{\mathbf{M}}$ has rank at most α relative to $\underset{\sim}{\mathbf{M}}$, then $\underline{\mathbf{M}}$ has height at most α relative to $\underset{\sim}{\mathbf{M}}$.*

Proof Assume that $\underline{\mathbf{M}}$ has rank at most α relative to $\underset{\sim}{\mathbf{M}}$, for some ordinal α. Then every finitary algebraic partial operation on $\underline{\mathbf{M}}$ has rank at most α relative to $\underset{\sim}{\mathbf{M}}$. Now let $\mathbf{A}$ be a subalgebra of a power of $\underline{\mathbf{M}}$. Using A.6.1, every $\mathrm{D}(\mathbf{A})$-fragment has a finitary algebraic partial operation on $\underline{\mathbf{M}}$ as an associate. So, by Lemma A.6.4, every $\mathrm{D}(\mathbf{A})$-fragment has height at most α in $\mathrm{Frag}_{\mathrm{D}(\mathbf{A})}$. Thus, by Theorem A.5.3, every homomorphism $x : \mathbf{A} \to \underline{\mathbf{M}}$ has height at most α in $\mathrm{D}(\mathbf{A})$. Hence, by definition, the algebra $\underline{\mathbf{M}}$ has height at most α relative to $\underset{\sim}{\mathbf{M}}$. ∎

We now have a finitary condition that guarantees that a duality can be upgraded to a strong duality, as an immediate consequence of Theorems A.4.6 and A.6.5.

A.6.6 Strong Dualisability via Rank Theorem *Let $\underline{\mathbf{M}}$ be a finite algebra and let $\underset{\sim}{\mathbf{M}}$ be an alter ego of $\underline{\mathbf{M}}$.*

(i) *If $\underline{\mathbf{M}}$ is dualised by $\underset{\sim}{\mathbf{M}}$ and has rank at most α relative to $\underset{\sim}{\mathbf{M}}$, for some ordinal α, then $\underline{\mathbf{M}}$ is strongly dualised by $\underset{\sim}{\mathbf{M}}$.*

(ii) *If $\underline{\mathbf{M}}$ is dualisable and has rank at most α, for some ordinal α, then $\underline{\mathbf{M}}$ is strongly dualisable.*

A.7 Some examples

We close this appendix by looking briefly at algebras of rank at most 2. First, we shall see that having enough algebraic operations guarantees rank at most 2—this strengthens Theorem A.4.11, where we showed that having enough algebraic operations guarantees height at most 2. We shall then prove that a finite algebra must have rank at most 1 if it is injective in the quasi-variety it generates, or if it generates a congruence-distributive variety. Finally, we shall characterise the dualisable algebras that have rank 0.

A.7.1 Lemma *Let $\underline{\mathbf{M}}$ be a finite algebra and let $\underset{\sim}{\mathbf{M}} = \langle M; G, H, R, \mathcal{T}\rangle$ be an alter ego of $\underline{\mathbf{M}}$. Assume that $h : \mathbf{B} \to \underline{\mathbf{M}}$ is an algebraic partial operation on $\underline{\mathbf{M}}$, where $\mathbf{B} \leqslant \underline{\mathbf{M}}^n$ for some $n \in \omega$. If h extends to a total operation in $[G \cup H]_n$, then h has rank at most* 1 *relative to $\underset{\sim}{\mathbf{M}}$.*

Proof Assume that h extends to a total operation $g : \underline{\mathbf{M}}^n \to \underline{\mathbf{M}}$ in $[G \cup H]_n$. We will show that h has rank at most 1 relative to $\underset{\sim}{\mathbf{M}}$, using the bound n. Assume we have $\xi(\mathbf{B}) \leqslant \mathbf{C} \leqslant \mathbf{D} \leqslant \underline{\mathbf{M}}^{n+t}$, for some $t \in \omega$ and coordinate embedding $\xi : \underline{\mathbf{M}}^n \hookrightarrow \underline{\mathbf{M}}^{n+t}$. Assume further that there is a homomorphism $h^+ : \mathbf{D} \to \underline{\mathbf{M}}$ with $h^+{\restriction}_{\xi(B)} = h \circ \xi{\restriction}_B^{-1}$.

As $\xi : \underline{\mathbf{M}}^n \hookrightarrow \underline{\mathbf{M}}^{n+t}$ is a coordinate embedding, there must be a surjective map $\sigma : n+t \twoheadrightarrow n$ such that

$$\xi(a_0, \ldots, a_{n-1}) = \big(a_{\sigma(0)}, \ldots, a_{\sigma(n+t-1)}\big),$$

for all $(a_0, \ldots, a_{n-1}) \in M^n$. As σ is surjective, there is a map $\tau : n \to n+t$ with $\sigma \circ \tau = \mathrm{id}_n$. For each $j \in n$, define the partial operation $h_j : \mathbf{D} \to \underline{\mathbf{M}}$ by $h_j := \pi_{\tau(j)}{\restriction}_D$. Then $h_0{\restriction}_C, \ldots, h_{n-1}{\restriction}_C$ all have rank 0 relative to $\underset{\sim}{\mathbf{M}}$, by definition.

We will use $g \in [G \cup H]_n$ as the required partial operation. The homomorphism $g^{\mathrm{D(C)}}(h_0{\restriction}_C, \ldots, h_{n-1}{\restriction}_C) : \mathbf{C} \to \underline{\mathbf{M}}$ is well defined, since g is a total operation. For all $b = (b_0, \ldots, b_{n-1}) \in B$, we have

$$\begin{aligned} g \circ \Big(\prod_{j\in n} h_j\Big)(\xi(b)) &= g \circ \Big(\prod_{j\in n} \pi_{\tau(j)}{\restriction}_D\Big)(\xi(b_0, \ldots, b_{n-1})) \\ &= g\Big(\prod_{j\in n} \pi_{\tau(j)}\big(b_{\sigma(0)}, \ldots, b_{\sigma(n+t-1)}\big)\Big) = g(b_0, \ldots, b_{n-1}) \\ &= g(b) = h(b) = h \circ \xi{\restriction}_B^{-1}\big(\xi(b)\big) = h^+{\restriction}_{\xi(B)}\big(\xi(b)\big). \end{aligned}$$

So $g^{\mathrm{D(C)}}(h_0{\restriction}_C, \ldots, h_{n-1}{\restriction}_C) : \mathbf{C} \to \underline{\mathbf{M}}$ is an extension of $h^+{\restriction}_{\xi(B)}$. Thus h has rank at most 1 relative to $\underset{\sim}{\mathbf{M}}$. ■

The previous lemma does not hold if we replace the assumption that h extends to a *total* operation in $[G \cup H]_n$ with the assumption that h extends to a *partial* operation in $[G \cup H]_n$. To see this, consider a finite algebra $\underline{\mathbf{M}}$ that is dualisable but not strongly dualisable. (See Chapter 4 for examples of such algebras.) By the Strong Dualisability via Rank Theorem, A.6.6, there must be a homomorphism $h : \mathbf{B} \to \underline{\mathbf{M}}$, for some $\mathbf{B} \leqslant \underline{\mathbf{M}}^n$ and $n \in \omega$, such that h does not have rank at most 1 relative to the strong brute-force alter ego $\underset{\sim}{\mathbf{M}}_\Omega$. Nevertheless, the algebraic partial operation h belongs to the type of $\underset{\sim}{\mathbf{M}}_\Omega$.

A.7.2 Theorem [45, 4.3] *Each finite algebra with enough algebraic operations has rank at most* 2.

Proof Let $\underline{\mathbf{M}}$ be a finite algebra, and let the map $f : \omega \to \omega$ witness the fact that $\underline{\mathbf{M}}$ has enough algebraic operations; see page 24. Now let $h : \mathbf{B} \to \underline{\mathbf{M}}$ be a homomorphism, where $\mathbf{B} \leqslant \underline{\mathbf{M}}^n$ for some $n \in \omega$. We must prove that h has rank at most 2.

Define $\ell := f(|B|)$. Assume that we have $\xi(\mathbf{B}) \leqslant \mathbf{C} \leqslant \mathbf{D} \leqslant \underline{\mathbf{M}}^{n+t}$, for some $t \in \omega$ and coordinate embedding $\xi : \underline{\mathbf{M}}^n \hookrightarrow \underline{\mathbf{M}}^{n+t}$. Assume further that there is a homomorphism $h^+ : \mathbf{D} \to \underline{\mathbf{M}}$ such that $h^+{\restriction}_{\xi(B)} = h \circ \xi{\restriction}_B^{-1}$. We can now apply the definition of enough algebraic operations, with $\mathbf{B}$, $\mathbf{A}$ and h replaced by $\xi(\mathbf{B})$, $\mathbf{D}$ and h^+, respectively. There exist algebraic operations $g_0, \dots, g_{k-1} : \underline{\mathbf{M}}^{n+t} \to \underline{\mathbf{M}}$, for some $k \leqslant f(|\xi(B)|) = f(|B|) = \ell$, and a homomorphism $p : \big(\prod_{j\in k} g_j\big)(\mathbf{D}) \to \underline{\mathbf{M}}$ such that

$$h^+{\restriction}_{\xi(B)} = p \circ \Big(\prod_{j\in k} g_j\Big){\restriction}_{\xi(B)}.$$

For every $j \in \{0, \dots, k-1\}$, define $h_j : \mathbf{D} \to \underline{\mathbf{M}}$ by $h_j := g_j{\restriction}_D$. Then, by Lemma A.7.1, the homomorphisms $h_0{\restriction}_C, \dots, h_{k-1}{\restriction}_C$ have rank at most 1. We have

$$\Big(\prod_{j\in k} h_j\Big)(C) \subseteq \Big(\prod_{j\in k} g_j\Big)(D) = \mathrm{dom}(p)$$

and

$$h^+{\restriction}_{\xi(B)} = p \circ \Big(\prod_{j\in k} g_j\Big){\restriction}_{\xi(B)} = p \circ \Big(\prod_{j\in k} h_j\Big){\restriction}_{\xi(B)}.$$

So $p^{\mathbf{D}(\mathbf{C})}(h_0{\restriction}_C, \dots, h_{k-1}{\restriction}_C) : \mathbf{C} \to \underline{\mathbf{M}}$ is a well-defined extension of $h^+{\restriction}_{\xi(B)}$, by the Diagram Lemma, A.3.4. Thus h has rank at most 2. ■

A.7.3 Remark The previous theorem can easily be generalised. If a finite algebra $\underline{\mathbf{M}}$ has enough algebraic operations relative to an alter ego $\underset{\sim}{\mathbf{M}}$, then $\underline{\mathbf{M}}$ has rank at most 2 relative to $\underset{\sim}{\mathbf{M}}$.

Because of the order of the quantifiers in the two definitions, it is not clear whether 'rank at most 1' implies 'enough algebraic operations'. We will introduce a new condition that is sufficient to guarantee both of these properties.

Let $\underset{\sim}{\mathbf{M}} = \langle M; G, H, R, \mathcal{T}\rangle$ be an alter ego of a finite algebra $\underline{\mathbf{M}}$. We say that $\underline{\mathbf{M}}$ has **enough projections relative to** $\underset{\sim}{\mathbf{M}}$ if $[G \cup H]_0$ contains all the nullary algebraic operations on $\underline{\mathbf{M}}$ and there is a map $f : \omega \to \omega$ for which the following condition holds:

> for all $n \in \omega\backslash\{0\}$, all $\mathbf{B} \leqslant \mathbf{C} \leqslant \underline{\mathbf{M}}^n$ and all non-constant homomorphisms $h : \mathbf{C} \to \underline{\mathbf{M}}$, there exists $k \in \omega$ with $k \leqslant f(|B|)$, projections $\pi_{i_0}, \dots, \pi_{i_{k-1}} : \underline{\mathbf{M}}^n \to \underline{\mathbf{M}}$, and a partial operation $p \in [G \cup H]_k$ such that $p^{\mathbf{D}(\mathbf{C})}(\pi_{i_0}{\restriction}_C, \dots, \pi_{i_{k-1}}{\restriction}_C) : \mathbf{C} \to \underline{\mathbf{M}}$ is a well-defined extension of $h{\restriction}_B$.

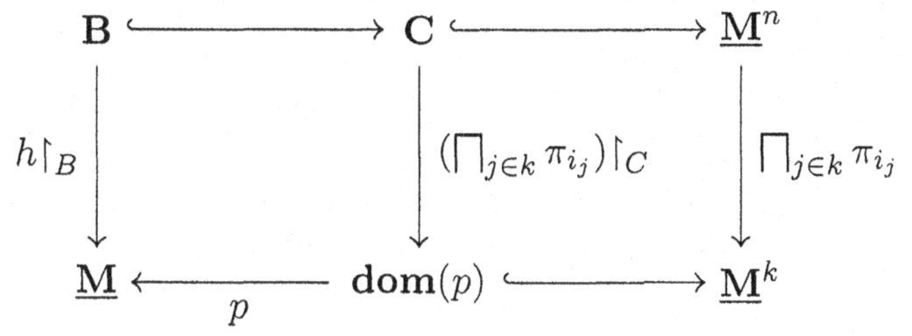

The algebra $\underline{\mathbf{M}}$ is said to have **enough projections** if it has enough projections relative to the brute-force alter ego $\underset{\sim}{\mathbf{M}}_{\Omega}$ of $\underline{\mathbf{M}}$.

Clearly, 'enough projections relative to $\underset{\sim}{\mathbf{M}}$' implies 'enough algebraic operations relative to $\underset{\sim}{\mathbf{M}}$'. Next we show that our new condition also implies 'rank at most 1'.

A.7.4 Theorem *Let $\underline{\mathbf{M}}$ be a finite algebra and let $\underset{\sim}{\mathbf{M}}$ be an alter ego of $\underline{\mathbf{M}}$. If $\underline{\mathbf{M}}$ has enough projections relative to $\underset{\sim}{\mathbf{M}}$, then $\underline{\mathbf{M}}$ has rank at most* 1 *relative to $\underset{\sim}{\mathbf{M}}$.*

Proof Assume $\underline{\mathbf{M}}$ has enough projections relative to $\underset{\sim}{\mathbf{M}} = \langle M; G, H, R, \mathcal{T}\rangle$, via the bounding function $f : \omega \to \omega$. Let $h : \mathbf{B} \to \underline{\mathbf{M}}$ be a homomorphism, where $\mathbf{B} \leqslant \underline{\mathbf{M}}^n$ for some $n \in \omega$. We want to show that h has rank at most 1 relative to $\underset{\sim}{\mathbf{M}}$. Define $\ell := f(|B|)$. Assume $\xi(\mathbf{B}) \leqslant \mathbf{C} \leqslant \mathbf{D} \leqslant \underline{\mathbf{M}}^{n+t}$, where $t \in \omega$ and $\xi : \underline{\mathbf{M}}^n \hookrightarrow \underline{\mathbf{M}}^{n+t}$ is a coordinate embedding, and assume there is a homomorphism $h^+ : \mathbf{D} \to \underline{\mathbf{M}}$ such that $h^+{\restriction}_{\xi(B)} = h \circ \xi{\restriction}_B^{-1}$.

First, consider the case where h is constant. As $\underline{\mathbf{M}}$ has enough projections relative to $\underset{\sim}{\mathbf{M}}$, all the nullary algebraic operations on $\underline{\mathbf{M}}$ belong to $[G \cup H]_0$. So there is some $p \in [G \cup H]_0$ that has the same value as h, and therefore the same value as $h^+{\restriction}_{\xi(B)}$. Thus $p^{\mathrm{D}(\mathbf{C})}(\varnothing) : \mathbf{C} \to \underline{\mathbf{M}}$ is a well-defined extension of $h^+{\restriction}_{\xi(B)}$. Hence h has rank at most 1 relative to $\underset{\sim}{\mathbf{M}}$.

Now assume that h is not constant. We can use the technical condition in the definition of enough projections relative to $\underset{\sim}{\mathbf{M}}$, with $\mathbf{B}$, $\mathbf{C}$ and h replaced by $\xi(\mathbf{B})$, $\mathbf{C}$ and $h^+{\restriction}_C$, respectively. There exists $k \in \omega$ with $k \leqslant f(|\xi(B)|) = f(|B|) = \ell$, projections $\pi_{i_0}, \dots, \pi_{i_{k-1}} : \underline{\mathbf{M}}^{n+t} \to \underline{\mathbf{M}}$ and a partial operation $p \in [G \cup H]_k$ such that $p^{\mathrm{D}(\mathbf{C})}(\pi_{i_0}{\restriction}_C, \dots, \pi_{i_{k-1}}{\restriction}_C) : \mathbf{C} \to \underline{\mathbf{M}}$ is a well-defined extension of $h^+{\restriction}_{\xi(B)}$. By definition, the homomorphism $\pi_{i_j}{\restriction}_C : \mathbf{C} \to \underline{\mathbf{M}}$ has rank 0, for each $j \in \{0, \dots, k-1\}$. Thus h has rank at most 1 relative to $\underset{\sim}{\mathbf{M}}$. So we have shown that $\underline{\mathbf{M}}$ has rank at most 1 relative to $\underset{\sim}{\mathbf{M}}$. ■

The completion of commutative diagrams inherent in the definition of rank is reminiscent of the definition of injectivity. A finite algebra $\underline{\mathbf{M}}$ is **injective** in the quasi-variety $\mathbb{ISP}(\underline{\mathbf{M}})$ if, for each set I and each $\mathbf{A} \leqslant \underline{\mathbf{M}}^I$, every homomorphism $x : \mathbf{A} \to \underline{\mathbf{M}}$ extends to a homomorphism $x^+ : \underline{\mathbf{M}}^I \to \underline{\mathbf{M}}$. Having rank α, for some ordinal α, seems to be a weak form of injectivity. For example, it will follow from Theorem A.7.8 that, if $\underline{\mathbf{M}}$ has rank 0, then $\underline{\mathbf{M}}$ is injective in $\mathbb{ISP}(\underline{\mathbf{M}})$. As we remarked in the introduction to Chapter 4, a simple application of Lemma 4.1.1 shows that, if $\underline{\mathbf{M}}$ is injective in $\mathbb{ISP}(\underline{\mathbf{M}})$, then $\underline{\mathbf{M}}$ has enough algebraic operations and so, by Theorem A.7.2, has rank at most 2. In fact, the injectivity of $\underline{\mathbf{M}}$ suffices to show that it has rank at most 1.

A.7.5 Theorem *Let $\underline{\mathbf{M}}$ be a finite algebra and assume that $\underline{\mathbf{M}}$ is injective in the quasi-variety $\mathbb{ISP}(\underline{\mathbf{M}})$.*

(i) *Let $\underset{\sim}{\mathrm{M}} = \langle M; G, H, R, \mathcal{T}\rangle$ be an alter ego of $\underline{\mathbf{M}}$ such that $[G \cup H]$ contains all the algebraic total operations on $\underline{\mathbf{M}}$. Then $\underline{\mathbf{M}}$ has enough projections relative to $\underset{\sim}{\mathrm{M}}$, and thus has rank at most 1 relative to $\underset{\sim}{\mathrm{M}}$.*

(ii) *The algebra $\underline{\mathbf{M}}$ has enough projections, and thus has enough algebraic operations and rank at most 1.*

(iii) *If $\underline{\mathbf{M}}$ is dualisable, then $\underline{\mathbf{M}}$ is strongly dualisable.*

Proof For claim (i), we will show that $\underline{\mathbf{M}}$ has enough projections relative to $\underset{\sim}{\mathrm{M}}$ using the bounding function $\mathrm{id}_\omega : \omega \to \omega$. Let $\mathbf{B} \leqslant \mathbf{C} \leqslant \underline{\mathbf{M}}^n$, where $n \in \omega\backslash\{0\}$, and let $h : \mathbf{C} \to \underline{\mathbf{M}}$ be a homomorphism. By Lemma 4.1.1, there are projections $\pi_{i_0}, \dots, \pi_{i_{k-1}} : \underline{\mathbf{M}}^n \to \underline{\mathbf{M}}$, for some $k \leqslant |B|$, such that

$$\varphi := \Big(\prod_{j\in k} \pi_{i_j}\Big){\restriction_B} : \mathbf{B} \hookrightarrow \underline{\mathbf{M}}^k$$

is an embedding. As $\underline{\mathbf{M}}$ is injective in $\mathbb{ISP}(\underline{\mathbf{M}})$, there exists an algebraic total operation $p : \underline{\mathbf{M}}^k \to \underline{\mathbf{M}}$ such that $p \circ \varphi = h{\restriction_B}$. So

$$h{\restriction_B} = p \circ \varphi = p \circ \Big(\prod_{j\in k} \pi_{i_j}\Big){\restriction_B} = p^{\mathrm{D}(\mathbf{C})}(\pi_{i_0}{\restriction_C}, \dots, \pi_{i_{k-1}}{\restriction_C}){\restriction_B}.$$

We must have $p \in [G \cup H]_k$, since $[G \cup H]$ contains all the algebraic total operations on $\underline{\mathbf{M}}$. Thus $\underline{\mathbf{M}}$ has enough projections relative to $\underset{\sim}{\mathrm{M}}$.

By Theorem A.7.4, it now follows that the algebra $\underline{\mathbf{M}}$ has rank at most 1 relative to $\underset{\sim}{\mathrm{M}}$. (We could have used Lemma A.7.1 to prove directly that $\underline{\mathbf{M}}$ has rank at most 1 relative to $\underset{\sim}{\mathrm{M}}$.)

We have shown that claim (i) holds. Claim (ii) follows easily, and claim (iii) uses the Strong Dualisability via Rank Theorem, A.6.6. ∎

Next we shall prove that every finite algebra that generates a congruence-distributive variety has enough projections, and therefore has enough algebraic operations and rank at most 1. We require the following definitions and lemma.

Consider a finite algebra $\mathbf{N}$. Every congruence on $\mathbf{N}$ is the meet of a finite set of meet-irreducible congruences. So we can define $\mathrm{irr}(\mathbf{N})$ be the least $n \in \omega$ such that the zero congruence on $\mathbf{N}$ is a meet of n meet-irreducible congruences. Thus an algebra $\mathbf{N}$ is subdirectly irreducible if and only if $\mathrm{irr}(\mathbf{N}) = 1$. The **irreducibility index**, $\mathrm{Irr}(\underline{\mathbf{M}})$, of a finite algebra $\underline{\mathbf{M}}$ is defined by

$$\mathrm{Irr}(\underline{\mathbf{M}}) := \max\{\, \mathrm{irr}(\mathbf{N}) \mid \mathbf{N} \text{ is a subalgebra of } \underline{\mathbf{M}}\,\}.$$

A variety $\mathcal{V}$ is **congruence distributive** if every algebra in $\mathcal{V}$ has a distributive congruence lattice. For example, an algebra with an underlying lattice generates a congruence-distributive variety [5, II.12.3].

The following lemma says that, for each finite algebra $\underline{\mathbf{M}}$ that generates a congruence-distributive variety, every algebraic partial operation on $\underline{\mathbf{M}}$ can depend on at most $\mathrm{Irr}(\underline{\mathbf{M}})$-many coordinates.

A.7.6 Lemma *Let $\underline{\mathbf{M}}$ be a finite algebra. Assume $\underline{\mathbf{M}}$ generates a congruence-distributive variety and define $k := \mathrm{Irr}(\underline{\mathbf{M}})$. Let $h : \mathbf{B} \to \underline{\mathbf{M}}$ be an algebraic partial operation, where $\mathbf{B} \leqslant \underline{\mathbf{M}}^n$ and $n \in \omega\backslash\{0\}$. Then there exist projections $\rho_{i_0}, \dots, \rho_{i_{k-1}} : \mathbf{B} \to \underline{\mathbf{M}}$ and a homomorphism $p : \big(\prod_{j\in k} \rho_{i_j}\big)(\mathbf{B}) \to \underline{\mathbf{M}}$ such that $h = p^{\mathrm{D}(\mathbf{B})}(\rho_{i_0}, \dots, \rho_{i_{k-1}})$.*

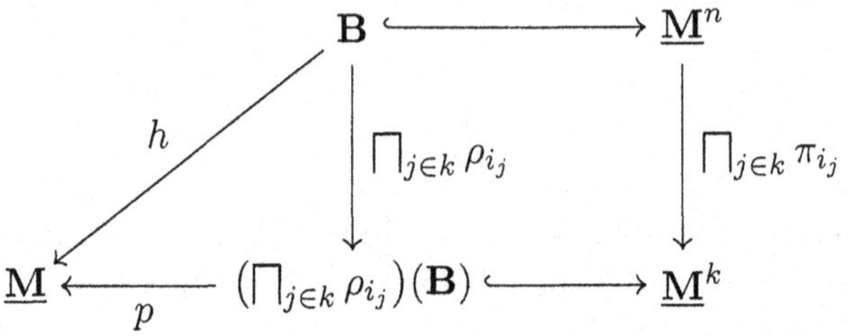

Proof We can assume h is not constant. So $\underline{\mathbf{M}}$ is non-trivial, whence $k \neq 0$. The lattices $\mathrm{Con}(h(\mathbf{B}))$ and $\uparrow_{\mathrm{Con}(\mathbf{B})}(\ker(h))$ are isomorphic. As $h(\mathbf{B})$ is a non-trivial subalgebra of $\underline{\mathbf{M}}$, this implies that $\ker(h) = \bigwedge_{j\in k} \theta_j$, for some (not necessarily distinct) meet-irreducible congruences $\theta_0, \dots, \theta_{k-1}$ on $\mathbf{B}$.

The zero congruence on $\mathbf{B} \leqslant \underline{\mathbf{M}}^n$ is equal to the meet of all the kernels of projections. Let $j \in \{0, \dots, k-1\}$. Then this gives us

$$\theta_j = \theta_j \vee \bigwedge_{i\in n} \ker(\rho_i) = \bigwedge_{i\in n} (\theta_j \vee \ker(\rho_i)),$$

as the lattice $\mathrm{Con}(\mathbf{B})$ is distributive. Since θ_j is meet-irreducible, there exists $i_j \in n$ such that $\theta_j = \theta_j \vee \ker(\rho_{i_j})$. So $\ker(\rho_{i_j}) \leqslant \theta_j$.

We now have

$$\ker\Big(\prod_{j\in k} \rho_{i_j}\Big) = \bigwedge_{j\in k} \ker(\rho_{i_j}) \leqslant \bigwedge_{j\in k} \theta_j = \ker(h).$$

Consequently, there is a homomorphism $p : \big(\prod_{j\in k} \rho_{i_j}\big)(\mathbf{B}) \to \underline{\mathbf{M}}$ such that $h = p \circ \big(\prod_{j\in k} \rho_{i_j}\big) = p^{\mathrm{D}(\mathbf{B})}(\rho_{i_0}, \dots, \rho_{i_{k-1}})$. ∎

We can now give a new proof that every dualisable algebra that generates a congruence-distributive variety is strongly dualisable [8, 3.3.7 and 3.3.8].

A.7.7 Theorem *Let $\underline{\mathbf{M}}$ be a finite algebra that generates a congruence-distributive variety and define $k := \mathrm{Irr}(\underline{\mathbf{M}})$.*

(i) *The algebra $\underline{\mathbf{M}}$ has enough projections, and thus has enough algebraic operations and rank at most 1.*

(ii) *Let $\underset{\sim}{\mathbf{M}} = \langle M; G, H, R, \mathcal{T}\rangle$ be an alter ego of $\underline{\mathbf{M}}$. Assume that $[G \cup H]_0$ contains all the nullary algebraic operations on $\underline{\mathbf{M}}$. Assume further that every k-ary algebraic partial operation on $\underline{\mathbf{M}}$ has an extension in $[G \cup H]_k$. Then $\underline{\mathbf{M}}$ has enough projections relative to $\underset{\sim}{\mathbf{M}}$, and thus has rank at most 1 relative to $\underset{\sim}{\mathbf{M}}$.*

(iii) *Let $\underset{\sim}{\mathbf{M}}$ be an alter ego of $\underline{\mathbf{M}}$ that yields a duality on $\mathbb{ISP}(\underline{\mathbf{M}})$. Define $\underset{\sim}{\mathbf{M}}'$ to be the alter ego of $\underline{\mathbf{M}}$ obtained from $\underset{\sim}{\mathbf{M}}$ by adding all the nullary algebraic operations on $\underline{\mathbf{M}}$ and all the k-ary algebraic partial operations on $\underline{\mathbf{M}}$. Then $\underset{\sim}{\mathbf{M}}'$ yields a strong duality on $\mathbb{ISP}(\underline{\mathbf{M}})$.*

Proof Let $\underset{\sim}{\mathbf{M}} = \langle M; G, H, R, \mathcal{T}\rangle$ be an alter ego of $\underline{\mathbf{M}}$ such that $[G \cup H]_0$ contains all the nullary algebraic operations on $\underline{\mathbf{M}}$ and every k-ary algebraic partial operation on $\underline{\mathbf{M}}$ has an extension in $[G \cup H]_k$. It follows from the previous lemma that $\underline{\mathbf{M}}$ has enough projections relative to $\underset{\sim}{\mathbf{M}}$, taking the bounding function $f : \omega \to \omega$ to be the constant map of value $\mathrm{Irr}(\underline{\mathbf{M}})$. So claims (i) and (ii) follow from Theorem A.7.4. Claim (iii) follows from claim (ii), in tandem with the Strong Dualisability via Rank Theorem, A.6.6. ∎

We now turn our attention to algebras of rank 0. Having rank 0 is clearly a very restrictive property on a finite algebra—every algebraic partial operation must be the restriction of a projection. Nevertheless, we will see that there are some well-known and intensively studied quasi-varieties that are generated by an algebra of rank 0: for example, the class of bounded distributive lattices.

The following theorem completely characterises the algebras of rank 0 within the class of dualisable algebras. While the proof of the theorem is quite short, it calls on some deep results from the general theory of natural dualities.

Our characterisation requires two definitions. First, an alter ego of a finite algebra is **purely relational** if it is of the form $\underset{\sim}{\mathbf{M}} = \langle M; R, \mathcal{T}\rangle$, for some set R of relations. Next, for all $n \geqslant 3$, a term $t(x_0, \dots, x_{n-1})$ in the language of an algebra $\underline{\mathbf{M}}$ is a **near-unanimity term of $\underline{\mathbf{M}}$** if the equations

$$t(x, \dots, x, y) \approx t(x, \dots, x, y, x) \approx \cdots \approx t(y, x, \dots, x) \approx x$$

are satisfied by $\underline{\mathbf{M}}$. The most commonly occurring examples of algebras with a near-unanimity term are those with an underlying lattice; in this case, we can use the ternary term $t(x, y, z) := (x \wedge y) \vee (y \wedge z) \vee (z \wedge x)$.

A.7.8 Theorem *Let* $\underline{\mathbf{M}}$ *be a non-trivial finite algebra. Then the following are equivalent:*

(i) $\underline{\mathbf{M}}$ *is dualisable and has rank* 0;

(ii) $\underline{\mathbf{M}}$ *is dualisable and has height* 0;

(iii) $\underline{\mathbf{M}}$ *is dualisable, and every duality for* $\mathbb{ISP}(\underline{\mathbf{M}})$ *is strong*;

(iv) $\underline{\mathbf{M}}$ *is strongly dualisable via a purely relational alter ego*;

(v) $\underline{\mathbf{M}}$ *has a near-unanimity term, every non-trivial subalgebra of* $\underline{\mathbf{M}}$ *is subdirectly irreducible, and the only homomorphisms from subalgebras of* $\underline{\mathbf{M}}$ *into* $\underline{\mathbf{M}}$ *are the inclusion maps.*

Proof Condition (i) implies (ii), by Theorem A.6.5. To prove that (ii) implies (iii), assume that (ii) holds and let $\underset{\sim}{\mathbf{M}}$ be an alter ego that dualises $\underline{\mathbf{M}}$. Since $\underline{\mathbf{M}}$ has height 0, it follows that $\underline{\mathbf{M}}$ has height 0 relative to $\underset{\sim}{\mathbf{M}}$. So $\underset{\sim}{\mathbf{M}}$ strongly dualises $\underline{\mathbf{M}}$, by Theorem A.4.6, and therefore (ii) implies (iii). Condition (iii) implies (iv), since every dualisable algebra is dualised by a purely relational alter ego.

To prove that (iv) implies (v), we call on a number of results that can be found in the text by Clark and Davey [8]. Assume that $\underline{\mathbf{M}}$ is strongly dualised by a purely relational alter ego $\underset{\sim}{\mathbf{M}} = \langle M; G, H, R, \mathcal{T}\rangle$, where $G = H = \varnothing$. Then every partial operation in $G \cup H$ is at most unary, vacuously. Using the Unary Structure Theorem [8, 6.2.2] (Clark and Davey [7]), it follows that $\underline{\mathbf{M}}$ has a near-unanimity term and that every non-trivial subalgebra of $\underline{\mathbf{M}}$ is subdirectly irreducible. (Both of these conclusions are far from obvious.)

Now let $\mathbf{N}$ be a subalgebra of $\underline{\mathbf{M}}$. There are various ways to prove that the inclusion map $\iota_{\mathbf{N}} : \mathbf{N} \to \underline{\mathbf{M}}$ is the only homomorphism from $\mathbf{N}$ into $\underline{\mathbf{M}}$. One of the simplest is to note, via the Dual Generation Theorem, A.2.3, that $\mathrm{D}(\mathbf{N})$ is the substructure of $\underset{\sim}{\mathbf{M}}^N$ generated by the projection $\iota_{\mathbf{N}} : \mathbf{N} \to \underline{\mathbf{M}}$. Since $G \cup H$ is empty, it follows that $\iota_{\mathbf{N}}$ is the only homomorphism from $\mathbf{N}$ into $\underline{\mathbf{M}}$, as required. Thus, (iv) implies (v).

Finally, we will prove that (v) implies (i). Assume that (v) is true. Since $\underline{\mathbf{M}}$ has a near-unanimity term, it follows that $\underline{\mathbf{M}}$ is dualisable, by the NU Duality Theorem [8, 2.3.4] (Davey and Werner [29]). It also follows that $\underline{\mathbf{M}}$ generates a congruence-distributive variety, by A. Mitschke [49]. As every non-trivial subalgebra of $\underline{\mathbf{M}}$ is subdirectly irreducible, we must have $\mathrm{Irr}(\underline{\mathbf{M}}) = 1$.

We want to prove that $\underline{\mathbf{M}}$ has rank 0. To this end, let $h : \mathbf{B} \to \underline{\mathbf{M}}$ be a homomorphism, where $\mathbf{B} \leqslant \underline{\mathbf{M}}^n$ and $n \in \omega$. The only endomorphism of $\underline{\mathbf{M}}$ is the identity. So we know that there are no constant homomorphisms into $\underline{\mathbf{M}}$, and therefore $n > 0$. By Lemma A.7.6, there is a projection $\rho_i : \mathbf{B} \to \underline{\mathbf{M}}$, for some $i \in n$, and a homomorphism $p : \rho_i(\mathbf{B}) \to \underline{\mathbf{M}}$ such that $h = p^{\mathrm{D}(\mathbf{B})}(\rho_i)$.

Our assumptions guarantee that p is the inclusion $\iota_{\rho_i(\mathbf{B})} : \rho_i(\mathbf{B}) \to \underline{\mathbf{M}}$, and thus $h = \rho_i$. Hence (i) holds. ∎

The conditions in (v) above are, of course, very restrictive on a finite algebra $\underline{\mathbf{M}}$. They imply that $\underline{\mathbf{M}}$ has no one-element subalgebras and that $\underline{\mathbf{M}}$ is injective in the quasi-variety $\mathbb{ISP}(\underline{\mathbf{M}})$ (via a Jónsson's Lemma argument, for instance). If $\mathbb{ISP}(\underline{\mathbf{M}})$ is closed under homomorphic images, they also imply that every subalgebra of $\underline{\mathbf{M}}$ is simple.

The previous two theorems tell us that the two-element bounded distributive lattice $\langle\{0,1\}; \vee, \wedge, 0, 1\rangle$ has rank 0, while its unbounded cousin $\langle\{0,1\}; \vee, \wedge\rangle$ has rank 1. Many other examples of algebras of rank at most 2 are known.

- Clark, Idziak, Sabourin, Szabó and Willard [14] proved that a finite commutative ring with identity is dualisable if and only if its Jacobson radical is self annihilating. They then showed that all these dualisable rings are strongly dualisable, by (implicitly) proving that they have rank at most 1 [65].
- Hyndman and Willard [38, 65] gave an example of a dualisable algebra whose rank is exactly 2: the four-element unar $\langle\{0,1,2,3\}; 0001\rangle$.
- The strongly dualisable three-element unary algebras are described in Chapter 4. They all have enough algebraic operations, and therefore have rank at most 2. Indeed, Hyndman and Pitkethly [40] proved that a three-element unary algebra has finite rank if and only if it has enough algebraic operations.
- In Chapter 7, we proved that all finite linear unary algebras are strongly dualisable by proving that they all have enough algebraic operations.
- Lampe, McNulty and Willard [45] have shown that the dualisable graph algebras and flat graph algebras are strongly dualisable, again by proving that they have enough algebraic operations.

We also have examples of finite algebras that do not have a height, and therefore have rank infinity. Any dualisable algebra that is not strongly dualisable must have rank infinity; see Chapter 4 for examples.

Between 2 and infinity, nothing is currently known. Is there a unary algebra of rank 3? A non-unary algebra? Are there any algebras of unbounded height? Is there a strongly dualisable algebra of rank infinity? Much remains to be discovered.

References

[1] M. E. Adams and W. Dziobiak (eds), *Studia Logica* **56**, Nos. 1–2, 1996.

[2] R. Balbes and A. Horn, Stone lattices, *Duke Math. J.* **37** (1970), 537–545.

[3] P. Berthiaume, The injective envelope of S-sets, *Canad. Math. Bull.* **10** (1967), 261–273.

[4] I. P. Bestsennyi, Quasiidentities of finite unary algebras, *Algebra and Logic* **28** (1989), 327–340.

[5] S. Burris and H. P. Sankappanavar, *A Course in Universal Algebra*, Springer, 1981. (Visit www.math.uwaterloo.ca/~snburris to download the Millenium Edition.)

[6] D. M. Clark and B. A. Davey, The quest for strong dualities, *J. Austral. Math. Soc. (Series A)* **58** (1995), 248–280.

[7] D. M. Clark and B. A. Davey, When is a natural duality 'good'?, *Algebra Universalis* **35** (1996), 265–295.

[8] D. M. Clark and B. A. Davey, *Natural Dualities for the Working Algebraist*, Cambridge University Press, 1998.

[9] D. M. Clark, B. A. Davey, R. S. Freese and M. Jackson, Standard topological algebras: syntactic and principal congruences and profiniteness, *Algebra Universalis* **52** (2004), 343–376.

[10] D. M. Clark, B. A. Davey, M. Haviar, J. G. Pitkethly and M. R. Talukder, Standard topological quasi-varieties, *Houston J. Math.* **29** (2003), 859–887.

[11] D. M. Clark, B. A. Davey, M. Jackson and J. G. Pitkethly, Inverse limits, ultraproducts, and the axiomatizability of topological quasi-varieties, La Trobe University preprint, 2005.

[12] D. M. Clark, B. A. Davey and J. G. Pitkethly, Binary homomorphisms and natural dualities, *J. Pure Appl. Algebra* **169** (2002), 1–28.

[13] D. M. Clark, B. A. Davey and J. G. Pitkethly, The complexity of dualisability: three-element unary algebras, *Internat. J. Algebra Comput.* **13** (2003), 361–391.

[14] D. M. Clark, P. M. Idziak, L. R. Sabourin, Cs. Szabó and R. Willard, Natural dualities for quasivarieties generated by a finite commutative ring, *Algebra Universalis* **46** (2001), 285–320.

[15] D. M. Clark and P. H. Krauss, Topological quasi varieties, *Acta Sci. Math.* **47** (1984), 3–39.

[16] B. A. Davey, Duality theory on ten dollars a day, *Algebras and Orders* (*Montreal, 1991*) (I. G. Rosenberg and G. Sabidussi, eds), NATO Advanced Science Institute Series C: Mathematical and Physical Sciences **389**, Kluwer, 1993, pp. 71–111.

[17] B. A. Davey, Dualisability in general and endodualisability in particular, *Logic and Algebra* (*Pontignano, 1994*) (A. Ursini and P. Aglianò, eds), Lecture Notes in Pure and Applied Mathematics **180**, Marcel Dekker, 1996, pp. 437–455.

[18] B. A. Davey, M. Haviar and T. Niven, When is a full duality strong?, La Trobe University preprint, 2004.

[19] B. A. Davey, M. Haviar, T. Niven and N. Perkal, Full but not strong dualities: extending the realm, La Trobe University preprint, 2005.

[20] B. A. Davey, M. Haviar and R. Willard, Full does not imply strong, does it?, *Algebra Universalis* (to appear).

[21] B. A. Davey, M. Haviar and R. Willard, Structural entailment, *Algebra Universalis* (to appear).

[22] B. A. Davey, L. Heindorf and R. McKenzie, Near unanimity: an obstacle to general duality theory, *Algebra Universalis* **33** (1995), 428–439.

[23] B. A. Davey, P. M. Idziak, W. A. Lampe and G. F. McNulty, Dualizability and graph algebras, *Discrete Math.* **214** (2000), 145–172.

[24] B. A. Davey, M. Jackson and M. R. Talukder, Natural dualities for semilattice-based algebras, La Trobe University preprint, 2005.

[25] B. A. Davey and B. J. Knox, Regularising natural dualities, *Acta Math. Univ. Comenianae* **68** (1999), 295–318.

[26] B. A. Davey and J. G. Pitkethly, Dualisability of p-semilattices, *Algebra Universalis* **45** (2001), 149–153.

[27] B. A. Davey and R. W. Quackenbush, Natural dualities for dihedral varieties, *J. Austral. Math. Soc.* (*Series A*) **61** (1996), 216–228.

[28] B. A. Davey and M. R. Talukder, Dual categories for endodualisable Heyting algebras: optimization and axiomatization, *Algebra Universalis* (to appear).

[29] B. A. Davey and H. Werner, Dualities and equivalences for varieties of algebras, *Contributions to Lattice Theory (Szeged, 1980)* (A. P. Huhn and E. T. Schmidt, eds), Colloquia Mathematica Societatis János Bolyai **33**, North-Holland, 1983, pp. 101–275.

[30] B. A. Davey and R. Willard, The dualisability of a quasi-variety is independent of the generating algebra, *Algebra Universalis* **45** (2001), 103–106.

[31] J. Dugundji, *Topology*, Allyn and Bacon, 1966.

[32] O. Frink, Pseudo-complements in semi-lattices, *Duke Math. J.* **29** (1962), 505–514.

[33] V. A. Gorbunov, *Algebraic Theory of Quasivarieties*, Consultants Bureau, 1998.

[34] G. Grätzer, *Universal Algebra*, 2nd edition, Springer, 1979.

[35] L. Heindorf, Some examples of non-dualizable algebras, *General Algebra and Applications (Potsdam, 1992)* (K. Denecke and H.-J. Vogel, eds), Research and Exposition in Mathematics **20**, Heldermann, 1993, pp. 143–151.

[36] D. Hobby and R. McKenzie, *The Structure of Finite Algebras*, 2nd printing with added appendix and bibliography, Cont. Math. **76**, Amer. Math. Soc., 1996.

[37] K. H. Hofmann, M. Mislove and A. Stralka, *The Pontryagin Duality of Compact 0-Dimensional Semilattices and its Applications*, Lecture Notes in Mathematics **396**, Springer, 1974.

[38] J. Hyndman, Mono-unary algebras are strongly dualizable, *J. Austral. Math. Soc.* **72** (2002), 161–172.

[39] J. Hyndman, Strong duality of finite algebras that generate the same quasivariety, *Algebra Universalis* **51** (2004), 29–34.

[40] J. Hyndman and J. G. Pitkethly, How finite is a three-element unary algebra?, *Internat. J. Algebra Comput.* (to appear).

[41] J. Hyndman and R. Willard, An algebra that is dualizable but not fully dualizable, *J. Pure Appl. Algebra* **151** (2000), 31–42.

[42] P. M. Idziak, Congruence labelling for dualizable algebras, preprint, 1994.

[43] M. Jackson, Dualisability of finite semigroups, *Internat. J. Algebra Comput.* **13** (2003), 481–497.

[44] J. L. Kelley, *General Topology*, Graduate Texts in Mathematics **27**, Springer, 1975.

[45] W. A. Lampe, G. F. McNulty and R. Willard, Full duality among graph algebras and flat graph algebras, *Algebra Universalis* **45** (2001), 311–334.

[46] S. Mac Lane, *Categories for the Working Mathematician*, 2nd edition, Graduate Texts in Mathematics **5**, Springer, 1998.

[47] A. I. Mal'cev, On the general theory of algebraic systems (Russian), *Mat. Sb.* **35** (1954), 3–20.

[48] R. N. McKenzie, G. F. McNulty and W. F. Taylor, *Algebras, Lattices, Varieties*, Vol. 1, Wadsworth & Brooks/Cole, 1987.

[49] A. Mitschke, Near unanimity identities and congruence distributivity in equational classes, *Algebra Universalis* **8** (1978), 29–32.

[50] J. G. Pitkethly, Strong and full dualisability: three-element unary algebras, *J. Austral. Math. Soc.* **73** (2002), 187–219.

[51] J. G. Pitkethly, Dualisability and algebraic constructions, *Acta Sci. Math. (Szeged)* **68** (2002), 571–591.

[52] J. G. Pitkethly, *Dualisability: Unary Algebras and Beyond*, PhD thesis, La Trobe University, 2002.

[53] J. G. Pitkethly, Inherent dualisability, *Discrete Math.* **269** (2003), 219–237.

[54] J. G. Pitkethly and B. A. Davey, A non-dualisable entropic algebra, *Algebra Universalis* **47** (2002), 51–54.

[55] L. S. Pontryagin, The theory of topological commutative groups, *Ann. Math.* **35** (1934), 361–388.

[56] H. A. Priestley, Representation of distributive lattices by means of ordered Stone spaces, *Bull. London Math. Soc.* **2** (1970), 186–190.

[57] H. A. Priestley, Ordered topological spaces and the representation of distributive lattices, *Proc. London Math. Soc. (3)* **24** (1972), 507–530.

[58] H. A. Priestley, Ordered sets and duality for distributive lattices, *Orders: Descriptions and Roles (L'Arbresle, 1982)* (M. Pouzet and D. Richard, eds), North-Holland Mathematics Studies **99**, North-Holland, 1984, pp. 39–60.

[59] R. Quackenbush and Cs. Szabó, Nilpotent groups are not dualizable, *J. Austral. Math. Soc.* **72** (2002), 173–179.

[60] R. Quackenbush and Cs. Szabó, Strong duality for metacyclic groups, *J. Austral. Math. Soc.* **73** (2002), 377–392.

[61] M. J. Saramago, Some remarks on dualisability and endodualisability, *Algebra Universalis* **43** (2000), 197–212.

[62] J. Sichler, Group-universal unary varieties, *Algebra Universalis* **11** (1980), 12–21.

[63] M. H. Stone, The theory of representations for Boolean algebras, *Trans. Amer. Math. Soc.* **40** (1936), 37–111.

[64] M. A. Valeriote, Decidable unary varieties, *Algebra Universalis* **24** (1987), 1–20.

[65] R. Willard, New tools for proving dualizability, *Dualities, Interpretability and Ordered Structures (Lisbon, 1997)* (J. Vaz de Carvalho and I. Ferreirim, eds), Centro de Álgebra da Universidade de Lisboa, 1999, pp. 69–74.

[66] L. Zádori, Natural duality via a finite set of relations, *Bull. Austral. Math. Soc.* **51** (1995), 469–478.

Notation

Notation is listed under every chapter in which it is used.

4

5

6

7

A

Index

Page references to definitions are given in italics.